SAFETY OF MARINE TRANSPORT

Safety of Marine Transport

Marine Navigation and Safety of Sea Transportation

Editors

Adam Weintrit & Tomasz Neumann
Gdynia Maritime University, Gdynia, Poland

CRC Press
Taylor & Francis Group
Boca Raton London New York

CRC Press is an imprint of the
Taylor & Francis Group, an **informa** business

A BALKEMA BOOK

Published by:
CRC Press/Balkema
P.O. Box 447, 2300 AK Leiden, The Netherlands
e-mail: Pub.NL@taylorandfrancis.com
www.crcpress.com – www.taylorandfrancis.com

First issued in paperback 2020

Typeset by V Publishing Solutions Pvt Ltd., Chennai, India

ISBN 13: 978-0-367-73821-1 (pbk)
ISBN 13: 978-1-138-02859-3 (hbk)

**Visit the Taylor & Francis Web site at
http://www.taylorandfrancis.com**

**and the CRC Press Web site at
http://www.crcpress.com**

Contents

List of reviewers

*Prof. Teresa **Abramowicz-Gerigk**, Gdynia Maritime University, Gdynia, Poland*
*Prof. Anatoli **Alop**, Estonian Maritime Academy, Tallin, Estonia*
*Prof. Ted **Bagfeldt**, Kalmar Maritime Academy, Linnaeus University, Sweden*
*Prof. Eugen **Barsan**, Constanta Maritime University, Romania*
*Prof. Angelica **Baylon**, Maritime Academy of Asia & the Pacific, Philippines*
*Prof. Christophe **Berenguer**, Grenoble Institute of Technology, Saint Martin d'Hères, France*
*Prof. Heinz Peter **Berg**, Bundesamt für Strahlenschutz, Salzgitter, Germany*
*Prof. Tor Einar **Berg**, Norwegian Marine Technology Research Institute, Trondheim, Norway*
*Prof. Vitaly **Bondarev**, Baltic Fishing Fleet State Academy, Kaliningrad, Russia*
*Prof. Neil **Bose**, Australian Maritime College, University of Tasmania, Launceston, Australia*
*Prof. Alfred **Brandowski**, Gdynia Maritime University, Poland*
*Prof. Zbigniew **Burciu**, Gdynia Maritime University, Poland*
*Prof. Doina **Carp**, Constanta Maritime University, Romania*
*Prof. Shyy Woei **Chang**, National Kaohsiung Marine University, Taiwan*
*Prof. Andrzej **Chudzikiewicz**, Warsaw University of Technology, Poland*
*Prof. German **de Melo Rodriguez**, Polytechnic University of Catalonia, Barcelona, Spain*
*Prof. Bolesław **Domański**, Jagiellonian University, Kraków, Poland*
*Prof. Eamonn **Doyle**, National Maritime College of Ireland, Cork Institute of Technology, Cork, Ireland*
*Prof. Branislav **Dragović**, University of Montenegro, Kotor, Montenegro*
*Prof. Daniel **Duda**, Polish Naval Academy, Polish Nautological Society, Poland*
*Prof. Billy **Edge**, North Carolina State University, US*
*Prof. Akram **Elentably**, King Abdulaziz University (KAU), Jeddah, Saudi Arabia*
*Prof. Alberto **Francescutto**, University of Trieste, Trieste, Italy*
*Prof. Jens **Froese**, Jacobs University Bremen, Germany*
*Prof. Masao **Furusho**, Kobe University, Japan*
*Prof. Wiesław **Galor**, Maritime University of Szczecin, Poland*
*Prof. Péter **Gáspár**, Computer and Automation Research Institute, Hungarian Academy of Sciences, Budapest, Hungary*
*Prof. Aleksandrs **Gasparjans**, Latvian Maritime Academy, Latvia*
*Prof. Avtandil **Gegenava**, Georgian Maritime Transport Agency, Maritime Rescue Coordination Center, Georgia*
*Prof. Andrzej **Grzelakowski**, Gdynia Maritime University, Poland*
*Prof. Marek **Grzybowski**, Gdynia Maritime University, Gdynia, Poland*
*Prof. Jerzy **Hajduk**, Maritime University of Szczecin, Poland*
*Prof. Esa **Hämäläinen**, University of Turku, Finland*
*Prof. Toshio **Iseki**, Tokyo University of Marine Science and Technology, Tokyo, Japan,*
*Prof. Marianna **Jacyna**, Warsaw University of Technology, Poland*
*Prof. Ales **Janota**, University of Žilina, Slovakia*
*Prof. Jung Sik **Jeong**, Mokpo National Maritime University, South Korea*
*Prof. Tae-Gweon **Jeong**, Korean Maritime University, Pusan, Korea*
*Prof. Mirosław **Jurdziński**, Gdynia Maritime University, Poland*
*Prof. Kalin **Kalinov**, Nikola Y. Vaptsarov Naval Academy, Varna, Bulgaria*
*Prof. Eiichi **Kobayashi**, Kobe University, Japan*
*Prof. Lech **Kobyliński**, Polish Academy of Sciences, Gdansk University of Technology, Poland*
*Prof. Serdjo **Kos**, University of Rijeka, Croatia*
*Prof. Pentti **Kujala**, Helsinki University of Technology, Helsinki, Finland*
*Prof. Shashi **Kumar**, U.S. Merchant Marine Academy, New York*
*Prof. Alexander **Kuznetsov**, Admiral Makarov State Maritime Academy, St. Petersburg, Russia*
*Prof. Bogumił **Łączyński**, Gdynia Maritime University, Poland*
*Prof. Andrzej **Lewiński**, University of Technology and Humanities in Radom, Poland*
*Prof. Mirosław **Luft**, University of Technology and Humanities in Radom, Poland*
*Prof. Zbigniew **Łukasik**, University of Technology and Humanities in Radom, Poland*
*Prof. Tihomir **Luković**, University of Dubrovnik, Croatia*
*Prof. Margareta **Lützhöft**, Australian Maritime College, Launceston, Australia*
*Prof. Melchor M. **Magramo**, John B. Lacson Foundation Maritime University, Iloilo City, Philippines*
*Prof. Michael Ekow **Manuel**, World Maritime University, Malmoe, Sweden*
*Prof. Jerzy **Matusiak**, Helsinki University of Technology, Helsinki, Finland*
*Prof. Jerzy **Merkisz**, Poznań University of Technology, Poznań, Poland*
*Prof. Jerzy **Mikulski**, University of Economics in Katowice, Poland*
*Prof. Daniel **Seong**-Hyeok Moon, World Maritime University, Malmoe, Sweden*
*Prof. Wacław **Morgaś**, Polish Naval Academy, Gdynia, Poland*
*Prof. Junmin **Mou**, Wuhan University of Technology, Wuhan, China*
*Prof. Rudy R. **Negenborn**, Delft University of Technology, Delft, The Netherlands*
*Prof. Nikitas **Nikitakos**, University of the Aegean, Chios, Greece*
*Prof. Gabriel **Nowacki**, Military University of Technology, Warsaw, Poland*
*Mr. David **Patraiko**, The Nautical Institute, UK*
*Prof. Vytautas **Paulauskas**, Maritime Institute College, Klaipeda University, Lithuania*
*Prof. Jan **Pawelski**, Gdynia Maritime University, Poland*
*Prof. Thomas **Pawlik**, Bremen University of Applied Sciences, Germany*

Safety of Marine Transport
Introduction

A. Weintrit & T. Neumann
Gdynia Maritime University, Gdynia, Poland
Polish Branch of the Nautical Institute

The contents of the book are partitioned into nine separate chapters: Human resource management and maritime crew manning (covering the subchapters 1.1 through 1.7), MET - Maritime Education and Training (covering the chapters 2.1 through 2.5), Sea ports and harbours (covering the chapters 3.1 through 3.6), Port facilities (covering the chapters 4.1 through 4.3), Ship's propulsion, main engine and power supply (covering the chapters 5.1 through 5.4), Maritime low and policy (covering the chapters 6.1 and 6.2), Piracy (covering the chapters 7.1 and 7.2), Ship's operations (covering the chapters 8.1 through 8.4), and Safety of transport (covering the chapters 9.1 through 9.5).

In each of them readers can find a few subchapters. Subchapters collected in the first chapter, titled 'Human resource management and maritime crew manning', describe: sample data from shipping companies: women in the Turkish seafarers registry and their employment situation, attractions, problems, challenges, issues and coping strategies of the seafaring career: MAAP seafarers perspectives, plights and concerns of Filipino seafarers on board vessels traversing Horn of Africa and Gulf of Aden: AMOSUP and other stakeholders responses, Swedish seafarers' occupational commitment in light of gender and family situation, web-based databank for assessment of seafarers' functional status during sea missions, implementation of CSR aspects in Human Resources Management (HRM) strategies of maritime supply chain's main involved parties, and analysis of factors influencing Latvian seafarers' outflow rate.

In the second chapter there are described problems related to Maritime Education and Training (MET): investigation of sea training conditions of deck cadets: a case study in Turkey, sleep quality, anxiety and depression among maritime students in Lithuania: cross-sectional questionnaire study, the use of the Portuguese Naval Academy navigation simulator in developing team leadership skills, paradigm shift in ship handling and its training, and experimental research with neuroscience tool in Maritime Education and Training (met).

Third chapter is about vessel's sea ports and harbours development. The readers can find some information about trends in environmental policy instruments and best practices in port operations, decreasing air emissions in ports – case studies in ports, port in a city – effects of the port, the influence of internalizing the external cost on the competiveness of sea ports in the same container loop, development of dry ports: significance of maritime logistics on improving the Iranian dry ports and transit, and a study on rapid left-turn of ship's head of laden cape-size ore carriers while using astern engine in harbour.

The fourth chapter deals with port facilities. The contents of the fourth chapter are partitioned into three subchapters: a geographical perspective on LNG facility development in the Eastern Baltic Sea; influence of "Suezmax" tankers size increase on mooring ropes at existing terminals; and the analysis of dredging project's effectiveness in the Port of Gdynia, based on the interference with vessel traffic.

The fifth chapter deals with ship's propulsion, main engine and power supply. The contents of the fifth chapter are partitioned into four: reliability of fuel oil system components versus main propulsion engine: an impact assessment study, impact of electricity generator on a small-bore internal combustion engine at low and medium loads, a comparative approach of electrical diesel propulsion systems, and method of determining operation region of single-transistor ZVS DC/DC converters.

In the sixth chapter there are described problems related to maritime low and policy: the implementation of a new maritime labour policy: the Maritime Labour Convention (MLC, 2006), and a new international law to protect abandoned seafarers: amendments to MLC, 2006.

The seventh chapter deals with piracy problem. The contents of the seventh chapter concerns effectiveness of measures undertaken in the Gulf of Guinea Region to fight maritime piracy, and counter piracy training competencies model.

Eighth chapter is about ship's operations. The readers can find some information about experimental study for the development of a ship hull cleaning robot, assessment of variations of ship's deck elevation due to containers loading in various locations on board, Tworty box to reduce empty container positioning, and consideration on dynamic modelling of ship squat.

The ninth chapter deals with safety in transport in general. The contents of the ninth chapter are partitioned into five: state of safety in the Polish land transport, surveys of the influence of telematics on the land transport safety, approaches and regulations regarding significant modifications in transportation and nuclear safety, optimization of the transport service of fishing vessels at ocean fishing grounds, and selected transport problems of dangerous goods in the European Union and Poland.

Each chapter was reviewed at least by three independent reviewers. The Editor would like to express his gratitude to distinguished authors and reviewers of chapters for their great contribution for expected success of the publication. He congratulates the authors and reviewers for their excellent work.

Human Resource Management and Maritime Crew Manning

Sample Data from Shipping Companies: Women in the Turkish Seafarers Registry and Their Employment Situation

H. Yılmaz, E. Başar & Ü. Özdemir
Karadeniz Technical University, Maritime Transportation & Management Engineering, Trabzon, Turkey

ABSTRACT: Throughout history, women have struggled to gain a place and establish their presence in social life. Mostly, they have continued their life in the position of carrying out family responsibilities. When women want to get involved in the business world, they are facing some problems arising from dogmatic thinking and prejudices, especially among the professions with "male-dominated" judiciary. Maritime profession, one of the oldest professions in the world, was also regarded as a single-gender area until the 20th century. However, in recent years, economical and political changes in the world, equal opportunity in education and incentive works of the International Maritime Organization have led women to work as seafarers. Although female seafarers constitute 2% of the world seafarers, this ratio is higher in developed countries than that of undeveloped or developing countries. In Turkey, women have played an active role and gained an apparent identity in the maritime sector since the 2000s. According to 2012 data, Turkey is ranks the 15th in the world maritime trade with more than 24 million deadweight tonnage and manages 1879 vessels. In Turkey, many studies related to the employment of seafarers are carried out as in the whole world. However, studies on the employment of women seafarers should be paid more attention. In this study, a questionnaire was carried out with the personnel department managers in Turkish shipping companies. Employment, career, educational status of female seafarers and the general difficulties they face were revealed.

1 INTRODUCTION

It is known that millions of women and men are pushed to certain job groups and worked with low wages because of their gender, skin colour, ethnic reasons or religion, without taking into account their abilities or qualifications, all over the world. Protection against discrimination is part of the fundamental human rights and equalizing the conditions of the employees in the workplace brings significant economic benefits. For the employers, this means more work force, higher quality and for the workers it means easier access to training and higher salaries. The benefits of a globalised economy are better allocated in an equalitarian society and they generate a higher social stability and a wider support from people in the favour of economic development (Popescu and Varsami, 2010).

The regulations of The International Labour Organization (ILO) about equality secure discrimination elimination from all fields of work and society. Since 1919, ILO has developed an international regulation work system focused on increasing men and women chances in obtaining a good and productive job in freedom, equity, security and dignity conditions regardless of the working domain. In the today worldwide economy, the international work regulations are an essential component of the international framework and they have as main purpose to ensure that everybody (men and women) profits from the world economy growth (Popescu and Varsami, 2010).

Women represent at least 1/3 of the world of labor. More than 50% of women are economically active in more than 90 countries. Despite the importance of women in the national economy and their income to the family, social protection is insufficient neither for them nor for their families. Due to the special operating conditions, marine transportation industry is one of the sectors which needs to be analyzed in protecting benefits of women. Female seafarers constitute 2% of the world seafarers (URL-1).

In Turkey, women have begun to show themselves in various branches of the maritime sector since the 1980s and their existence in the sector has showed a rapid increase especially after 2000. Women's employment in commercial vessels and the acceptance of female students to maritime schools have been realized.

According to data from the year 2012, Turkey ranked the fifteenth in the world maritime trade with more than 24 million deadweight tonnage management (URL-4). Management of 1879 ships is performed in Turkey (Maritime Trade 2012 Statistics, 2013). They are registered in both international and national record (150 GT and over) and have total capacity of more than 10 million DWT (URL-2,3). The total number of seafarers registered in Turkish Seafarers' Registry is 178134 and 2246 of them are female seafarers. 45677 of them are the officers and 132457 of them are ratings. However, the number of active employees are 36254 as officers and 83316 as ratings (e-Maritime Database, 2013).

2 FEMALE SEAFARERS REGISTERED IN TURKISH SEAFARERS REGISTRY

According to the 2013 year-end data, there are 2246 women seafarers registered in Turkish Seafarers' Registry. Turkish female seafarers' mean ages and distribution of competences are shown in Table1.

Competencies of female seafarers and the mean age data are shown in Fig 2 separately for each proficiency level. In Fig 3, the mean age data of the registered female seafarers are classified by departments. The number of people in each proficiency level are shown in Fig 4.

Table 1. Female seafarers recorded in Turkish Seafarers' Registry

Rank	Age Period	Mean Age	Nr.
Oceangoing Master	30-40	32,5	22
Oceangoing Chief Officer	26-39	32	55
Oceangoing Watchkeeping Officer	22-47	27,76	124
Master	41-57	48,17	6
Chief Officer	36-73	50	3
Watchkeeping Officer	24-36	26,92	13
Restricted Master	32-41	35,5	4
Restricted Watchkeeping Officer	27-61	40,6	10
Deck Cadet	17-54	22,16	423
Shipping Clerk	35-68	53,29	14
Boatswain	55	55	1
Ableseaman (Ableseafarer deck)	20-68	45,35	123
Ordinary Seaman (Ordinary seaferer deck)	17-67	36,46	367
Deck Boy	22-73	39,68	284
Cook	22-67	40,46	97
Steward	20-69	33,88	520
Doctor	37-55	44,67	3
Nurse	29-53	39	5
Oceangoing Chief Engineer	31-34	32,33	6
Oceangoing Second Engineer	30-34	32,33	3
Oceangoing Watchkeeping Eng.	27-34	28,79	19
Chief Engineer	41-71	59	2
Engineer Officer	24-25	24,5	2
Engine Cadet	18-30	22,78	18
Electrician/Electronic Officer/ Electronic Operator	22-41	31,75	4
Donkeyman	54	54	1
Wiper	27-65	41,17	12
Oiler/Motorman (Ableseafarer engine)	29-54	39,55	11
Yacht Master	21-69	37,85	40
Unknown record	21-60	39,87	54
Total			2246

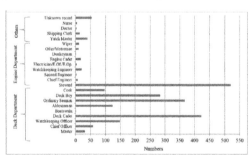

Figure 1. Female seafarers recorded in Turkish Seafarers' Registry
E.Of.: Electric Officer, E.Op.: Electric Operator

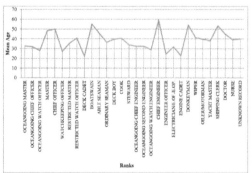

Figure 2. Registered female seafarers' mean age by all competencies
WATCH.: Watchkeeping, E.OF.: Electric Officer, E.OP.: Electric Operator

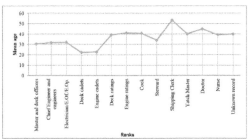

Figure 3. Registered female seafarers' mean age in categories.
E.OF.: Electric Officer, E.OP.: Electric Operator

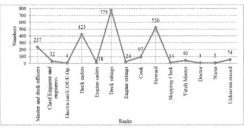

Figure 4. Registered female seafarers' numbers in categories

3 METHODS

In this study, a questionnaire was carried out with the personnel department managers in Turkish shipping companies. Employment, career, educational status and the general difficulties of female seafarers were revealed.

40 corporate firms were interviewed. In 19 of these cases, it was determined that there was no female seafarer employment. It was reported that female personnel were currently working on board in the remaining 21 firms at the time of the survey. Study assessments have been performed with the data of 16 companies who participated in the survey.

In the survey, 24 multiple choice questions were directed to the participants. The personnel department managers' opinions were also mentioned. A part of the survey questions aimed to determine the service years of companies, types, tonnages and numbers of vessels, engine powers, number of maritime-related employees, competence and education level of seafarers in the fleets. In the other section, it was intended to identify the number and competencies of female seafarers, education levels, types of ships they are employed, age and marital status, employment and wage policies. The survey also tried to reveal the conditions that complicate the work of female seafarers.

4 FINDINGS AND CONCLUSIONS

4.1 Company and Fleet Profiles

The graphics in Fig 5 and Fig 6 show the profiles of the companies and fleets of the surveyed firms. All of the surveyed firms are experienced corporate companies with large tonnage vessels in the transportation sector.

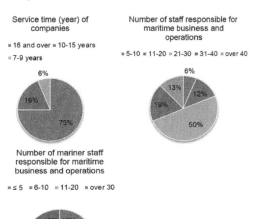

Figure 5. Company profiles of surveyed firms

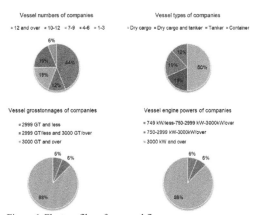

Figure 6. Fleet profiles of surveyed firms

4.2 Findings Related to Fleet Personnel

The total number of fleet personnel (5648) and the distribution of competences for 16 surveyed firms are shown in Fig 7. In Fig 7, "others" group means seafarers working as fitter and pumpers (117) and crew's family members with ordinary seafarer and steward competencies (14).

122 of all detected crew is consists of female seafarers. The distribution of female seafarers' competencies is shown in Fig 8. "Others" group of women seafarers have been reported as relatives of the crew on board with the competencies of ordinary seafarer and steward. They are not active seafarers.

Fig 9 shows the number of male and female employees for each competency. When Fig 9 is analyzed, it is observed that women are working in tasks that require many operational capabilities like captain, officers and engineers. Also, the presence of female cadets on deck and engine departments shows that the number of female seafarers have increased in competencies that require technical and operational skills. In spite of the fact that the belief "Women can do better" is accepted by society for public service jobs such as food preparation and cleaning, female employment is not seen as cook and steward. However, there are 97 female cooks and 520 female stewards already registered in Turkish Seafarers' Registry (e-Maritime Database, 2013). Similarly, although a large number of other rating proficiencies registered, no female seafarers are seen as rating.

The officer seafarers are subjected to classification as management level and operational level, and are shown in Fig 10a for deck department and in Fig 10b for engine department. For deck department, master and chief officers are evaluated at management level and watchkeeping officers are evaluated at operation level. For engine department, chief engineer and second engineers are evaluated at management level and watchkeeping engineers and electricians are evaluated at operation level.

In Fig 11., the crew employed on vessels of surveyed firms are categorized in four classes. Deck class is composed of bosun, ableseaman and ordinary seaman ranks. Engine class is composed of donkeyman, ableseafarer engine and oiler/motorman ranks. "Ableseafarer engine" term means the experienced oiler/motorman. Hotel class is composed of cooks and stewards. No female seafarers are employed at deck, engine and hotel departments. "Others" class is composed of pumpers and fitters of males. Although 14 female seafarers in "others" class have ordinary seafarer and steward ranks, they are family members of the crew and they are on board like passengers.

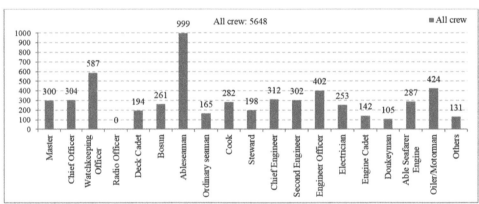

Figure 7. Number of all crew employed on vessels of surveyed firms

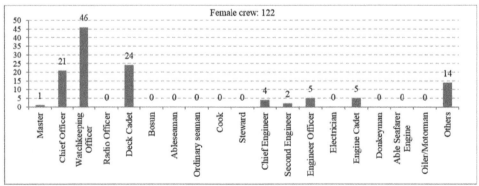

Figure 8. Number of female crew employed on vessels of surveyed firms

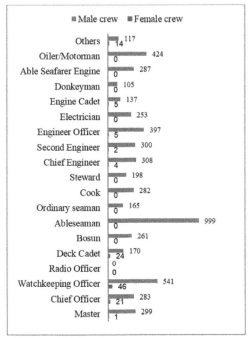

Figure 9. Seafarers by genders on surveyed firms

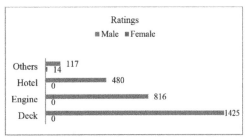

Figure 11. Ratings employed on vessels of surveyed firms by genders and departments

In Fig 12., cadets employed on vessels of surveyed firms are shown by genders and departments. According to academic calender of maritime schools in Turkey, the traniee is getting more on ships in the summer term. The numbers of traniee (cadets) in Fig 12 were identified in the fall term during which the survey was conducted.

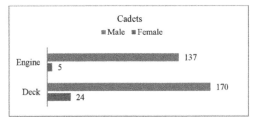

Figure 12. Cadets by genders and departments

Fleet personnel of surveyed firms were also examined in terms of educational status. Educational status of 2186 of 2381 men in the officer class has been determined (1033 of 1123 men in the deck officer, 1153 of 1258 men in the engine officer). The total number of female seafarers in the officer class is 79 and educational status of all female seafarers have been reported. Educational status of the officer class are given separately in Fig 13 by genders.

When the educational status of ratings of surveyed firms were analyzed, it was seen that the educational status of 2551 ratings out of 2838 ratings were reported. 1730 of them are deck ratings and 821 of them are engine ratings. All of them are male seafarers. There are no female seafarers as rating in the surveyed fleets. Detected 14 female ratings are family members on board with seafarer documents. Educational status of ratings is shown in Fig 14. The qualifications of cook, steward, fitter and pumper have been included in the deck and engine class.

Deck and engine cadets' educational status is shown in Fig 15. The numbers belong to the fall term and they increase in the summer term.

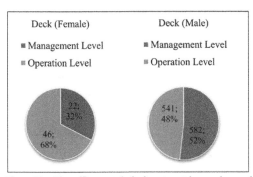

Figure 10a. The officers at deck department by genders and levels

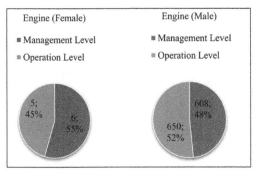

Figure 10b. The officers at engine department by genders and levels

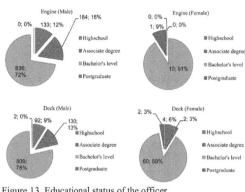

Figure 13. Educational status of the officer

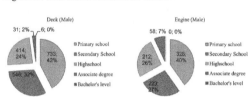

Figure 14. Educational status of ratings

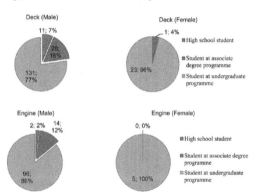

Figure 15. Deck and engine cadets' educational status

Ship types were also investigated in terms of female employment in surveyed fleets. Percentage of 16 firms is given in Fig 16 according to the type of vessel in which female personnel are employed. In three companies from these, operating both types of tanker and dry cargo vessels and female seafarers are employed on tankers. One of the three companies is also operating oil and fuel barges, but there is not female employment in these barges.

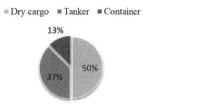

Figure 16. Vessel types in which female personnel are employed

In addition, it is determined that there is not a restriction on women's employment in the Turkish-flagged and foreign-flagged vessels of surveyed firms. Female seafarers were found to be employed in both classes. When the age range of female seafarers is investigated, Table 2 has come to light.

Table 2. Female seafarers' age ranges on surveyed fleets

| Company | 18 and less | Age period of female seafarers | | | |
		19-25	26-30	31-35	36 and over
1			X		
2			X		
3		X	X	X	
4		X			
5		X			
6		X			
7			X		
8		X	X		
9		X	X		
10		X	X		

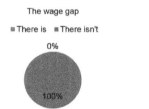

Figure 17. The wage gap between male and female seafarers

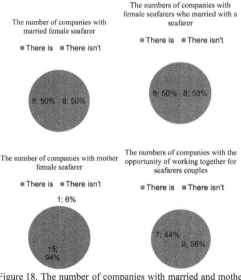

Figure 18. The number of companies with married and mother female seafarers

The companies asked about "married female seafarers"; "female seafarers who have children"; "female seafarers who married with a seafarer" and "the opportunity of working together for seafarers

18

couples". It was determined as "there is" or "there isn't" in Fig 18.

The companies also asked about what reasons female seafarers who left the job had for quitting work. In 3 companies, there were no female seafarers who left job. In other companies, the reasons of female seafarers who left job was found to be as follows;
– Workplace dissatisfaction
– Marriage
– Contract end
– Internships end
– Leave work at sea
– Prefer to work on oceangoing voyage
– Excessive alcohol consumption
– Study abroad
– Tempo on tanker and transportation of chemical cargo
– Her own request

When the existence of a written policy regarding female seafarers was investigated, in two of the 16 companies it was found that there were written employment policies. 14 companies did not have a written policy regarding the female seafarers. The past of companies with female seafarers is shown in Fig 19.

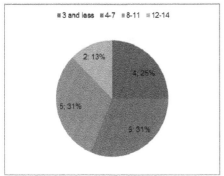

Figure 19. The number of companies by years with female seafarers

The companies asked about the situations that complicate the work of female seafarers. According to the personnel department managers' answers, the results are categorized as follows:
– Conditions of accommodation
– Physical differences
– Superior-subordinate relationship
– Difficulties in destination ports (personnel changes, etc.)
– Ship's personnel not to accept female staff
– Family life and marriage in later years
– Emotionality

The personnel department managers were also asked to indicate their own opinions about female seafarers. 13 managers clearly said that they were satisfied to work with female seafarers and that they wanted to work with female seafarers for many years. However, other managers made some other comments. One of them was also very interesting. Following are the comments made by them;
– Female seafarers tend to go ashore early with the prejudice of "I'm a woman. They do not give me the captain or chief engineer position."
– Usually, women married with sailors stay at sea longer. In each case, female seafarers' professional marine life is ending after birth. They can not go to sea because they can not leave children to their husbands.
– Female officers have no problems about documentation and in compliance with company procedures. However, it was observed that they are experiencing difficulties in the parts of the physical needs of the operational procedures and technical skills, according to their male counterparts.
– "The biggest problem of female employment is still female employee herself. Female staff's ego is fed unnecessarily due to the small number and being in the limelight. Over time, they are having a request to establish an unfounded superiority against other crew. Also, they're being selective among the crew and they're getting disproportionate force in directing the crew behaved well. This situation often leads to imbalance in the ship. For example; the crew member behaved well by female seafarer, feels himself in a privileged position than others. Other people are taking sides against this situation and begins to hate. This hatred often explode in smaller events."

5 RESULTS

As a result of this study, for governmental record, following items were detected:
– The total number of seafarers registered on Turkish Seafarers' Registry (2013 year-end) is 178134 and 2246 of them are female seafarers. It means *1,26% of Turkish seafarers are female.*
– 45677 of all seafarers are officers and 132457 of them are ratings. It means 25,6% of Turkish seafarers are officers and 74,3% of them are ratings.
– The number of active employees are 36254 for the officer and 83316 for ratings. It means 79.3% of the officer and 62,9% of ratings are active seafarers.
– The percentage of 2246 registered female seafarers is as follows; 14% the officer, 20% cadets, 64% ratings and 2% unknown.

The following items were detected as the results of our questionnaire study with 16 companies,:
– Total crew number is 5648, and 5526 of them are male, 122 of them are female seafarers. It means

97,8% of all fleet personel of surveyed firms are male and *2,2% of them are female seafarers*. However, the active female seafarers on surveyed firms are 108 people. In this case, the actual rate of seafarers is *98,1% for male and 1,9% for female*. It also shows us that in these Turkish ships, the atmosphere is suitable for families and away from hostility.

- In surveyed firms, the percentage of women on board by ranks is as follows: master 1%, chief officer 21%, 46% watchkeeping officer, deck cadet 24%, chief engineer 3%, second engineer 2%, watchkeeping engineer 4%, engine cadet 4% and others (families) 11%.
- In surveyed firms, the percentage of men on board by ranks is as follows: master 5%, chief officer 5%, watchkeeping officer 10%, deck cadet 3%, chief engineer 6%, second engineer 5%, watchkeeping engineer 7%, electrician 5%, engine cadet 2%, bosun/boatswain 5%, ableseaman 18%, ordinary seaman 3%, cook 5%, steward 4%, donkeyman 2%, ableseafarer engine 5%, oiler/motorman 8% and others (fitter and pumper) 2%.
- The ratio of women on board for all these 16 companies is; for the deck officer 5,7%, for the engine officer 0.8%, for active ratings 0%, for deck cadet 12% and for engine cadet 3,5% at the time of the survey.
- The percentage of educational status for 2186 of 2381men in the officer class is as follows; high school 10%, associate degree 14%, bachelor's level 75%, postgraduate 0,1%.
- The percentage of educational status for all women in the officer class is as follows; high school 5%, associate degree 2,5%, bachelor's level 88,5%, postgraduate 4%.
- The percentage of educational status for 2551 of 2838 men in the rating class is as follows;

primary school 41,6%, secondary school 30,2%, high school 24,5%, associate degree 3,5%, bachelors level 0,2%.
- There is no restriction related with ship's flag for female seafarers' employment.
- In some companies, tankers are more available for female seafarers' employment and there are no women on barges in one company.
- For female seafarers employed in surveyed fleets, main age range is 19-25, max age range is 31-35.
- There is no wage gap between female and male seafarers.
- Married and/or mother female seafarers can be employees on vessels.
- Experience of surveyed firms with female seafarers reaches to 12-14 years.

REFERENCES

e-Maritime Database, 2013. Transport, Maritime Affairs and Communications Ministry of Turkish Republic, General Directorate of Marine and Inland Waters, December 2013

"Employment and Career Situations of Women Seafarers at Turkish Fleet" Interview Documents, 2013

Maritime Trade 2012 Statistics (2013), Ministry of Transport, Maritime Affairs and Communications of Turkish Republic,

http://www.kugm.gov.tr/BLSM_WIYS/DTGM/tr/Kitaplar/201
30514_102843_64032_1_64480.pdf December 2013

Popescu, C.,Varsami, A.E. (2010), Latest Trends on Engineering Education. In: Dondon P. and Martin O. (ed), The Place of Women in a Men's World from a Maritime University Perspective, Corfu Island, Greece,pp. 182-186

URL-1, http://www.itfseafarers.org/ITI-women-seafarers.cfm December, 2013

URL-2, http://www.denizcilik.gov.tr/tr__/istatistik/Guncel_Filo.asp?rf=ybs December 2013

URL-3, https://atlantis.denizcilik.gov.tr/istatistik/istatistik_filo.aspx December 2013

URL-4, http://www.virahaber.com/haber/turk-armatore-ait-gemi-tonaji-3-kat-artti-28990.htm December, 2013

Attractions, Problems, Challenges, Issues and Coping Strategies of the Seafaring Career: MAAP Seafarers Perspectives

A.M. Baylon & E.M.R. Santos
Maritime Academy of Asia and the Pacific Mariveles Bataan, Philippines

ABSTRACT: This is an exploratory study that aims to develop a critical understanding and prepare the basis of improvement for seafaring life in its interaction with and within the changing global environment and to explore alternative approaches to the prevailing trends in seafaring lifestyle from the perception of the seafarers themselves affected by the challenges and opportunities for "life that is worth living ". This study presents the seafaring work patterns, lifestyle and how this affects their well being and that of their family from the MAAP seafarers' perspective. The paper presents the points of views of the MAAP seafarers in terms of: how they and their family cope up with these work patterns; the attractions/benefits of the seafaring career; the potential difficulties and problem areas and some coping strategies and sources of support from the family in ascertaining a healthy, secure and happy environment to the seafarer whether at home or at sea. This is a basic research driven by the curiosity or interest of the researcher thru a scientific exploratory study that makes use of the following methods of data collection: observation, a questionnaire, documentary analysis from internet and library research and structured interview The respondents were seafarers on vacation, and were teaching at MAAP in 2012- 2013 who had voluntarily agreed to a structured interview intended for the study. Demographic details of respondents were collated by means of a questionnaire and verbatim quotes are also included from the interviews, providing a vivid account of how respondents think, talk and behave. Previous studies on seafaring work patterns; on health research; on ship-shore communication opportunities; on role of the family as support system; on family life and its impact on work performance and safety, had all been validated thru the responses by the MAAP seafarers who are on vacation and currently teaching in 2013-2014. Of the 19 respondents, their profile shows that half (10) are married while the other half (9) are single. Majority of them are Catholics (84%) and most of them are from provinces outside Bataan (73%). On educational attainment, 58% of the respondents are BSMarE graduates while 42% are BSMT graduates. The identified appealing aspects of seafaring career are: Money; Tax benefit ; Flexibility ; excitement with being reunited with family and travel around the world whereas the identified problems and difficulties of seafaring career are: Impact of Children; intensity of family relationship; communications ; routenary of activities ; social isolation; role displacement; transitions between ship and shore and health and sexuality The family particularly the wife and the children have an important role to play to successfully cope with seafaring lifestyle by making the necessary adjustments and deal with the periodic presences and absences of the seafarers through the following suggestions from the seafarers themselves : Trust (in the context of sexual fidelity); regular communications; support social networks; confident and competent wife and keeping busy. The paper ends with concluding remarks.

1 INTRODUCTION

The world of seafarers is truly the first known international workforce representing different nationalities and they work on a range of varied vessels, operating on different routes with diverse range of working conditions. Their common denominator is that their work necessitates prolonged separation from their home and families.

This is in addition to numerous trainings they have to take while on vacation as a result of the revised STCW Convention and Code collectively named as the 2010 Manila Amendments to ensure competency of the seafarers.

Seafaring is therefore seen not just an occupation but a way of life. A typical lifestyle characterized by a regular cycle of goodbyes and hellos , parting and reunions with family, coupled by a transition from a shore-based home environment to a unique work environment of the vessel . This lifestyle would definitely impact on seafarers, their partners, their family, home and work lives which are inextricably linked.

Forty one years ago , the Rochdale Report (1970) pointed to the role of the family in the seafarers decision to cut short their sea career and more recent research suggest that stands true today (Telegraph 1999 , SIRC 200). This points the significance of seafarer's family in sustaining the seafarer's participation as seagoing members of the maritime industry.

The problems of the retention and recruitment of well-trained seafarers has been highlighted as a matter of global concern both in the BIMCO Report 2005 and 2010 respectively. Attention to seafarers families and attempts to reduce the negative consequences of seafaring on family life may have considerable beneficial implications for the retention of seafarers. The seafarers from developing country like the Philippines with less economic power and weaker economic positions within the global seafarer's labor market typically have less favorable employment contracts and power work conditions than their western European counterparts. This condition would amplify the detrimental impact of seafaring work patterns on family life with the subsequent implication for both the seafarers' retention and seafarer's family well- being.

This paper aims to present seafaring lifestyle and work patterns to the well being of the seafarers and their family with exploration of some relevant literatures in this area, attractions and benefits of the seafaring career, the potential difficulties and problem areas and some coping strategies and sources of support from the family.

2 REVIEW OF RELATED LITERATURE AND STUDIES

A review of related literature and studies suggests that seafaring lifestyle and work patterns impact on both seafarers and their families. Managing the work and home has been found to be the largest cause of stress of seafarers (Horbulewiccz, 1978; Parker et al. 1997). In particular, transition between the work environment (vessel) and home are the most likely problematic periods for seafarers and their family as they have to adjust to new situation, work patterns, that impact on the health and well-being, causing in some cases of anxiety and depression. Communications may sometimes be slow, unreliable or very expensive. Seafarers' family or wives may also be socially isolated from their contemporaries and may not be offered specialized support and services frequently offered to military personnel and families (Brown-Decker, 1978). Seafarers may also

be similarly isolated without sufficient support networks, both while at sea and while ashore.

2.1 *Studies on Seafaring Work Patterns*

Seafaring work patterns will vary based on a number of factors such as nationality and rank, employers' policies, type of trade and route sailed.

A typical length of contract for a Filipino rating is 9 months whereas senior western officers (British) typically work contracts between 3-4 months (SIRC, 1999) Further seafarers form the Philippines may take no more than 2 months leave before returning to sea for periods of 9 months or more, whereas senior western officers may enjoy "back to back status "working 4 months on and then having a corresponding 4 month leave period with pay. Employment contracts may differ ranging from permanent employment with paid vacation leave and other associated contracts to single contracts with no income during the leave periods and no assurance of employment when the seafarer wishes to return to sea. However, regardless of these considerable differences, all seafarers share the same situation. Their seafaring career takes them away from home and their family and loved ones for appreciable period of time.

There is also irregularity in work schedules and uncertainties surrounding anticipated dates for joining and leaving the vessels. These are due to unforeseen events like finding relief for seafarers who are due on leave for personal situations such as birth, death and sickness etc. The seafaring schedules involve a constant process of change, readjustment and transition, making the work patterns harder to manage for the seafarers, their partners and families.

2.2 *Studies on Health Research*

Most of the researches on seafarers were focus on accidents and injuries (Hansen , 1996; ILO/WHO ,1993; Mayhew, 1990) or occupational physical illness (Hansen et al, 1996; Nilsson et al., 1997) and not much on the psychological and emotional well being of the seafarers except on studies conducted by the Seafarers and International Research Center (Bloor et al. 2000 ; Lane , 2002 ; Thompson 2002)

However, physical health is more simply the prevention of injury or disease , which is very much reflected by the seafarers' annual medical examination which includes physical check-ups (sight, hearing etc) and laboratory tests (diabetes , drugs , HIV etc) and explored psychological tests in a limited way {MCA, 1999}

Seafaring is a psychologically demanding occupation that requires long work of hours in socially isolated conditions. Separation from family and home is one of the most significant factors contributing to stress (Southerland and Finn, 1998). There are three main psychological problems identified among seafarers: loneliness, homesickness and "burn out" syndrome "which were primarily caused by long periods away from home, the decreased number of seafarers per ship and by increased automation (Aagterberg and Passcschier, 1998).

Investigations into suicide at sea have identified the role of marital and family problems that resulted to symptoms of depression and severe mental illness with some having disappeared at sea and thought to have taken their own lives by jumping over board (Roberts, 1998). Seafarers find their periodic absence from home problematic. Likewise transition from home to work (vessel) is reported to be a source of stress among seafarers (Parker et al., 1997)

2.3 *Studies on Ship-Shore Communication Opportunities*

Studies have proven that opportunities to communicate ship-shore can potentially have a considerable impact on the experience of separation for seafarers and their partners and family. Regular contact is crucial to maintain relationships with the family and shore-based life (Davies and Parfelt, 1998) with a reduced frequency of contact potentially leading to relationship decline and eventual breakdown (Argyle, 1990). Contact with home can be particularly important at times of illness or health problems in the family at which stress levels at sea can rise dramatically (Parker et al 1997).

It has been found that in general, the shipping industry has been very slow to utilize computers and telecommunications facilities particularly on board the vessels (Davies and Parfelt, 1998). Research in seafarer's communications patterns and opportunities has shown that much of the ship-shore communications occurs via Inmarsat satellite communications services that are often prohibitively expensive (Davies and Parfelt, 1998). Although, email can significantly increase opportunities for communication at greatly reduced costs, however, email access to seafarers continues to be limited, often restricted to officers and is impeded by the fact that many seafarers may be computer illiterate and that family and friends ashore may not have access to email facilities. (Davies and Parfelt, 1998).

Developments in telephone technology mean that seafarers can phone home using mobile phones in national and international waters as long as ship is in port or within close range of land, however, for those on deep sea routes, this services is limited and access is also restricted due to cost.

2.4 *Studies on the Role of the Family as support system*

Dependence on intimate partnerships and immediate family as a support system may be particularly important for seafarers who are potentially an extremely socially isolated group. While on vacation, seafarers are often geographically removed from their work mates and even they have been closed, however, they may have different work patterns which result in few occasions where vacation periods overlap. Current crewing patterns and strategies often result to seafarers working with different nationalities each time they sail thus preventing or hindering the establishment of a work-place friendships and encouraging more of an on-board acquaintances. This isolation on board may be amplified by the military style or structure of the ship where officers and ratings and different departments (deck, engine etc) are separated both socially and physically (Forsyth and Bankston, 1984). Furthermore, the nature of seafaring work patterns may make it harder to initiate and maintain shore–based friendship networks resulting in isolation (Forsyth and Bankston, 1984). It is in this context that the importance of family and marital relationship may be amplified.

2.5 *Studies on Family Life and Its Impact on Work Performance and Safety*

While it may be argued that any person may bring with them their domestic problems and stresses at work, however, it is opined that for those who are occasionally absent like in the case of seafarers may take on more importance as a result of concurrent lack of opportunities to communicate with home along with limited opportunities for leisure and socialization at work place. This may amplify the impact of psychological well being and work performance of the seafarers. Indeed even when there are no perceived problems in family relations , the emotional deprivation associated with prolonged absence from partner and loved ones can lead to psychological deterioration and increased rate of emotional tension which in turn may lead to increase in stress, emotional alertness and aggression, threatening individual and workplace health and safety (Horbulewicz, 1978).

3 METHODOLOGY

This is a basic research driven by the curiosity or interest of the researcher thru a scientific exploratory study that makes use of the following methods of data collection: observation, a questionnaire, documentary analysis from internet and library research and structured interview with the 10 MAAP alumni aged 25- 30 (2 already married) and 5 MAAP maritime professionals. All the respondents who were teaching at MAAP and ASTC Training Center AY 2013-2014 were interviewed individually. An interview method was used as it provides:

"The opportunity for the researcher to probe deeply to uncover new clues, to open up new dimensions of a problem and to secure vivid, accurate, inclusive accounts that are based on personal experience." (Burgess, 1982:107).

The topic guide for the interview was drawn up based on the review of literature and studies and by pilot interviews to two seafarers and their wife. Topics covered included: the attractions or benefits and problems or difficulties associated with the seafaring work patterns, perceived impact on emotional and well-being, sources of support and coping strategies. The interview was conducted face to face either at MAAP staff house assigned to them, at the MAAP premises or at their house. The interviews were tape recorded and later transcribed prior to analysis. The respondents are aware that their participation is voluntary. All of the respondents were positive during the interview and felt the importance of the study.

Demographic details of respondents were collated by means of a questionnaire and verbatim quotes are also included from the interviews, providing a vivid account of how respondents think, talk and behave. The interview method resulted in a total of 17 interviews, 15 seafarers with all officers ranged in rank from operational to management level (captains and chief engineers) and the two wives of seafarers. Married duration varied from a few months to over thirty five years. The 15 seafarers are either single (8) or married (7) with dependent children, with adult children with the wife or from previous marriages and with no children.

The previous study had been validated with the seafarers on vacation who are teaching at MAAP this 2013-2014. The 19 MAAP alumni who are seafarers on vacation graduated the IMO Instructors' Course or IMO 6.09 at MAAP campus with the researcher as the keynote speaker and inducting officer on 22 November 2013. After their graduation ceremonies, they were requested to answer the questionnaires that have been explained and distributed to them and the same were retrieved after they have accomplished on the same day. The objective is to validate the insights, comments and concerns generated last year as regards various human security issues like: appealing aspects, problems and coping mechanisms on their seafaring work patterns and their suggested solutions.. The respondents who were interviewed last AY 2012-2013 were different from the respondents who answered the questionnaire this year November 2013- 2014.

4 RESULTS AND DISCUSSIONS

Of the 19 respondents, 53% (10) are married while 47% (9) are single. Majority of them are Catholics (84%) and most of them are from provinces outside Bataan (73%). On educational attainment, 58% of the respondents are BSMarE graduates while 42% are BSMT graduates.

Table 1.Profile of the Respondents

Profile		Frequency	Percent (%)
Civil Status:	Single	9	47
	Married	10	53
Religion:	Catholic	16	84
	Others	4	16
Province:	Bataan	5	26
	Manila	1	5
	Others	13	68
Degree Course	BSMT	8	42
	BSMarE	11	58
Rank on Board:	2E/1E	4	21
	OIC-EW/NW/3E	5	26
	2M/3M	7	37
	CHIEFMATE	2	11
	MASTER	1	5
PRC License:	2E/1E	8	42
	OIC-EW/NW/3E	5	26
	CHIEFMATE	5	26
	MASTER	1	5

4.1 *The Appealing Aspects of a Seafaring Career*

Table 2. Appealing Aspects of Seafaring Life

Aspects of Seafaring Life	Mean	SD	Remarks
1. Money (High Salary)	4.26	0.99	Agree
2. Tax benefit	4.11	0.94	Agree
3. Flexibility and autonomy seafarers experience in their work	3.89	0.76	Agree
4. Excitement and pleasure associated with seafarers' reunions with their partners and/or family	3.89	1.05	Agree
5. Travel the World	3.53	1.17	Agree
Average	3.94	0.98	Agree

From the perspective of the 19 maritime instructors, money still ranks as the most appealing aspect of seafaring career with mean of 4.26. This is followed by tax benefit, and then flexibility and autonomy at work and excitement and pleasure associated with reunion with partners and/or family. Travel is the least appealing aspect cited by the respondents.

4.1.1 *Money (Seafarers as the Breadwinner)*

To work at sea is certainly a personal choice made by the seafarer as the breadwinner. The most common reason working on board is Money. Salaries are significantly higher than one could possibly earn ashore. The following were also mentioned namely: financial security with the money that goes directly into the bank, a comfortable and nice house, a nice car, education for children, brothers , sisters and relatives , pay debts, a property at Kamaya Coast, decent lifestyle and able to buy whatever I want with no second thoughts. Everything is free on board - the accommodations, the food, laundry, toiletries etc. Tax benefits are also extended to seafarers and provided with risk pay if aboard the tanker vessel. Although one of the respondents says:

"I have saved more while at MAAP teaching rather than on board. I was able to buy a motorcycle "(Operational level officer)

Although still single, one of the reason cited was that while on board , families and relatives rely on him hence he had sent all his hard earned money to help relatives and family in the Province . Majority of the young respondents intend to sail after the teaching contract to secure the mortgage for the property and the house and to save enough money to start a family and business.

4.1.2 Benefits to Relationships
Married seafarers talked about the intense pleasure and happiness they experience when they returned home feeling excited and happy, using the term "honeymoon" to describe the relationship during reunion. For the single seafarers on the hand, the long absence encouraged them and their girlfriends to appreciate and value each other and their time together. As much as possible, they avoid arguments and consider everyday a holiday while they are on vacation. One even commented:

"It's the quality of time being together not the quantity or number of hours".

4.1.3 Other Benefits
Common popular responses are the "opportunity to travel and see the world for free".

Also mentioned are: trainings, airfare and hotel accommodations at the company's expense. The nature of work also provides them the flexibility and autonomy that they would not find on shore based employment. It also saves time from long drive of going to work and vise versa as they are within the confines of the workplace (vessel) unlike with shore-based employment, wherein you consume a lot of time going from home to the office and vise versa due to heavy traffic particularly in Manila On the other hand, it also provides the family some opportunities and freedom during the seafarer's absence. Two of the retired seafarers disclosed:

"My wife particularly took advantage to pursue her interests like higher studies and hobbies like cooking and business in my absence which are all good for the family."

" I am happy with my wife that while being left alone , she has been skilled in managing household and family issues and events from the routines of household bills and finances to unexpected occurrences or other problems at home and had saved me from all worries "

4.2 Problems and Difficulties of Seafaring career
Table 3. Problems and Difficulties of Seafaring Career

What are the Difficulties and Problems of Seafarers?	Mean	SD	Remarks
1. Impact of children on the family life of seafarers	4.32	0.75	Agree
2. Intensity of couple or family relationship	4.26	0.65	Agree
3. Communications	4.16	0.90	Agree
4. Maintenance of Routenary Activities Upon return of the Seafarer	4.11	0.94	Agree
5. Social Isolation Aboard ship	4.05	0.78	Agree
6. Role Displacement and Redundancy in the family	3.95	0.97	Agree
7. Transitions between Ship and Shore	3.95	0.85	Agree
8. Health and Sexuality	3.88	0.78	Agree
Average	4.08	0.83	Agree

On problem associated with being a seafarer, the respondents provided the highest mean rating of 4.32 on the impact of children on the family life of seafarers and then followed by intensity of couple or family relationship with mean of 4.26. Communications ranked fourth. Maintenance of Routenary activities upon return of the seafarer is fifth. The respondents also believed that social isolation aboard ship, role displacement and redundancy in the family, transitions between ship and shore, and health and sexuality are difficulties experienced by the seafarers.

On the average, the respondents agreed with the problems enumerated having a mean value of 4.08.

4.2.1 Transitions between Ship and Shore
For married seafarers, transitions between ship and shore were reported to be the most difficult period of the work cycle as they struggled to adjust to shore life and their partners to being part of a "couple" again. It is as if the seafarers have two lives, single (on his own) while on board and married (with partner) while at home. (Baylon, 2012) , however on this study , it ranked number 7 .

Seafarers reported taking considerable time to unwind and recover from the stress associated with their shipboard life and many reported difficulties adjusting sleeping patterns to fit in with those of their families. As returning home for the seafarers was problematic so too was the return to work is also characterized by stress and sadness. Likewise the uncertainty on the exact date of departure and arrival particularly on postponements on date's added stress to the married seafarers who were awaiting the end of their trip with some anticipations and longing to be back home with the family soon. Due to dates being sometimes unrealizable, one of strategies being adopted by married seafarers is

" I usually therefore inform my family on my return date only when I had left the ship and was safely on dry land" (Management level officer)

4.2.2 *Communications*

Couples reported using a wide range of means of communication, from satellite phone calls to letters, mobile telephone call and email. . Frequent communication eased the transition at home and ensured that a seafarer had a role within the family whether at sea or ashore. However, rates of ownership of cell-net phones and personal computers were high. The financial cost associated with purchase and on-going uses of telecommunication equipment were significant and sometimes prohibited.

Opportunities to communicate are also sometimes seen problematic. In particular, when a seafarer heard about the difficulties and sad news at home while at sea, and he has no power to address or be of assistance, caused him emotional stress.

"When I learned that my family had a car accident, with my brother in critical condition, I was very much worried and stressed and decided to go home "

Likewise telephone communication in particularly could be emotionally upsetting due to the fact that the seafarer might miss the girlfriend or wife all the more. As a seafarer commented:

"The more I miss my girlfriend and family , hence when on board , I prefer not to call anyway , times run fast and sooner , I will be home and see them " (Operational Level Officer)

4.2.3 *Social Isolation Aboard ship*

Social isolation was found to be a significant issue for seafarers. Changes in crewing patterns meant that relationship at sea were often limited to 'on-board acquaintances', and, where relationships did developed, geographical separations and unsynchronized leave periods made the maintenance of these relationship problematic.

The family of seafarers also has little opportunity to visit the workplace and be involved in the workplace social events on ship unlike with shore-based work. Seafarer's irregular presences at home along with the cost of ship-shore communication appeared to make it difficult for some seafarers to sustain friendships at home.

Shipboard culture or male culture seemed to reflect the belief that one's problem should be kept to oneself. As one seafarer noted

"On board you are by your own good self and you do not have the time to talk about personal and emotional problems at sea "

Captains talked about his responsibility and social distance necessitated by the position prohibiting emotional disclosure. Most of the seafarers disclosed of their being dependent on their family or wive's friendship networks for social contact and support.

4.2.4 *Intensity of couple or family relationship*

When asked the seafarers who they would turn to if they felt down, all the married seafarers said they would turn to their wife. This was regardless of the nature of the problem (be it professional or personal) and whether the problem occurred while at sea or ashore.

The single seafarers on the other hand made mention of their family members (mother, , brother, sisters or father) or best friends .Seafarers seems to be very much dependent on their partners , family and loved ones to provide emotional support and maintain social networks . Such dependence could leave seafarers particularly vulnerable should the relationship breakdown.

4.2.5 *Impact of children on the family life of seafarers*

Seafarers may be fathers and husbands and their wives may have dual roles of both a wife and a mother. The presence of children would have an impact on the family life of the seafarers

For seafarers, children could make separations more manageable as children were a source of company for the wife and a means of 'keeping her busy' and hence making time pass more quickly. Seafarers reported difficulties leaving children for long periods and sometimes found the changes in their children upon their return as young children often did not recognized their seafaring parent and this could be upsetting for families. The seafarers also missed significant events like birthdays, first communion, school activities etc. One of the retired married seafarer shared:

"During my early married years while still active in sailing , I felt my child's indifference and seems not affected by my leaving, hence made me decide to work while my child was still growing " (Management Level officer)

On the other hand, most of the married seafarers said that:

"Once the children are grown, they could managed well with the lifestyle and accepted my continuous absence and presence, this is probably attributed to the fact that the children were used to it and it was what they knew and accustomed to , hence my family has already cope up with this particular kind of lifestyle".

4.2.6 *Role Displacement and Redundancy in the family*

A significant issue for seafarer was their sense of redundancy ((unnecessary and unwanted) upon their return home. Most of the seafarers spoke of the importance of having a partner who was independent and capable of managing the home and family in their absence. A retired seafarer shared:

"My wife characteristically was not independent before our marriage, but she eventually learned to be as she has adjusted to the realities of the seafaring lifestyle"

However those very abilities that allowed the seafarers to work aboard ship without undue worry about home often also resulted in problem for the seafarers when they returned home on vacation. As one of the retired seafarer complained: "Too much ability from my wife who doesn't call for my help. Sometimes left me feeling displaced, unimportant and unnecessary both practically and emotionally upon my return home "(Management level officer)

Upon return at home, household routines were complicated and involved several people in the routines who are unknown to them or activities with which they were unfamiliar. Despite their best efforts, seafarers often found it difficult to participate in the day to day events. Hence, sometimes felt to be an outsider at home and felt that the family will still survive even in their absence.

4.2.7 *Maintenance of Routenary Activities Upon return of the Seafarer*

One seafarer said "life must go on ". Keeping busy was an often used strategy to manage the seafarer's absence and to avoid feelings of loneliness and loss. However keeping busy would require routenary activities that require commitments and could not abruptly be cancelled upon return of the seafarer. The wife is also aware that such activities need to be continued so they could cope up with their husband's next call of duty and be of help to the family once the seafarer so decides to discontinue working at sea.

While the established routines was important to the well being of the seafarers family, however the continuation of such practices could leave the seafarers feeling that his return at home was unimportant and his presence at home was irrelevant to the day to day existence and well being of his wife and family, and felt that his worth and position in the family was related solely to his ability to bring in the support money.

Research with wives of those in the military has found that as naval wives becomes more experience at coping alone, problems move from coping with partner absence to attempting to reintegrate them into household upon their return (Chandler 1991). The mot practices wives in Chandler's study reported pretending to be helpless when their husbands were home in order to minimize tension caused by husband's feeling emasculated by their wives competence. Another research reflects a dilemma where husbands wanted wives to be independent and good managers but if they become

to self reliant, men felt that they were loved only for their pay packets (Tunstall 1962)

4.2.8 *Health and Sexuality*

Contrary to popular images of seafarers as sexually promiscuous (a girl in every port), however, previous researches found that in general majority are monogamous , resigned to the long period of sexual abstinence associated with their sea voyages and are faithful to partners at home. Most of the seafarers did not feel that pro-longed separations affected their sexual relationship with partners upon their return home , although a small number reported difficulties and sexual frustration associated with long separations and re-establishing intimacy and that the high work loads and work- related stress sometimes temporarily affected their sexual function (Thomas, 2003). In general, most of the married seafarers in this study are one in saying that:

"Sexual abstinence is considered an inevitable aspect of the seafaring lifestyle that has to be accepted"

4.3 *Successfully Coping with a Seafaring Lifestyle*

The family particularly the wife and the children have an important role to play to successfully cope with seafaring lifestyle by making the necessary adjustments and deal with the periodic presences and absences of the seafarers.

4.3.1 *Trust (also in the context of sexual fidelity)*

Trust was regarded as vital to successfully managing the seafaring lifestyle. This is referred to in the context of sexual monogamy. Trust is also extended to include a belief or faith that the wife and family could successfully manage home while the seafarer is at sea. This includes the management of the household financial affairs to allow the seafarer to work aboard ship without the burden of anxieties and concern over other people and events at home.

4.3.2 *Communication*

Communication between ship and shore is very important in managing separations and reunions. Regular communications allowed the seafarers and their partners/family to remain emotionally close and be updated of good news and happenings at home. It also allowed the seafarers to continue to participate in family decisions and events while at sea. Modern communication technologies like email are extremely positive and the arrival of regular post on board continued to be a matter of importance. Frequent communications could help bridge emotional gaps and provide the seafarer with sense of belongingness that he still continues to actively participate in the family affairs while at sea.

4.3.3 Support Social Networks

The difficulties associated with the seafarer absence are both practical (household problems) and emotional (loneliness, anxieties and depression). In both these domains, the existence of local support network of family and friends could be vital both for the family and the seafarer who could continue his work at sea without worrying about his family's well being.

Contact with other seafaring families would be of help. A good example is a Seafarers Association Wives of the Philippine Inc (SWAPI), which the wife of seafarers may opt to join. It would be an opportunity for the wife to meet up with other women with seafaring partners who shared like experiences and lifestyles and would understand particular issues that faced seafaring families. This would also be good for children as they share similar situations wherein the Dad worked on a ship and therefore promote understanding that the absence of the father, being a seafarer was normal.

Support of the family would also involve a conscious effort to protect the seafarer's well-being The family particularly the wife as recipients of problems by the seafarers both when ashore and at sea, should provide emotional support and comfort for the seafarers personal and professional problem alike . The family must take considerable efforts to protect the seafarer from news or events that might induce negative emotions or feelings while at sea. If ever problems arise, the family must jointly try to sort all things out, until the matters have been addressed and the crisis had passed and to just report belatedly once everything is fine. Anyway, the seafarer can not do anything and has no power to help as he is at sea.

4.3.4 Confident and Competent Wife

The wife must not show any signs of signs of unsatisfaction or discontentment. She must avoid nagging or complaining at all times for the seafarer to be at peace and happy as work at sea is already problematic and full of tensions due to the responsibility of keeping the ship sea worthy at all times for the safety of life and cargo. The wife must also be independent with an ability to cope up with day to day events coupled with the ability to manage household, finances and family affairs alone. It would also be good to maintain her identity and interest outside their marital relationship.

Those with a family member (having a seafarer in the family like a father or brother for example), felt better able to cope with the lifestyle due to positive role models and an awareness of the potential problems associated with the occasional absence of the seafarer husband

4.3.5 Keeping Busy

This is a strategy to deal with the 'seafarers' absence. This may be done by involvement and immersion in domestic labor or increase social contacts with family and friends. Paid employment is also a great means to help time pass, getting out of the house and making social contacts. Although during the seafarer's vacation, women's work is seen as interfering with plans of the husband; the wife should take the imitative to file for a vacation leave while the husband is on vacation to take short breaks and holidays together to prevent the seafarer spending considerable portions of his vacation period alone.

Children also provide a source of company and therefore combat loneliness although their needs may be quite time– demanding and sometime all consuming for the mother to fulfill the need and to keep busy. Parental activities and school activities also provide extra sources of social contact for the woman while the seafarer is away. Children should do well in schools so that they have something good to report, that the seafarer Dad will be proud of.

4.3.6 Coping with Seafaring Lifestyle

Table 4. Coping with Seafaring Lifestyle

Aspects of Seafaring Life	Mean	SD	Remarks
1. Communication	4.42	0.61	Agree
2. Trust (also in the context of sexual fidelity)	4.32	0.82	Agree
3. Confident and Competent Wife	4.26	0.73	Agree
4. Keeping Busy	4.21	0.54	Agree
5. Coping with Tension from the Seafarer	4.21	0.63	Agree
6. Support Social Networks	3.95	1.03	Agree
Average	4.23	0.73	Agree

While communication is one of the difficult aspects of seafaring lifestyle, it is also the number one (1) coping mechanism of seafarers in dealing with problems and difficulties. Moreover, the respondents agreed that trust, also in the context of sexual fidelity, is important for seafarers. They also believed that confident and competent wife is a key in coping with seafaring lifestyle. Further, keeping busy, coping with tension from the seafarer and having support social networks help in dealing with the lifestyle of seafarers.

It is interesting to note that the average level of agreement on the coping mechanisms on seafaring lifestyle is 4.23, which indicate positivity on the part of the respondents who are seafarers themselves.

In the context of the work environment, the seafarer being in position of command and authority carries with him at home his experience and rank.

Upon return to the family with wife used to being in charge of the family and household , may experience tension from an officer husband bringing with him the ship-board –status at home, making the family feel as junior officers or ratings. Although the wife is independent and capable in his absence , she must not deter responsibility to the husband upon his return home. Hence at home, the woman should take the second place while the seafarer is having a vacation

The family must adjust to the seafarer's presence at home. Wives must took steps to accommodate their husband within household routines and to integrate them into the day to day running of the home and the family

5 CONCLUSIONS

The result of this study is not to make a generalization but rather , this was an exploratory study of an under-researched area that attempts to explore the views and experiences of the seafarers which somehow had provided valuable insights on the various human security issues on the seafaring work patterns, the various challenges on maritime and education (MET) into the lives of the seafarers and their families and the suggested solutions to cope up with the seafaring lifestyle thru the support of the family.

ACKNOWLEDGEMENT

The National Research Council of the Philippines for the funding of the 6-month MAAP Research Project on challenges, issues and concerns in seafaring (March -September 2012)

REFERENCES

Aagterberg, G. and Passchier J (1998) "Stress among seamen", Psychological Reports, 83: 708-710.

Agryle, M (1990) "Social Relationship 'M, Hewestone, W Stoebe, J Codol and G Stephenson (eds) Introduction to Social Psychology. Oxford: Blackwells

Baylon, AM and Santos, EMR (2012) " Impact of the Seafaring Lifestyle to the Family : MAAP Seafarers Viewpoints " presented during the 2012 World Research Congress held at Marco Polo Plaza Hotel , Cebu City Philippines , August 22- 25,2012

BIMCP /ISF (2010) 2010 Manpower Updates: The world-wide Demand for the Supply of Seafarers – April 2010

Bloor, M, Thomas, M and Lane, AD (2000) "Health risks in the global shipping industry: an overview" Health, Risk and Society 2: 329-340

Burgess, R.G. (1982) "The unstructured interview as a conversation in R Burgess (ed) Field Research: a Sourcebook and Field Manual. London: Allen and Unwin.

Brown-Decker, K. (1978) 'coping with sea duty: problems encountered and resources utilized during periods of family separation in Military Families Adaptation to Change. E. J

Hunter and D Stephens Nice (Eds) New York and London: Praeger Publishers, Praeger Special Studies.

Chandler J. (1991) Women without Husbands: An Exploration of the Margins of Marriage. London: Macmillan.

Davies, A.J. and Parfett, M.C. (1998) Seafarers and the Internet: Email and Seafarers' welfare. Cardiff: Seafarers International research Centre, Cardiff

Forsyth, C. J. and Bankston, W.B. (1984) 'the social psychological consequence of a life at sea: a causal model', Maritime Policy and management, 11 (2):123-134 retrieved on July 23, 2014 http://goo.gl/5oKzZU.

Hansen, H. L (1996) 'Surveillance of deaths on board Danish merchant ships 1986-93: implications for prevention, Occupational & Environmental Medicine, 53 (4): 269-75.

Hansen, H. L , Hansen H .K., Andersen P.L. (1996) ' Incidence and relative risk for hepatitis A, hepatitis B and tuberculosis and occurrence of malaria among merchant seamen', Scandinavian Journal of Infestations Disease, 28: 107-110.

Horbulewicz, J. (1978) 'The parameters of the psychological autonomy of industrial trawler crews' in P. Frickie (Ed) Seafarer and Community: Towards a Social Understanding of Seafaring. London: crooms Helm.

ILO/WHO Committee on the Health of Seafarers (1993) Occupational Accidents among Seafarers Resulting in Personal Injuries , Damage to their General Health and Fatalities , Working paper No 1 . Geneva: ILO.

Lane, A.D. (Ed) (2002) the Global Seafarer. Geneva: International labour Organization Maritime & Coastguard Agency (1999) MC/18/3/069/17. Southampton. UK

Nilsson , R. , Nordliner , R , Hogstedt, B., Karlsson, A and Jarvholm, B. (1997)' Symptoms , lung and liver function , blood counts and genotoxic effects in coastal tanker crews', International Archives of Occupational and Environmental Health, 69: 392-398.

Parker, A.W. Hubinger. L. M., Green, S., Sargaent, L. and Boyd, R. (1997 A Survey of the Health, Stress and Fatigue of Australian Seafarers. Canberra: Australian maritime Safety Authority.)

Rochdale Report (1970), Committee of Inquiry into Shipping CMND 4337 British parliamentary papers, London: HMSO. " Roberts, S... (1998) Occupational Mortality among Merchant Seafarers in the British, Singapore and Hongkong Fleets, Cardiff: Seafarers International Research Centre, Cardiff University.

SIRC (1999) The Impact on Seafarers 'Living and Working Conditions from Changes in the Structure of the Shipping Industry. Report prepared for the International Labour Organization by the Seafarers International Research Centre, Cardiff University, UK .retrieved on July 23, 2014 from http://goo.gl/8MqsSv.

Sutherland, V. J and Flin, R.. (1989) 'Stress at sea: a review of working conditions in the offshore oil and fishing industries', Work and Stress, 3 (3): 269-285.

Telegraph (1999)' Seafarer Suicide Linked to Fatigue' NUMAST Telegraph, July 1999

Thomas , M (2003) Lost At Sea and Lost at Home : The Predicament of the Seafaring Families, Seafarers International Research Centre, Cardiff University, UK pp118

Thompson, L. and Walker, A. (1989)' Gender in families: women and men in marriage, work and parenthood', Journal of Marriage and the Family 51: 845-71.

Tunstall J. (1962) the Fishermen. London: MacGibbon and Kee.

Plights and Concerns of Filipino Seafarers on Board Vessels Traversing Horn of Africa and Gulf of Aden: AMOSUP and other Stakeholders Responses

A.M. Baylon, E.M.R. Santos & J.W. Vergara
Maritime Academy of Asia and the Pacific, Bataan, Philippines

ABSTRACT: This paper presents the suffering of Filipino seafarers and their family as a result of Somali piracy, with an aim to better prepare seafarers for this unfortunate eventuality should it befall them. With the presentations of latest piracy case data and reports, the paper aims to renew interest from the maritime stakeholders (government, industry and international organization) involved in maritime trade in the issue of violence against the hijacked seafarers particularly on their plight and concerns while on board the vessels traversing the horn of Africa and Gulf of Aden. The Association of Marine Officers and Seamen's Union of the Philippines (AMOSUP) and some international measures have also been cited as terms of reference for all concerned to address the piracy threat for the protection and promotion of the seafarers' and their family's welfare. Everybody is encouraged to support and reflect on various possible ways or means in providing care for the seafarers before, during and after the incident of piracy.

1 INTRODUCTION

It is an accepted fact that over 90% of world trade is carried by the shipping industry and the seafarers are playing a significant role in moving products around the world. Also, 25% of these seafarers worldwide are Filipinos manning international vessels of various types plying international routes to carry products from port to port. Over the last 40 years or so, the Philippines has evolved into the leading maritime providing nation of the world. Mr. Gerardo Borromeo of Philippine Transmarine Carriers, Inc (PTC) in his speech during the 2014 Filipino National Seafarers Convention emphasized, "*a Filipino seafarer is referred to as a global maritime professional enjoying the preference of many leading maritime nations for many reasons such as fortitude, commitment, attitude, values and service orientation/the can do promise/and the perpetual smile.*" He further stated that Filipino seafarers are given the opportunity to continue to be of service in this lucrative yet challenging industry with quality and competency coupled with the right values and attitude.

While Filipino seafarers are clearly doing a significant role for the world trade as well as for the Philippine economy because of their substantial remittances of about 4.3 Billion USD a year, it cannot be denied that they are faced with challenges, difficulties and hardships ranging from fatigue, anxiety, and communication to risks associated with passing through piracy-infested routes such as Somalia and the Gulf of Aden.

Piracy and armed robbery is one of the foremost threats facing the international shipping community of Somalia, Gulf of Aden and other international shipping routes. Recognizing the economic cost and most especially human cost of piracy, the international maritime community– government, flag state, ship owners, and others have collaborated, debated and coordinated to develop plans and programs to counteract violence against seafarers. In fact, these challenges are part of the discussions during international maritime conferences, conventions, meetings and other fora. Though many measures such as naval escorting of ships, installation of citadel as refuge on ship, best management practices/guidelines were undertaken to combat piracy, seafarers remain concerned about their welfare and protection for themselves and respective loved ones, especially as pirates become more violent and heavily armed frequently with automatic weapons, rocket-propelled grenades (RPGs), and explosives.

2 PIRACY CASE REPORTS

Preparing The Gulf of Aden, off the Coast of Somalia, remains one of the most notorious pirate-infested waters in modern times. Somali piracy along with the associated hijackings has prevailed for a number of years. Not only did the Somali pirates attack an unprecedented number of vessels they also continue to threaten a vast sea area as well as some of the busiest trade and energy routes. There are a total of 80 crew fatalities from 2007-2012 (Jan-May).

There are a total of 210 ships from 2007-2012 (Jan-May) attacked by Somali pirates. Attacks leveled off in 2010, after having experienced a significant jump in 2008. While the Gulf of Aden remains a high-threat area, attacks fell there in 2009-2010 due to the presence of foreign navies, while incidents are occurring with increasing frequency off Somalia's eastern/southern coast. Attacks by Somali pirates increased only minutely in 2010, with the last significant jump occurring in 2008. A total of 219 attacks were attributed to Somali pirates in 2010, compared to 215 attacks in 2009 (the number in 2008 was 551 after 44 in 2007). International monitoring group Ecoterra has recorded at least 551 incidents of sea jacking since 2008. It is in 2008, that piracy of the coast of Somalia escalated to levels previously unheard in modern times.

Many more attacks remained undocumented, with vessels disappearing off the radar, never to be found again.

Table 1 shows the threat of Somali piracy is still present; as long as this threat remains, the seafarer will always feel at risk and under stress while sailing through these high risk waters. The numbers shown for 2012 cover all incidents through Jan 2008- May 2012) as summarized on Table 1:

Table 1. Somali Piracy Ships Attacked/Hijacked 2008-2012

	2008	2009	2010	2011	2012 (Jan-May)	Total
Ships Attacked	511	215	219	237	61	843
Ships Hijacked	42	47	49	28	12	178
% success	38%	22%	22%	12%	20%	21%
Hostages	815	867	1016	470	188	3356

From the reports submitted to the IMB Flag States, ship owners or operators, seafarers and by the Maritime Piracy Humanitarian Response Program (MPHRP) that details the experience aboard 23 of the 77 vessels hijacked in 2010 and 2011, revealed that all of the captive seafarers were subject to violations of their basic human rights, the most significant being the right to life, liberty and security of persons. Seafarers aboard these 23 vessels were held hostage for an average of six and half months (196 days). The International Transport Workers' Federation (ITF), the largest alliance of unions in the transport sector, had sounded the alarm about the use of hostages as "*anti-piracy attack shields*". From 2007 to 2011, the fatalities among hostages have reached to 62, and increasing, notes the ITF.

As of July 16, 2012, the International Maritime Bureau (IMB) Piracy Reporting Centre reported a worldwide piracy and armed robbery incidents of 180 attacks and 20 hijackings. From this, 69 incidents, 13 hijackings with 212 hostages are reported in Somalia. Also, IMB reported 11 vessels and 174 hostages currently held by Somalia pirates. The International Maritime Bureau (IMB) reported 769 Filipino seafarers who have fallen in the hands of pirates' between 2006-2011. But one no longer needs to look farther than 2012 to understand the gravity of Somali piracy.

Based on reports in 2012, there are still 50 Filipinos who are held captives of the pirates. These people, says Vic, in his early 40s and a survivor of piracy and had been in the "*enemy's nest*" for more than five months, are the most insane peoples that they have met. "*Either they will hit us, or wake us up in the middle of the night, to torture us; the foods that they are giving us are inedible, but we eat it to be able to survive,*" he said in an afternoon chat with Charia Semabrano of ABS CBN in a law firm in Quezon City in November 2010. Because of the bitter and painful experience, he refused to elaborate and further said "*I am now going home, to our province, to take plenty of rest. When I got here, after our abduction, I was so sick, my mind is not working fine, and I am like crazy! I was very angry, we're all angry with them (pirates). I am thankful that I got home here, safe. But I don't want to talk about it again,*"

Filipino Seaman Luther Amaro was once chief mate of a vessel that often took the 3- to 5-day voyage across the Gulf of Aden. He recalls the fear that would grip their hearts as they approached, and the haunting distress calls that would bombard their ship's VHF radio. "*They would say 'mayday, mayday....,'*" recounts Amaro. "*We are under pirate attack.' Sometimes, they would say, 'they are firing their guns!' and the coordinates they would give are just a few miles ahead of us.*" When asked if he would sail to the Gulf of Aden again, Amaro replies, "*Yes. When we became sailors, we knew there were risks involved. I've accepted these risks.*"

Engr. Jun Valmonte, president of a manning agency that deploys sailors to the various ships at sea, explained that nearly all of the seamen he has encountered share the sentiments of Amaro. Even sailors-in-training regard passage into the Gulf of Aden as potential bragging rights. Ship owners themselves prefer to risk the Gulf of Aden, rather than take the long way round to Europe, which would add at least 10 more sailing days to the voyage. Valmonte commutes a ship's consumption to amount to $20,000 dollars per day. "*The Gulf of*

Aden is the most economical route. It's also the correct route, the fastest route. You can avoid it, but that would add 2 to 3 thousand miles on your expense". When asked whether it would consider imposing a ban on Filipino sailors from entering the Gulf of Aden, considering the circumstances, the Department of Foreign Affairs merely answered *"what is important is we prepare our seamen for the risks they have to take. All professions have risks, and our seamen know this. Manning companies and shipping lines should train their crew and prepare them for such incidents."*

Instead of discouraging sailors from entering the Gulf of Aden, the Philippine Overseas Employment Agency has ordered all sea-based employers to pay double the amount to any sailor who enters "pirate alley." Pirate Alley is a colloquialism for the area immediately surrounding and including the Gulf of Aden, a small body of water that begins at the foot of the Red Sea at the Bab el-Mandeb Strait (Arabic for Gates of Grief), and ends at the Arabian Sea and the Indian Ocean. This, more than pride and honor, convinces the Philippines's sailors to push on. Ship captains would say *"Okay, after this port, we enter the Gulf of Aden. All those who do not wish to enter, can go home at the shipping line's expense. Those who want to continue, get twice the normal pay."*

Seldom of the Filipino seafarers would chose to go home. They even like the Gulf of Aden more because they get paid more. *"So, with no one to give a definitive order to stop sailing onto the Gulf of Aden, ships now come up with their own ways to defend themselves from pirates."* Countries like France, the United States, Italy, Japan, Korea, and Great Britain have sent armed naval vessels to accompany any ship to and back from the Gulf of Aden. Ships who wish to seek their protection only need to meet the designated schedule of the convoy's departure, and they are assured safe passage.

But not all ships arrive in time for the convoy, and many are still left to fend for themselves. Some crew members have established a *"panic room"* within their ships—a room stocked with food, water, and communication devices, where the crew can lock themselves in the moment pirates reach their ship. Panic rooms are usually situated near the engine, so the crew members can disable the vessel and prevent the pirates from taking them away. Seaman Amaro, as ship captain on his next voyage, showed ABS-CBN photographs of his vessel completely wrapped in barbed wire and explained *"The point is to slow the pirates down while we radio for rescue or just to injure them a bit."*

But the torture and the unpalatable, if not "sickening" meals are not the only things that are associated with pirates. Lately, the pirates had been killing their hostages, for two apparent reasons: the slow release of the ransom money; and as self-defense against possible attacks from the armed forces of the countries of origin of their hostages and against litigation.

The Philippines had its first casualty on January 26, 2012). He is identified as Farolito Vallega, 48, a Filipino crew of MV Beluga Navigation (German cargo ship). He was shot dead by Somali pirates, of the coast of Seychelles, about 1,500 kilometers or 932 miles off the Continent of Africa.

The victim seafarers attacked by pirates face physical dangers from weapon fire and explosives, smoke or chemical inhalation from pirates as well as abuse or violence if pirates gain access. They are also psychologically endangered with the experience of threat during attack, fear and uncertainty about success in attack and potential exposure to combat experience of exchanged gunfire (Hurlburt, et.al, 2011).

Physical Abuse, as defined in the '*Human Cost of Piracy Report*, includes deprivation of food and water, beating (often with the butt of a gun), shooting at hostages with water cannons, locking hostages in ship's freezer, tying hostages up on deck exposed to scorching sun and hanging hostages by their feet submerged in the sea. Some **physical abuse** reported were: severe bruising and swelling of arms and legs, extreme pain in back and spine, severe sun burn, broken teeth and eye damage to such a degree that surgery was required upon release, humiliation, giving up on life and the desire to commit suicide. Nearly all the reports indicate that the majority of the physical abuse towards the hostages was due to the pirates' lack of understanding or misperceptions of operations. The abuse was also used to show the hostages that the pirates had absolute power over the hostages' lives. The abuse increased when pirate guards and negotiators mistrusted or became paranoid about hostages' actions. This most often occurred when guards or negotiators suspected hostages of trying to communicate with the outside world without their permission, of providing inaccurate information on the quantities of fuel oil and fresh water remaining on board, or when the hostages had difficulties starting the main engines due to lack of pump-able fuel. Increases in abuse also occurred when negotiations appeared to be breaking down or had come to a complete halt. The reports suggest the best way for hostages to avoid abuse was to avoid pirates and comply with their orders. One example of an extreme abuse or torture was the report on one hostage that was hung over the side of the vessel by his legs so that his head was just above the water line. On another vessel, the pirates forced two hostages to stand in a cargo net with the intention of lowering them into the water using the crane. Luckily the crane did not work and the crew was spared. In yet another incident, hostages were forced at gun point to choose and appoint who would go underwater to clean the propeller. Hostages were regularly tied up and made to stand in the sun and on one occasion made to strip and stand in the ships' freezer compartments, kept at -18°C, for 40 minutes. There are also incidents that describe the ignorance and inflexibility in the pirates' attitudes toward the hostages. In one case, the pirates required the vessel to be moved to another anchorage for a short duration. When informed that there would not be sufficient fuel to return, the hostages were beaten and then forced to operate the vessel. On the return journey when there was insufficient pump-able fuel the engines would not start. The pirates severely beat a few hostages and, out of fear of losing their lives, the hostages finally opened the fuel tanks and manually scooped out, collected and transferred the fuel to start the engines. On another vessel, pirates demanded that the hostages run the diesel generators using crude oil. When the hostages refused and tried to explain that it would spoil the generators, they were slapped, kicked, tied up and made to stand in the sun, stripped and beaten with wooden sticks and iron rods. In this incident, the pirates also squeezed the hostages' fingers using pliers.

Psychological Abuse, as defined in the 'Human Cost of Piracy Report', include firing weapons as an intimidation tactic, solitary confinement, calling family members while threatening hostages, parading hostages naked around the vessel, and taking hostages ashore to see the hostages' supposed graves. Reports say that the psychological abuses appeared to increase when the pirates or negotiators felt that the progress of the negotiations was not going well or when it had stalled. The pirates repeatedly told the hostages that their office and families had abandoned them and that no one cared for them. The pirates then often tried to apply psychological distress to the families of hostages' who were told that their loved ones would be killed or beaten. Weapons were also fired close to or directed at a hostage's head and then fired on an empty chamber. Seafarers report that this terrified them and served as a constant reminder that they might not survive the experience. There were reports in which crews were divided and some taken ashore, sometimes for nearly a month. There was an incident in which a senior crewmember was isolated for nearly a month while the remaining crew was informed that he was dead. There were a couple of occasions when certain senior ranks were relieved of their duties and junior ranks were made to replace them. One report also suggests that as part of the pirate tactics, a senior pirate or, as stated in the report, a "psychologist" would come onboard the vessel and talk to certain hostages for hours to try to make them turn on their fellow crewmembers. In that case, one hostage succumbed to the pirate's pressure and this created considerable friction and misunderstanding among the hostages that resulted in extreme physical and psychological abuse to the remaining crewmembers.

Hygiene, Living, and Sanitary Conditions on the Vessels: All reports indicated that the living, hygiene and sanitary conditions onboard the vessel deteriorated as the length of captivity increased. This exacerbated the hostages' challenge of maintaining hope and morale. Pirates ransacked and looted the cabins early on after the hijacking, with subsequent groups of pirate guards continuing to steal goods until the vessel was stripped of nearly all amenities. Most of the hostages' possessions, including their clothing, were stolen. This led to the seafarers having to survive their period of captivity with one or two sets of clothes and no facilities for washing except sometimes in salt water and on deck.

The pirate's favorite location to keep hostages is the bridge. Some reports also suggest that hostages were made to share cabins (four to a cabin) or on some occasions the engine room. The overall cleanliness and hygiene in the accommodation rapidly deteriorated. Pirates spat and left food around and did not use toilets properly.

Most reports indicated the pirates did not allow the hostages to clean the accommodation regularly. The hostages were not allowed access into all areas of the accommodation. The conditions in which they were forced to live for months grew unhygienic and were completely different from what they were used to normally at sea.

The lack of hygiene or adequate food and exercise contributed to a number of skin and health issues, both major and minor, which remained untreated for the duration of their captivity. A report indicates that some of the pirate guards appeared to have skin diseases and had open wounds which looked unpleasant and infectious.

Normal ship routines and watches were not allowed by the pirates. Bridge and engine room watches were not maintained except when essential and required. At times the engine room crewmembers were forced to work on a 12-hours-on, 12-hours-off routine, during which time they were not allowed to leave the engine room, even to use the toilets, forcing the hostages to use the sludge tanks to relieve themselves.

The hostages were deprived of basic freedom of movement. All movement was subject to the permission of the pirate guards and ensured that the hostages were totally dependent and subjugated to the pirates moods and wishes. Hostages even had to ask permission to use the toilets, the duration of which was also dictated by the pirates. While there were no restrictions on the number of times the hostages could use the toilets, there are indications that sometimes the pirate guards became "difficult". The bridge toilet remained one of the most frequently used because the bridge was the preferred location to keep hostages and it would sometimes clog up because the pirates threw cigarette butts and other items into it.

Purely depending on the availability of fresh water, and at the pirates' discretion, the hostages were allowed to shower. The frequency varied from once in two or three days or a week or a month, sometimes more. One report also indicated that in the initial six months of captivity the hostages were only allowed to shower twice due to lack of water. This only changed when the hostages had collected rain water to use. The time allowed to each hostage varied from between two to 15 minutes. The only saving grace was that privacy was provided during this time.

Food and Water: It appears from reports that the policy of the Somali pirates was to use ship stores until nearly depleted and after that to provide the vessel with a combination of meat (goats which were slaughtered onboard usually by the pirates and then cooked by ship's cook), rice, flour, cooking oil, sugar, beans and noodles. Sometimes the food provided was UN aid food and some pirates allowed the hostages to fish to supplement their meals. As the negotiations prolonged, the quantity of the food reduced resulting in rationing. The reports indicate that the hostages were allowed to cook and eat one to three times a day every day, depending on the availability of stores and the supplies received from shore which depended mainly on the weather and sea conditions. At times the food cooked was mainly comprised of meat, the vegetarian hostages remained hungry until they learned to adapt.

Reports indicate that at times the pirates used to eat in front of the hostages and waste a lot of food but not give them any. The meal times varied from being inconsistent to following a regular pattern. Additionally, the place where the meals were served also varied from the mess room, to the cabin (when hostages were kept isolated), to the bridge which appeared to be the preferred place to keep all the hostages.

Depending on the availability of fresh water, the utensils were allowed to be either cleaned in fresh water or first cleaned in salt water and then rinsed in fresh water. Over time, as the vessel remained under negotiation the hygiene of the galley deteriorated as the hostages were not allowed to clean the galley regularly. In one case as the ship was extremely low on fuel it was not possible for them to use the galley to cook. They had to improvise a kitchen on the poop deck where wood and other combustible materials from the accommodation (cabins, cabinets, bunks, etc.) were used as fuel for the fire. This resulted in accommodation being completely destroyed.

This report also indicated that the quantity and quality of the food was relatively good during the initial negotiation period. However, once negotiations broke down, the hostages were allowed only one meal a day of reduced quantity and quality, and fresh water was limited to one bottle per hostage each day. This led to stomach infections and other problems for nearly all the hostages. There were times when this caused the hostages to fight over the limited provisions with the stronger ones getting more. During the monsoons, as the seas became choppy, the supply boat was not always able to make the journey to the ship; in these situations the hostages had to go hungry and collect rain water for drinking

Survival Techniques used by Hostages: International convention prohibits seafarers from arming themselves with weapons, so standard self-defense measures include the sounding of alarms, the use of fire hoses as repellants, and occasionally, the use of scarecrows that look like soldiers.

Nearly all the reports indicated that the hostages used prayer, meditation, and reading books to pass their time and maintain their sanity. If allowed by the pirate guards, the hostages watched videos or played games. In most cases the hostages stayed together and avoided any confrontation with the pirate guards. On one vessel, all communication was conducted via a designated hostage to reduce the chance of miscommunication. In most cases, the hostages encouraged and gave hope to each other. Overall upon release the hostages were mentally tired, exhausted and stressed. Most of them had suffered significant weight loss.

None of the hostages felt sympathetic towards the pirates. There was only one case wherein the pirate guards asked one hostage to join them. There was one case in which, after the first few weeks in captivity, the hostages started confronting and abusing each other and fighting among themselves. The pirates used this to their advantage and the hostages of this vessel suffered extreme physical abuse.

Vessel Used as Mother Ship: In cases when the pirates forced the hostages to sail the vessel and use it as mother ship either all hostages were kept on the bridge or those required for navigation were kept on bridge while the rest were kept in the mess room. The vast majority of attacks are aimed at holding crews and vessels for ransom. Pirates attack all types of ships and are capable of operating farther from the coast than in the past due to the increased use of pirate mother ships. All factors that tend to fuel piracy are very pronounced in Somalia and will likely remain so in the near future.

The presence of foreign navies cannot be a comprehensive solution to the problem of Somali piracy. As the figures show, the pirates have adapted to the presence of the navies by using mother ships to shift more of their attacks to the Arabian Sea and the Indian Ocean, an area that because of its size is much more difficult to patrol.

3 PLIGHTS, CONCERNS AND EFFECTS OF PIRACY

All hostages were subjected to confinement and loss of privacy; loss of self-esteem and dignity; hunger and malnourishment; psychological stress and abuse; and the threat of physical violence.

Physical abuse towards hostages included slapping, punching or pushing in 50% of the cases.

Some 10% were extremely abused and had experienced torture. Nearly all hostages were affected psychologically and few seafarers were so affected by the captivity that they chose not to return to their seafaring profession.

Other plights and concerns are the stress faced for their lives and in having no regular contact with their families and not knowing how the families are coping in their absence including in some case, the lack of financial support during their long months in captivity, fears or the deterioration in standards of living of the hostages family members.

Pirate negotiators and security guards rarely differentiated between nationalities and ranks in their treatment of hostages. However, pirates appear to give Muslim hostages marginally better treatment regardless of their nationality.

Hurlburt, et. al. (2011), in the report "*Human Cost of Somali Piracy*", mentioned that both successful and unsuccessful attacks expose seafarers to dangerous experiences, with the potential for long-tern physical and psychological trauma. It was also noted in this reported that the trauma caused by violent crimes at sea adds significant stress to the jobs of seafarers transiting through pirated waters. Moreover, trauma to seafarers is undervalued and misunderstood, but have lasting negative implications both for seafarers and their families. Further, the report highlighted that families and others who depend on seafarers are faced with stress and fear from the time a seafarer enters high risk regions until the seafarers' returns home. Also, in case of hijacking, families may be subjected to psychological manipulation from pirates.

The previous government ban for seafarers to board ships transiting high-risk zones created oppositions from ship operators and manning agencies as this would significantly affect the recruitment industry. Seafarers themselves also opposed this move as their means of livelihood would be affected.

Compensation premium for seafarers' transversing Somalia and Gulf of Aden --- The doubling of hazard pay and benefits has elicited mixed reactions from Filipino seamen.

Filipino seafarers, despite the dangers of plying through Somalia and Gulf of Aden, continue to board ships considering the benefit of providing quality life for their families. They are hoping for an improved not from the Philippine government but

from a US-led coalition of 10 countries, including Russia, that is working to secure sea lanes beset by pirates off the Eastern African coast.

On the other hand, what makes the Filipino seafarers' lives more difficult is the **existence of the flags-of-convenience (FOCs)**. According to Entero (2011), most ocean-going vessels in operation are sailing under an FOC where the nationality of the ship owner is different from the nationality of the flag [it carries]. The ITF, in 1974, has given this working definition of an FOC: "*Where beneficial ownership and control of a vessel is found to lie elsewhere than in the country of the flag the vessel is flying, the vessel is considered as sailing under a flag of convenience.*" The FOC had been used by the ship-owners to enable them to register their fleets at the very low cost, to operate tax-free and to employ cheap labor to man their usually substandard ships. Based on the ITF list, there are 32 countries being used as FOCs: Antigua and Barbuda; Bahamas; Barbados; Belize; Bermuda (UK); Bolivia; Burma (Myanmar); Cambodia (Kampuchea); Cayman Islands; Comoros; Cyprus; Equatorial Guinea; French International Ship Register (FIS); German International Ship Register (GIS); Georgia; Gibraltar (UK); Honduras; Jamaica; Lebanon; Liberia; Malta; Marshall Islands (USA); Mauritius; Mongolia; the Netherlands; Antilles; North Korea; Panama; São Tome and Príncipe; St Vincent; Sri Lanka (Laos); Tonga; and Vanuatu.

The burden of the piracy costs, the other costs of shipping operations and the effects of the international financial crunch in 2007–2008, is often reflected in the Filipino seafarers' salaries which were frozen if not reduced. As Atty. Edwin S. De la Cruz, President of the International Seafarers' Action Center (ISAC) disclosed: "*While there are standard salaries being imposed, supposedly, based on the existing collective bargaining agreements (CBAs) between the ship-owners and the existing unions (ITF and the International Bargaining Forum or IBF) and as stipulated in the Philippine Overseas Employment Administration-Standard Employment Contract (SEC) for Seafarers, there are times that this is not being implemented.*" Ideally, an ordinary seaman could get as much as $1,250; but there are many instances, this is far from what is supposedly to receive by the seaman. Boarded in a Saudi-flagged ship, a concerned seafarer says that he can only send back home around $235.738 (P10, 000.00). He is boarded on a Saudi-flagged ship, which he did not disclose in fear that his employer would no longer renew his contract. Atty de la Cruz further cited that: "*Wages of the crew are always kept at a minimum to maintain the level of profit. This is coupled with the high risks, extreme conditions and hardships entailed by work at sea. Despite the strategic importance of seafarers to world commerce, they are among the most exploited and oppressed of workers*". He further disclosed that: "*The merchant ship-owners altogether earn an average annual income of US$380 billion, which is five percent of the world economy. The average annual profit of an individual merchant ship-owner is $5.42 million, but only about US $ 636,000 goes to the seafarers as wages, showing a clean 800 percent profit in relation to wages.*" In 2009, during the onslaught of the global economic crunch that resulted from the collapse of the Wall Street, there had been a decision of freezing the wages of seafarers, regardless of nationality and position inside the ship. This wage freeze, until now, exists. Furthermore, there is also discrimination on salaries between Filipinos and non-Filipino officers and crew members.

4 INTERNATIONAL RESPONSE

4.1 *24-Maritime Security Hotline*

In 2007 the International Maritime Bureau Piracy Reporting Centre launched a dedicated hotline for seafarers, port workers, shipping agents, shipyard personals, brokers, stevedores and all concerned parties to report any information that they may have seen/heard/ known of relating to maritime crime and/or security, including terrorism, piracy and other illegal activities.

4.2 *SaveOurSeafarers Program*

Launched in March 2011, this is an international, not-for-profit, anti-piracy campaign which was established by a group of five influential maritime associations. In the past year, the number of maritime associations, trade unions and P&I insurers supporting the campaign has risen to thirty- three (33); The supporting organizations (33) are as follows: ASF (Asian Shipowners' Forum), BIMCO (international shipping association), Britannia (P&I insurance), CEFOR (Nordic Association of Marine Insurers), DNK (Den Norske Krigsforsikring for Skib), GARD (P&I insurance), GSF (Global Shippers' Forum), IMEC (International Maritime Employers' Committee), ICS/ISF (International Chamber of Shipping/International Shipping Federation) , IMB (International Maritime Bureau), IMO (International Maritime Organization), Intercargo (International Association of Dry Cargo Shipowners), INTERTANKO (Int'l Association of Independent Tanker Owners), InterManager (International Ship Managers' Association), IG (International Group of P&I Clubs), IPTA (International Parcel Tanker Association), ITF (International Transport Workers' Federation), IUMI (International Union of Marine Insurance),LMA (Lloyd's Market Association), The

London (P&I insurance), North of England (P&I insurance), SIGTTO (Society of International Gas Tanker and Terminal Operators), SKULD P&I club, The Seamen's Church Institute, The Standard (P&I insurance), UGS (Union of Greek Shipowners), UK Chamber of Shipping, UK P&I Club (P&I insurance), WISTA UK (Women's International Shipping & Trading Association),WSC (World Shipping Council), IFSMA (International Federation of Ship Masters Associations), Nautical Institute, the Association of Marine Officers and Union of the Philippines (AMOSUP) and the Maritime Academy of Asia and the Pacific (MAAP).

This is the largest number of maritime organizations ever to unite behind a single cause. The campaign has succeeded in putting Somali piracy on the public and political agenda in numerous maritime nations during the past years. There are clear signs these nations' political resolve to defeat piracy is strengthening. This aims to eradicate piracy around the world; in particular Somalia- based piracy in the Gulf of Aden and Indian Ocean, by;

1 Increasing the strength of naval forces patrolling the Gulf of Aden and the 2,000,000 square nautical miles of the Indian Ocean.
2 Ensuring that when pirates are arrested by security forces, evidence will be gathered and that they will face trial, sentencing and punishment.
3 Endorsing the UN principle of financing, building and operating courts and jails in the cooperating autonomous regions of Somalia and neighboring states.
4 Seeking a sustainable political solution to the underlying problems in Somalia.
5 Supporting the introduction of counter measures and a criminal information database.

4.3 *Maritime Piracy-Humanitarian Response to Piracy (MPHRP)*

The objectives of the MPHRP are to implement a model for assisting and responding to seafarers and their families with regards to the humanitarian aspects of a traumatic incident caused by piracy, armed robbery or being taken hostage. In this regard the MPHRP has published good practice guidelines for Shipping Companies and Manning Agents for the Humanitarian Support of Seafarers and their Families involved in Piracy incidents. This programme has been built around: a task group of multi-disciplined, international experts; extensive fact finding and the feedback gained from first-hand meetings and interviews with seafarers and families worldwide, including many with firsthand experience of attacks and hijackings; advisory groups on industry practices and procedures, pre-deployment piracy training and the skills required of responders, and; the advice and assistance of a project steering group.

In its first phase the programme is developing: "*good practice*" guides for use by shipping companies, manning agents and welfare associations to support both seafarers and seafarers' families through the three phases of a piracy incident from pre-departure, during the crisis and post release/post incident, associated training modules, an international network of trained first-responders with appropriate skills within Partner and associated organizations, access to a network of professional aftercare, and a 24 hour seafarer's international telephone helpline.

4.4 *Best Management Practice Ver. 4: (dated 18 Aug 2011).*

Industry bodies, along with the navies, have developed guidelines to assist masters in transiting the dangerous waters in the Gulf of Aden and off the east coast of Somalia. The guideline, called the Best Management Practice (BMP), is a fluid document which gets updated regularly as new lessons are learned.

4.5 *Citadel (with communication facilities)*

Citadel is assigned at ship to serve as the refuge of crew for several days, protected by double-layer security doors during piracy attack. The facility is also equipped with communication devices usable even in case of blackout, ship maneuvering equipments such as stop main engine and steering gear, and is able to gather information of the ship's data, including video picture and sound. The accommodation windows are bulletproof, and water cannons are placed on the upper deck to prevent pirates' entry into the accommodation.

Imabari Shipbuilding has developed the Aero-Citadel, a newly designed superstructure which brings to realization both energy efficiency and anti-piracy measures as well as the improvement of safety and living condition of crew members. http://www.marinelink.com/news/aerocitadel-antipiracy344469.aspx

4.6 *Other Measures*

Private security personnel allowed on board; UN allowed military boats to attack boats of pirates ashore Somalia [Maritime Safety Convention]

5 AMOSUP/MAAP ACTIONS

AMOSUP, as part of its welfare program, continue to assist its members in terms of acquiring immediate support from their shipping companies.

An example was the M/V IZUM which was hijacked at Mobasa Kenya from October 10, 2010 – February 5, 2011 (6 months) with twenty AMOSUP member seafarers on board master mariners to ratings. Aside from meetings on April 1,2011(26 consist of crews of M/V Izumi and their family, three officials from Marsun Shipping Corporation , NRCO-OWWA representative and two AMOSUP lawyers), the AMOSUP have made representation for and on behalf of the Filipino officers and crew of MV Izumi thru follow-up letter communications dated June 16, 2011 with the President/Gen Manager of Marsun Shipping Corporation to urge his Japanese principal to settle the agreed account as per compensation mandated by the POEA Rules and Regulations /Board Resolutions # 04 Series of 2008 which must be complied unconditionally. This was for the immediate pay-out to alleviate the AMOSUP seafarers from their financial difficulties as they were still unemployed at that time. AMOSUP also reminded the company that they are in coordination with the AJSU-Manila Coordinating Office and if no positive action from their side, AMOSUP would be constrained to make necessary recommendations to effect immediate payment and requested the concerned company to take preferential attention to avoid unnecessary legal complications both local and overseas.

On July 14, 2011, AMOSUP have been informed that Marsun's Japanese principal have agreed on the settlement of the obligation(double compensation) on the total amount claimed by the crews who have agreed on staggered or installment payments during their meeting as the principal is short of funds due to the quite big ransom money paid to the pirates With the assistance of AMOSUP, on December 23,2011, AMOSUP have been informed that the 20 AMOSUP seafarers have completely been paid the crews double compensation during the period of their captivity by the hijackers from the Marsun Shipping Corporation thru their principal Fair Field Shipping Co Ltd of Japan. The first part was the double payment (regular pay on board USD 35,000.00) and the other part of the double payment which consists of the basic pay, the overtime pay and the leave pay, all multiplied by two which is the high–risk compensation (USD 159,835.59) in accordance to the POEA, SEC and the CBA, which at least had compensated the trauma and stress that the AMOSUP seafarers have obtained from the piracy incident at sea while they were performing their duties. The request of AMOSUP for the re-employment of the seafarers who have signified their availability for service, have also been granted, as AMOSUP was also been informed that 15 of them were already employed back on board other ships manned by Marsun and some with other company. On the other hand, MAAP has supported the international campaign "SaveOurSeafers" Program.

6 CONCLUDING REMARKS

Filipino Seafarers are very much aware that their profession requires challenges, difficulties and hardships to improve quality of life not only of their own but of the world because of their services– moving products which are very vital in the world trade. The economic cost of piracy may be estimated, but the human cost is difficult to estimate and even misunderstood by public. They are concerned, first and foremost, to their personal welfare and protection on board vessels especially transversing in high risk zones like Somalia and Gulf of Aden. Filipino seafarers are required by the government to undertake the Anti-piracy training before going onboard. Also, there are a number of international measures and some ships are escorted by naval ships. However, the anxiety and fear still does not recede especially as they travel in Somalia and Gulf of Aden. As one seafarer says after passing through that area, "Thanks God, we have passed through the area safely, now we can relax a bit". Unless government steps in, Filipino sailors will keep on sailing. The risk, for the sailors, is nothing compared to the money they can bring home to their families. However, if attacked and held hostage, what support would they expect? Who would come to their aid? Hence, the following must be reflected to form part of the government measures:
- Welfare and Protection from Ship-owners and Government
- Welfare for Families
- Pre-passage of Somalia and Gulf of Aden
- If Attacked by Pirates
- After Attack

Seafarers move more than 90% of world trade. Their invaluable role needs to be acknowledged, and their rights be protected by pursuing all possible means to bring Somali Piracy to an end.

ACKNOWLEDGEMENT

The National Research Council of the Philippines for the funding of the 6-month MAAP Research Project on challenges, issues and concerns in seafaring (March -September 2012)

REFERENCES

Baylon, AM and JK Vergara (2013) Unpublished Proceedings of the 2012 Philippine Seafarers Convention transcribed by Department of Research and Extension Services (DRES, MAAP)

Chiara Zambrano, ABS-CBN News (2010) Why Pinoy seamen risk going to 'pirate alley' Posted at 11/12/2010 5:03 PM | Updated as of 11/14/2010 4:01 PM

International Maritime Bureau Reporting Center Hotline. http://www.icc-ccs.org/piracy- reporting-centre/24-hour-maritime-security-hotline

Lowe, Mark. Aero-Citadel Published on May 11, 2012 retrieved on September 15, 2014 at http://www.marsecreview.com/2012/05/aero-citadel/ and http://www.marinelink.com/news/aerocitadel-antipiracy344469.aspx

Maritime Piracy-Humanitarian Response to Piracy (MPHRP). Peter Swift Programme Chair, September 2011 retrieved on September 14, 2014 at http://www.mphrp.org/publication.ph

SaveOurSeafarers retrieved on September 14 2014 (http://www.saveourseafarers.com)

The Human Cost of Somali Piracy by Kaija Hurlburt, et. al. June 6, 2011 retrieved on September 12, 2014 at www.oceansbeyondpiracy.org.

Human Resource Management and Maritime Crew Manning
Safety of Marine Transport – Marine Navigation and Safety of Sea Transportation – A. Weintrit & T. Neumann (eds.)

Swedish Seafarers' Occupational Commitment in Light of Gender and Family Situation

C. Hult & C. Österman
Kalmar Maritime Academy, Linnaeus University, Sweden

ABSTRACT: The present study focuses on the pattern of Swedish seafarers' occupational commitment relative to gender and family situation. Statistical analyses are employed, using a survey material of Swedish seafarers randomly collected from a national register in 2010. It was hypothesized that the effect from having children at home should be negative on commitment to seafaring occupation. However, the effect was found to be strongly positive and statistically significant for women and close to significant for men. Another important family effect was, as expected, the positive effect of having a relative working, or having worked, at sea. This effect was, however, only significant for male seafarers in the age group below the early 40's. When controlling for possible mediating effects due to gender distribution in the onboard departments, it became clear that working in the catering department comes with a strong negative effect on commitment to the seafaring occupation, but only so for women. This effect did not, however, alter much of the already observed gender patterns. In the concluding discussion, the findings are discussed in more details and recommendations put forward.

1 INTRODUCTION

1.1 Background to the study

There has been a lingering problem with high turnover of onboard staff in the Swedish shipping industry. A common estimate is that the average time a Swedish ship's officer remains in the occupation is only eight years (Arbetsförmedlingen 2010:10, SMA 2010:17). According to a survey made in 2010, over 20% of seafarers aged 30 or younger, and close to 18% of those between 31-42 years of age, declared they were likely to leave their occupation within a few years (Hult 2012a). One reason for this problem may be found in the long periods of time when the seafarer has to be separated from home, family and friends.

The combination of high turnover and long periods away from home, calls for an attempt to shed some light over the possible correlation between seafarer's family situation and their occupational commitment. It is most likely that these recurring separations from home may be emotionally harder for seafarers who have a spouse or partner, compared to those who are single; and particularly difficult if they have young children at home. There may, however, be family circumstances which have the potential to strengthen commitment to the seafaring occupation. One such circumstance would be to have a close relative who works, or has worked, at sea and as such acts as a source of inspiration.

1.2 The objectives of the study

The purpose of the study presented in this paper is to investigate Swedish seafarer's occupational commitment. Specifically, the objectives of the study are to investigate (i) whether family situation have any net-effects on occupational commitment for Swedish seafarers, (ii) whether the patterns of effects are similar for both men and women, (iii) whether the effects are similar in deck, engine and catering departments.

2 THEORETICAL BASIS FOR THE STUDY

2.1 Earlier research on attitudes to work

Quantitative research on seafarers' attitudes to work is not abounded. There are, however, a number of studies from different parts of the world (e.g. Guo et al. 2005, Guo et al. 2010, Pan et al. 2011, Sencila et al. 2010, Turker & Er 2007). A particular interest can be seen within the cruise sector, probably due to the link between employee job satisfaction and customer satisfaction in service occupations (e.g. Larsen et al. 2012, Testa 2001, Testa & Mueller 2009, Testa et al. 2003). Moreover, there are four attitude studies on Swedish seafarers with quantitative approaches. Two older studies that focus on job satisfaction on board merchant ships (Olofsson 1995, Werthén 1976), and two more recent studies that focus on commitment to work and occupation (Hult 2012b, Hult & Snöberg 2013).

2.2 Occupational commitment

Succinctly, commitment to an occupation has to do with perceptions of the generalities for that occupation. In that respect, the prospect of a decent income is, of course, one important factor. More interestingly, there is a qualitative and emotional driver for this type of commitment, which has drawn most attention in earlier research (c.f. Lee et al. 2000). It is, for example, primarily within an occupation that people can develop a sense of social status and identity.

Earlier research show that the duration of education, age, and years invested in the occupation have positive effects on occupational commitment (Nogueras 2006). Returning to the high turnover among Swedish seafarers, it has been shown that time spent on the same ship has a negative effect for younger seafarers (Hult & Snöberg 2013). It has been reported that perceptions of social quality and leadership quality in the work organization influence occupational commitment (van der Heijden et al. 2009). A positive correlation between occupational commitment and perceived autonomy at work has also been reported (Giffords 2009). Research has further shown that strong occupational commitment restrains decisions to leave a job (Hult 2012b, Nogueras 2006).

2.3 Occupational commitment and family situation

The first assumption in this study is that some social circumstances may have negative effects on seafarers' occupational commitment. It is plausible that having to endure long periods of separation from home and family is emotionally harder for those seafarers who have a spouse or partner, compared to those who are single. The existence of problems related to family separation among seafarers also finds support in earlier research (e.g. Thomas et al., 2003). Moreover, it is likely that the separation would be particularly difficult if there are young children at home.

On the whole, this rather intuitive assumption concerning effects on occupational commitment strikes close to Becker's (1960) side bet theory

which suggests that continuance in a job does not solely depend on the degree of affiliation with it. Rather, it is a result of a more holistic calculation including impacts on other aspects of life. In this case, this may assume a possible situation of conflicting commitments; one commitment directed towards the family and another towards the occupation.

The next assumption is firmly anchored in the tradition of socialisation and social capital theory. Here, we assume that the existence of a relative who are, or have been, working at sea will have a positive effect on commitment to the seafaring occupation. The social relationship to the term 'capital' has been neatly explained by Portes (1998:7) as follows:

Whereas economic capital is in people's bank accounts and human capital is inside their heads, social capital inheres in the structure of their relationships. To possess social capital, a person must be related to others, and it is those others, not himself, who are the actual source of his or her advantage.

Studies on social capital may deal with the effects of social contacts, social ties and social networks, on occupational choices; the likelihood of success on the labour market and on satisfaction with job related aspects (Bentolila et al. 2010, Flap & Völker 2001, Mouw 2003, Requena 2003, Seibert et al. 2001). Another strand of research, in the realm of social capital, focuses on family effects on people's career outcome (e.g. Egerton 1997). This effect has metaphorically been labelled *career inheritance* (Goodale & Hall 1976, Inkson 2004). It has been suggested that this effect can arrive from parenting practise during childhood, from reinforcement of work values and vocational interests during adolescence, and from more tangible support later on (Aldrich & Kim 2007). According to Gottfredson (2002:139):

...people tend to glean information about their options from people in close proximity and who thus populate their birth niche, which constitutes a recipe for minor adjustment rather than major change. It should come as no surprise, then, that people's adult niches tend to resemble their birth niches, that children re-create the society...

Attempts to incorporate inheritance in a more genetic sense have also been made (Aldrich & Kim 2007, Gottfredson 2002). Here, it is argued that socialisation theory have difficulties explaining individual differences in career choices, especially for people originally from similar social niches. Instead, a combination of socialisation and individually driven search for a person–environment fit in the social world has been suggested. From the adolescent's perspective, the career formation may here be described as a struggle of circumscription and compromises of preferences on a socially chaired map of gender and prestige differences in occupations (Gottfredson 2002).

The debate of nurture *vs.* nature lingers on, however, astray from the topic of this study. In sum, the bulk of research gives us reason to believe that job relevant social capital is positive for individual perceptions and satisfaction within a specific occupation. When it comes to the maritime sector, however, it has been pointed out that the chances for post-adolescent building of sustainable social capital is increasingly challenged because of the world wide dispersion of shipping companies and frequent changes of crews (Grøn & Svendsen 2013, Sampson 2013). Thus, the parental or family related social capital may be of particular importance for seafarers' perceptions of the occupation. In fact, the assumption that seafarers who have close relatives working, or having worked, at sea are likely to express greater commitment to their occupation than others finds some support from an earlier study (Hult 2012b). Thus, family situation entails social capital that matter for seafarers.

2.4 *The study's rationale*

If career choice largely is a compromise based on information from people in close proximity, and preferences towards positions on a socially chaired map of gender and prestige in occupations, we may expect different patterns of commitment due to gender. The seafaring occupation has historically been a traditionally male occupation. A close relative at sea, often a man, may therefore work better as a role model. And as such, may be re-created with less effort by young men than by women. Thus, we expect that the positive effect from close relatives at sea on occupational commitment will be stronger for male seafarers than for female. Likewise, because the catering department traditionally has been relatively female dominated, we expect a lower positive effect here, than in deck and engine departments.

As pointed out earlier, we also expect that occupational commitment will be lower for seafarers who have a spouse or partner, and for those having young children at home.

3 METHOD

3.1 *The sample*

This study is based on a sample taken from the Swedish Register of Seafarers using unrestricted random selection of deck and engineering personnel for the men, and of catering personnel for both men and women. Because women still are strongly underrepresented among deck and engineering

personnel, all women from these departments were drawn into the framework. The only effect of this decision would be that it gives more, of very few, women in these departments the opportunity to participate.

The data were collected via postal surveys during the period of March 8 to September 8 in 2010. The questionnaire as a whole was based on pre-existing questionnaires from the International Social Survey Programme, Work Orientations III study (ISSP 2005).

The final material consists of 1309 respondents with an answering rate of 54%, which must be considered sufficient given the general trend of shrinking answering rates. More important, the control of different aspects, such as gender, age, onboard position, trade area, and type of ship, found the material representative for Swedish seafarers. Although sufficient demographic and work-related representativity, it is always difficult to estimate the likeliness of non-response effects on the attitudinal representativity. An educated guess would be that people who take great interest in their work may be more likely than others to complete this type of questionnaire and therefore be over-represented in the sample. If so, the attitudinal patterns found in the analysis would still be correct, but the levels of commitment would be slightly overestimated (e.g. Hult & Svallfors 2002).

3.2 *Processing of data and analysis*

The Statistical Package Social Science (SPSS) were used throughout the analysis. The dependent variable of occupational commitment was constructed as an index using Principal Component Analysis (PCA) and internal reliability control. In order to control for the influence of different and competing variables, multiple regression analysis (OLS regression), allowing adjusted effects, were used in several steps of the analysis.

The dependent variable of occupational commitment was developed with theoretical connections to the Porter scale (Porter et al. 1974) and the three-component measurement (Meyer & Allen 1991) and carefully adapted to the specifics of the seafaring occupation. The variable is based on five attitude questions expressed as statements on which respondents were asked to take a position by selecting a fixed option on a five-point Likert Scale, from *strongly agree* to *strongly disagree*. After mapping the pattern of latent factors underlying a number of different indicators, using PCA, the appropriate indicators for occupational commitment came out as shown in Table 1.

The indicators were recoded so that 0 denotes the option that entails the lowest commitment and 4 the highest. The indicators were then summarized in the index of occupational commitment. To facilitate interpretation of the results, the index was divided by its maximum value and multiplied by 100. The index is thus permitted to vary between 0 and 100.

Table 1. Indicators of occupational commitment for seafarers

Please agree or disagree with the following:
There are qualities to the seafaring occupation that I would miss in another occupation.
The seafaring occupation is part of my identity
The seafaring occupation is not just a job, it is a lifestyle.
I feel proud of my occupation as a seafarer
I would prefer to remain in the seafaring occupation even if I were offered a job with higher pay on land

Table 2 presents the mean value and standard deviation for the index. The high mean value and the low standard deviation indicates that seafarers are quite united in their high commitment to, and identification with, the seafaring occupation. Cronbach's Alpha is a test of the internal correlation among the indicators in each index – the higher the value (between 0 and 1), the more reliable the index. Table 2 shows that the index turn out very stable.

Table 2. Occupational commitment index – Swedish seafarers in 2010

	Occupational commitment
Mean value (0-100)	71.7
Standard deviation	16.78
Cronbach's Alpha	0.82

4 RESULTS

4.1 The full sample

In Figure 1, the effects of the family situation on occupational commitment are presented. It is obvious that having family with connection to the seafaring occupation is the only family aspect that is significantly positive for commitment.

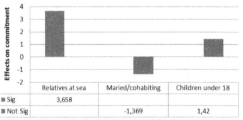

Figure 1. Occupational commitment and family situation

Figure 2 illustrates how the main strength of this effect originates from the two youngest age categories, which understandably indicates that older relatives inspire occupational identification among younger seafarers.

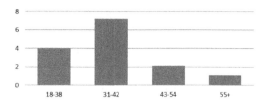

Figure 2. Effects of a relative at sea to commitment by age group

Moreover, the older the respondents are, the more likely they are having a younger relative in mind when answering the question. Thus, for the remainder of the study we will concentrate on seafarers not older than 42 years of age.

4.2 Seafarers aged 18-42

Table 3 and 4 display the effects of family situation on occupational commitment for men and women separately. Both tables display the results of multiple analyses in four steps (I-IV). New variables and their effects are entered into a stepwise increasing model. The mean value of the total comparison group is displayed for each step at the top of the table. Statistical significance is denoted in bold type. Three asterisks (***) denote the strongest significance and one asterisk (*) the weakest. Numbers shown in bold with no asterisk indicate that the effect is close to the lowest significant level.

Table 3. Occupational commitment and family situation, male, age 19-42

	I	II	III	IV
Mean values for the comparison group, no relatives at sea, not married/partnered, and no children under 18	67,79	69,01	68,66	69,01
Having a relative at sea (compared with not having one)	**7,41***	**7,35***	**7,29***	**7,41***
Being married/partnered (compared to being single)	-	-1,90	**-3,17**	-3,07
Having children under 18 in the household (compared to having none)	-	-	**3,21**	**3,31**
Catering (compared to other departments)	3,80	-	-	-
Explained variance (%)	*0,05*	*0,06*	*0,06*	*0,7*
Number of respondents	*360*	*360*	*359*	*358*

Significance levels: **Bold** and *** = 0.001 level, ** = 0.01 level, * = 0.05 level, **bold** only = 0.1 level.

Table 3 displays the effects on commitment for men. Step 1 shows the effects from having relatives at sea compared to having no such relatives. In step II to IV, the effects of being married or cohabiting, having younger children, and finally a control for onboard department are stepwise introduced to the model. Here we can see that these variables do not alter the significant effect from having relatives at sea. The effect from having children is positive and close to significant. The effect from being married or cohabiting is negative and close to significant in step III. The effect of working in the catering department is not significant.

In Table 4 the effects on commitment for women are displayed in the same stepwise manner as in Table 3.

Table 4. Occupational commitment and family situation, female, age 19-42

	I	II	III	IV
Mean values for the comparison group, no relatives at sea, not married/partnered, and no children under 18	69,36	70,47	70,40	73,12
Having a relative at sea (compared with not having one)	1,34	2,36	1,93	2,03
Being married/partnered (compared to being single)	-	-3,37	-4,42	-4,40
Having children under 18 in the household (compared to having none)	-	-	6,66	**8,81** *
Catering (compared to other departments)	-	-	-	**-7,95** **
Explained variance (%)	*0,002*	*0,01*	*0,03*	*0,09*
Number of respondents	*135*	*134*	*134*	*134*

Significance levels: **Bold** and *** = 0.001 level, ** = 0.01 level, * = 0.05 level, **bold** only = 0.1 level.

Here, the attitudinal pattern for women is shown to be totally different from that of men. Having relatives at sea has no impact at all on occupational commitment for women. The effect from having children is positive, strong and significant, when department is controlled for. The effect of working in the catering department is however negative, strong and significant. The only effect resembling the male pattern is that from being married or cohabiting, which is negative and rather weak.

5 CONCLUDING DISCUSSION

5.1 *The results*

It was initially hypothesized that the effect from having children at home would be negative on commitment to seafaring occupation. Contrarily, this effect was found to be strongly positive and statistically significant for women and close to significant for men. This finding begs for explanation although mainly give rise to new questions. Does the finding indicate that the seafaring occupation in some ways liberates people from the everyday attempt to solve the life puzzles

of work and home? Does the fact that women's occupational commitment is more strongly positively affected by having young children at home, indicate that the occupation works as some sort of cooping strategy for otherwise being caught in a double work role at home? The role of strategies has recently been highlighted concerning countries where policies promoting the dual-earner families and 'have-it-all" aspirations (Grönlund & Javornik 2014). It takes more research, however, to know whether this could offer explanations to our findings.

Another important family effect was, as expected, the positive effect of having a relative working, or having worked, at sea. This effect was however, only significant for male seafarers in the age group below the early 40s, indicating that it is the older relatives that convey and encourage occupational identification among younger seafarers. The gender difference can be explained by the male dominant occupation and that a relative at sea often is a man and therefore work better as a role model for young men than for women.

The results further show that working in the catering department comes with a strong negative effect on commitment to the seafaring occupation, but only so for the women. This effect did not, however, alter much of the already observed gender patterns.

5.2 *Future work and recommendations*

The findings in this study give rise to further questions that calls for future research. The only justified recommendation we can give at this point is that strategies are needed to strengthen occupational commitment for the personnel working in the catering department. It is reasonable to believe that any efforts in that respect would be met with reduced labour turnover, increased job satisfaction, and thus most likely increased customer satisfaction on passenger ships. That would in turn have a positive effect on overall business performance as well as employee health and wellbeing.

As a rule, statistical results can only be generalized to the population from which the sample is drawn. Given the large differences in working and living conditions for seafarers worldwide, as well as the diverse institutional and cultural settings of their backgrounds, that rule indeed counts for this study. However, this diversity calls for future cross national comparative research, studying the correlation of family situation and commitment to seafaring life over time. Such an approach would make it possible to continuously evaluate the effects of a developing global and uniform compliance and enforcement of international conventions, such as the Maritime Labour Convention (ILO 2006) and the

STCW convention regarding training, certification and watch-keeping (IMO 2011).

ACKNOWLEDGEMENT

We like to thank all the Swedish seafarers who completed their questionnaires, one anonymous reviewer, and the Swedish Mercantile Marine Foundation for funding the data collection.

REFERENCES

Aldrich, Howard E, & Kim, Phillip H. (2007). A life course perspective on occupational inheritance: Self-employed parents and their children. *Research in the Sociology of Organizations*, 25, 33-82.

Arbetsförmedlingen. (2010). Prognos Arbetsmarknad Sjöfart 2010 och 2011. (In Swedish). Stockholm: Swedish Employment Service Maritime

Becker, Howard S. (1960). Notes on the concept of commitment. *American journal of Sociology*, 32-40.

Bentolila, Samuel, Michelacci, Claudio, & Suarez, Javier. (2010). Social contacts and occupational choice. *Economica*, 77(305), 20-45.

Egerton, Muriel. (1997). Occupational Inheritance: the role of cultural capital and gender. *Work, Employment & Society*, 11(2), 263-282.

Flap, Henk, & Völker, Beate. (2001). Goal specific social capital and job satisfaction: Effects of different types of networks on instrumental and social aspects of work. *Social networks*, 23(4), 297-320.

Giffords, Elissa D. (2009). An examination of organizational commitment and professional commitment and the relationship to work environment, demographic and organizational factors. *Journal of Social Work*, 9, 386-404.

Goodale, James G, & Hall, Douglas T. (1976). Inheriting a career: The influence of sex, values, and parents. *Journal of Vocational Behavior*, 8(1), 19-30.

Gottfredson, Linda S. (2002). Gottfredson's theory of circumscription, compromise, and self-creation. *Career choice and development*, 4, 85-148.

Grøn, Sisse, & Svendsen, GunnarLindHaase. (2013). "Blue" social capital and work performance: anthropological fieldwork among crew members at four Danish international ships. *WMU Journal of Maritime Affairs*, 12(2), 185-212.

Grönlund, Anne, & Javornik, Jana. (2014). Great expectations. Dual-earner policies and the management of work–family conflict: the examples of Sweden and Slovenia. *Families, Relationships and Societies*, 3(1), 51-65.

Guo, Jiunn-Liang, Gin-Shuh, Liang, & Ye, Kung-Don. (2005). Impact of Seafaring Diversity: Taiwanese Ship-officers' Perception. *Journal of the Eastern Asia Society for Transportation Studies*, 6, 4176-4191.

Guo, Jiunn-Liang, Ting, Shih-Chan, & Lirn, Taih-Cherng. (2010). The Impacts of Internship at Sea on Navigation Students. *Maritime Quarterly*, 19(4), 77-107.

Hult, Carl. (2012a). Seafarers' attitudes and perceptions towards age and retirement. In C. Hult (Ed.), *Swedish Seafarers and the Seafaring Occupation 2010 – A study of work-related attitudes during different stages of life at sea*. Kalmar: Kalmar Maritime Academy, Linnaeus University.

Hult, Carl. (2012b). Seafarers' commitment to work and occupation. In C. Hult (Ed.). *Swedish Seafarers and the Seafaring Occupation 2010 – A study of work-related*

attitudes during different stages of life at sea. Kalmar: Kalmar Maritime Academy, Linnaeus University.

Hult, Carl, & Snöberg, Jan. (2013). Wasted Time and Flag State Worries. *Marine Navigation and Safety of Sea Transportation*, 121-128: CRC Press.

Inkson, Kerr. (2004). Images of career: Nine key metaphors. *Journal of Vocational Behavior*, 65(1), 96-111.

ISSP. (2005). ZA4350: International Social Survey Programme: Work Orientation III - ISSP 2005. Retrieved January 29, 2010, from http://zacat.gesis.org/webview/

Larsen, Svein, Marnburg, Einar, & Øgaard, Torvald. (2012). Working onboard – Job perception, organizational commitment and job satisfaction in the cruise sector. *Tourism Management*, 33(3), 592-597.

Lee, Kibeom, Carswell, Julie J, & Allen, Natalie J. (2000). A meta-analytic review of occupational commitment: relations with person-and work-related variables. *Journal of Applied Psychology*, 85(5), 799.

Meyer, John P, & Allen, Natalie J. (1991). A three-component conceptualization of organizational commitment. *Human resource management review*, 1(1), 61-89.

Mouw, Ted. (2003). Social capital and finding a job: Do contacts matter? *American sociological review*, 868-898.

Nogueras, Debbie J. (2006). Occupational commitment, education, and experience as a predictor of intent to leave the nursing profession. *Nursing economics*, 24(2), 86-93.

Olofsson, Martin. (1995). *The work situation for seamen on merchant ships in a Swedish environment*: Chalmers University of Technology.

Pan, Yu-Huei, Lin, Chi-Chang, & Yang, Chung-Shang. (2011). The Effects of Perceived Organizational Support, Job Satisfaction and Organizational Commitment on Job Performance in Bulk Shipping. *Maritime Quarterly*, 20(4), 83-110.

Porter, Lyman W, Steers, Richard M, Mowday, Richard T, & Boulian, Paul V. (1974). Organizational commitment, job satisfaction, and turnover among psychiatric technicians. *Journal of applied psychology*, 59(5), 603.

Portes, Alejandro. (1998). Social capital: Its origins and applications in modern sociology. *Annual Review of Sociology*, 24(1), 1-24.

Requena, Felix. (2003). Social capital, satisfaction and quality of life in the workplace. *Social indicators research*, 61(3), 331-360.

Sampson, Helen. (2013). Globalisation, Labour Market Transformation and Migrant Marginalisation: the Example of Transmigrant Seafarers in Germany. *Journal of International Migration and Integration*, 14(4), 751-765.

Seibert, Scott E, Kraimer, Maria L, & Liden, Robert C. (2001). A social capital theory of career success. *Academy of Management Journal*, 44(2), 219-237.

Sencila, V, Bartuseviciene, I, Rupšiene, L, & Kalvaitiene, G. (2010). The Economical Emigration Aspect of East and Central European Seafarers: Motivation for Employment in Foreign Fleet. TransNav: International Journal on Marine Navigation and Safety of Sea Transportation, 4(3), 337-342.

SMA. (2010). Handlingsplan för ökad rekrytering av personal till sjöfartssektorn (In Swedish). Norrköping: Swedish Maritime Administration.

Testa, Mark R. (2001). Organizational commitment, job satisfaction, and effort in the service environment. *The Journal of Psychology*, 135(2), 226-236.

Testa, Mark R, & Mueller, Stephen L. (2009). Demographic and cultural predictors of international service worker job satisfaction. *Managing Service Quality*, 19(2), 195-210.

Testa, Mark R, Mueller, Stephen L, & Thomas, Anisya S. (2003). Cultural fit and job satisfaction in a global service environment. MIR: *Management International Review*, 129-148.

Thomas, M., Sampson, H., & Minghua, Z. (2003). Finding a balance: companies, seafarers and family life. *Maritime Policy and Management*, 30(1), 59-76.

Turker, F, & Er, ID. (2007). Investigation the root causes of seafarers' turnover and its impact on the safe operation of the ship. *International Journal on Marine Navigation and Safety of Sea Transportation*, 1(4), 435-440.

van der Heijden, Beatrice IJM, van Dam, Karen, & Hasselhorn, Hans Martin. (2009). Intention to leave nursing: the importance of interpersonal work context, work-home interference, and job satisfaction beyond the effect of occupational commitment. *Career Development International*, 14(7), 616-635.

Werthén, Hans-Erik. (1976). *Sjömannen och hans yrke: en socialpsykologisk undersökning av trivsel och arbetsförhållanden på svenska handelsfartyg.* (Doctoral thesis), University of Gothenburg.

Web-based Databank for Assessment of Seafarers' Functional Status During Sea Missions

G. Varoneckas, A. Martinkenas, J. Andruskiene, A. Stankus, L. Mazrimaite & A. Livens
Klaipeda University, Klaipeda, Lithuania

ABSTRACT: This paper introduces the basic concept of a new developed web-based databank for an assessment of seafarers' functional status during the sea missions. The Web system is based on client-server architecture and the international open source technologies including Apache web server, PHP scripts and MySQL database. The paper focuses on the aspects and first results of the initial practical realization of the web-based databank. The main operational advantage of the developed system is the capability to on-line handle up to a dozen users at the same time. The system includes administrative data and questionnaires. Electronic data entry saves the time and material resources.

1 INTRODUCTION

In spite of the development of the modern technology the sea transport depends largely on the human action. Crews of the cargo and passenger ships' have high responsibility, which requires good health, to perform their duties. A tight ship working environment, limited sleep duration, working time by shifts, difficulties in communicating with colleagues, nostalgia, isolation feeling; they all are considered to be factors of fatigue and stress that can lead to various consequences [1].

In order to prevent human caused accidents and to increase the safety awareness among the crews it is obligatory to measure, control and manage functional status of seafarers in maritime work-environment. The sea transport leaders drew attention to the factors causing fatigue and stress, since only by reducing or eliminating them the safety of navigation could be ensured and more workers attracted. The leaders also pointed out the need to initiate a new crew views toward maritime education and training. Such measures would help to create better working and living environment, increase efficiency, productivity and safety for crew members. Assessment of seafarers' functional status during marine missions requires tele informatic system, which should be based on the online access [2].

The most modern online database management systems (DBMS) were designed to work in a global network, using Web technologies. Database management systems are multifunctional enabling to store structured data, edit the existing data, and to perform a quick search for the desired criteria. The main feature of such systems is integrity – a possibility to collect data from different users (databanks) and to get access to the database independently from other users. Database management tools not only allow at any time quickly to get complete and accurate information, keep it in compact format, but also provide an opportunity to deal with automated data overload, consistency, integrity and standardization security problems.

One of the most widely used types of DBMS – relational model, based on the mathematical theory of relational algebra. The relational model is special because it provides data tables and performs specific operations with these tables. Another important feature of this model - structural independence, i.e., the data can be changed without changing the structure of the software or data processing procedures.

Such DBMS becomes available to consumers from virtually any location and requires minimum cost, but gives a lot of comfort, especially when working for many users simultaneously. The data of research are collected from the variety of tests and questionnaires; because of that the central storage of data is more efficient and less costly, especially using them for statistical analysis [3].

This paper concentrates on the development of data bank for seafarers' functional status assessment.

In section 2 and 3, which are the main parts of the paper, the initial practical realization of the Web-based databank for assessment of seafarers' functional status during sea missions is presented.

2 WEB-BASED DATABANK CONCEPT

Database was created using MySQL, Apache, PHP, Java scripts (Fig. 1) and placed on the server. To connect to the server the user can use any web browser. Apache web server – the most widely used web server software. Apache performs a key role in the initial growth of the World Wide Web function.

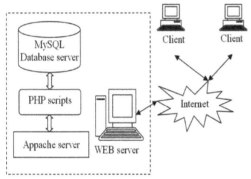

Figure 1. Web-based tool for the databank management

Apache Web server applications communicate to the data bank using PHP. The main advantage of the Apache server is the system's independence from the operating system.

Website was developed using PHP – dynamic programming language. This open-source language supports number of relational databases and runs on most operating systems and with most web servers. This DBMS runs on MS Windows operating system platform. PHP code (command) inserted into the HTML code generates mapping results in the same HTML page and displays it in a Web browser.

Database administration intended for running MySQL GUI (graphical user interface) tool – dbForge Studio for MySQL [4]. The database system's block diagram is presented in Figures 2 and 3.

JavaScript – an object oriented scripting programming language, which, like PHP, added to HTML pages, extends the static HTML pages. Java scripts are used for control of the parameters of the questionnaires.

Algorithms based on SQL language were developed for analysis and interpretation of the standard questionnaires. The data bank was developed taking into account that it should work as quickly as possible, expanding system's functionality.

A Web-based tool is easy to maintain and guarantees user access to the latest, most recent version. The user can connect to the Web server using convenient Web browser. For user interface interaction between the system's components and data flow the Web server uses PHP, Java and other programming techniques.

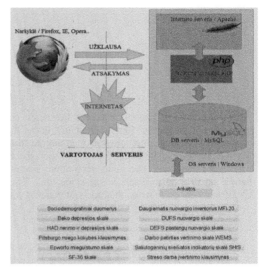

Figure 2. Database management system

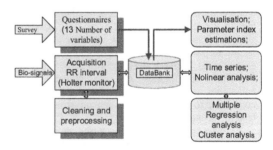

Figure 3. Database system and data management of functional and physiological testing, signal inputs and data analysis

The data depending on the security level and user's role are accessible at different communication levels only to physicians and specialists of strongly defined field and group. The system recognizes user's four roles:
1. Information system administrator;
2. Databank manager;
3. Reviewer;
4. Seafarer (interviewed).

Each role has a particular set of transaction capabilities. Information system administrator's capabilities are following: input, update, delete, review system users and patient's administrative and biomedical data. The user, having manager role of person's data can upload, modify and delete data.

The user having prescribed reviewer role can review data and generate reports. Individual interviewer can review his own data of the different treatment stages.

The elaborated database is in compliance with national (Lithuanian) and EU regulations on safety and data protection. Data are kept strictly confidential. Electronic transfers of data precisely follow the data protection guidelines.

3 DATA OF SEA-FARERS' FUNCTIONAL STATUS

3.1 Questionnaires

Data of the specialized questionnaires (n=13) and physiological recordings of the heart rate during active orthostatic test as well as during Holter monitoring were used for assessment of sea-farers' functional status. The data of following questionnaires are stored in the databank for assessment of:

- Anxiety and depression using *Hospital Anxiety and Depression Scale*, and *Beck Depression Inventory* [5, 6];
- Subjective sleep quality using *Pittsburgh Sleep Quality Questionnaire* [7-8];
- Daytime sleepiness using *Epworth sleepiness scale* [9];
- Health-related quality of life using SF-36 questionnaire "*Short Form Medical Outcomes Study Questionnaire*" [10, 11];
- Fatigue using *Multidimensional fatigue inventory*, fatigue scale DUFS (*Dutch Fatigue Scale*), and effort fatigue scale DEFS (*Dutch Exertion Fatigue Scale*) [12, 13];
- Work experiences and health indicators using WEMS (*Work Experience Measurement Scale*) and SHIS (*The Salutogenic Health Indicator Scale*) [14, 15];
- Encountered stress and stress at work place using special questionnaire (V. Reigas, 2012);
- Subjectively estimated by operator workload intensity at work place using special questionnaire (J. Andruškienė, 2012, General data form);
- Harassment at the workplace using WHS (*Work Harassment Scale*) [16].

Attribute groups and number of variables in each group are presented in the Table 1.

In database the raw data of physiological signals obtained during functional testing are collected and prepared for further analysis using self-developed statistical tools.

The possibility to provide convenient statistical analysis as well as sophisticated HR analysis using power spectra or Poincare plots is foreseen and might be performed remotely.

Table 1. Attribute groups and number of variables in web-based databank for assessment of seafarers' functional status

Attribute groups	Number of variables
1. Administrative data	80
2. Hospital Anxiety and Depression Scale	14
3. Beck Depression Inventory	23
4. Pittsburgh Sleep Quality Questionnaire	24
5. Epworth sleepiness scale	9
6. SF-36 Short Form Medical Outcomes Study Questionnaire	47
7. Dutch Fatigue Scale	10
8. Dutch Exertion Fatigue Scale	10
9. The Multidimensional Fatigue Inventory, MFI-20	22
10. WEMS	34
11. SHIS	14
12. Stress at Work	47
13. Heart rate data during active orthostatic test	52
14. Heart rate data during Holter monitoring	62
Total	448

3.2 Heart rate data

Assessment of operator's (seafarer's) functional status is based on objective evaluation of human autonomic heart rate (HR) control which determines the adaptation of cardiovascular function to different stressors, such as mental and physical work. HR recording during active orthostatic test or Holter monitoring is used for assessment of autonomic HR control.

The parameters of HR variability are stored in the data base: the average value of the R-R interval (RR, ms) and its standard deviation (σRR, ms); root mean square difference between the value of the successive RR interval duration (RMSSD, ms); percentage of the adjacent RR intervals differing from each other by more than 50 ms (pNN50%) and the number of successive RR intervals differing by more than 50 ms, (NN50, n); very low frequencies component (VLFC, from 0.003 to 0.04 Hz), low frequency component (LFC, from 0.04 to 0.15 Hz) and high frequency component of heart rate (HFC, from 0.15 to 0.4 Hz), absolute values (VLFC ms, LFC ms HFC ms) and normalized values (LFC norm, HFC norm) and percentage values (VLFC%, LFC% and HFC%), and LFC and HFC ratio (LFC/HFC, %) [17].

Poincare´ plot indexes of heart rate variability (HRV). Minor axis (SD1, ms), major axis (SD2, ms) and the SD1/SD2 ratio indices; RR histogram triangular interpolation (TINN, ms) and triangular index (TrI); heart rate level in A, B and C points (RR_A, ms, RR_B, ms, RR_C, ms); maximum heart rate response to active orthostatic test), absolute values (dRRB, ms) and percentage values (dRRBpr, %), the duration of the transition process (T_AB, s, T_BC, s) [18-20].

3.3 *Data extraction from web-based database*

Interviewer (user) before starting to fill in the questionnaires for the first time must register and identify himself in the information system. After completing a questionnaire in the database, the management system generates the conclusion and submits results to the user. Response model from one of the 13 questionnaires, i.e., Anxiety and Depression Scale (HAD), is shown in Figure 4.

A problem of feature extraction from heart rate data related to the diagnosis of sleep disorders and diseases is considered. Raw data of heart rate interval (RR, ms) sequences are taken from web-based database. The standard methods for time series analysis, e.g. statistical inference, hypotheses testing, correlation analysis, spectral analysis, etc., are included in the developed tool.

Hospital Anxiety and Depression Scale

Test No.1	Time & Date: 06/05/2013		Code: 100
		Score	Assessment
	Anxiety	12	Moderate
	Depression	4	Non-case

Estimation	Grading: Non-case (0-7), moderate (8-10) and severe (10-21)

Figure 4. The example of the conclusion from the test using Anxiety and Depression Scale.

Algorithm for diagnostics of sleep quality and disturbed sleep is based on evaluation of sleep structure (sleep stages) using sophisticated analysis of parameters of RR interval time series recorded during Holter monitoring.

Different statistical, spectral and non-linear dynamic parameters of RR interval time series are stored in the databank.

After HR data analysis, a verbal conclusion regarding the functional state of autonomic HR control reflecting seafarer's functional status is presented.

The results of analysis of questionnaires' data and physiological signals from the heart (RR intervals) provide useful information of seafarer's functional stateus, including mental and physical fatigue, stress and ability to perform the daily duties as well as changes of functional status to extreme environmental and labour factors. The developed system allows detecting the crew working capacity and proposes measures that can be faster and more efficient in restoring function status and improving management of the human factors during sea missions.

4 PILOT STUDY

The pilot study, in which 25 seafarers were investigated during their sea mission, was performed.

A database testing - the testing of 25 seafarers. The system's testing results confirmed that Web based databank for practical use is convenient; users give priority to electronic data entry.

All the test data collection in one database allows them to quickly export data and to perform statistical analysis.

The study results demonstrated that sea farer's functional status mostly depends on the physical and mental fatigue during operational activity. The functional status is also influenced by gender, age, physical fitness, pain, boredom and emotions.

The limitation of the pilot study is a relative small number of investigated persons, including monitoring of HR during sea missions.

5 CONCLUSIONS

The developed Web system was based on the client-server architecture and international open source technologies, including Apache web server, PHP scripts and MySQL database. The system is implemented in client server architecture. The server stores the data and controls system's basic functions.

The main operational advantage of the developed system is the access to the data bank for several users at the same time. Users give priority to electronic data entry against filling questionnaires on paper. Electronic data entry saves operational time and material resources.

ACKNOWLEDGMENTS

Presented research was carried out in Klaipeda University and funded by a European Social Fund Agency grant for national project "Lithuanian Maritime Sectors' Technologies and Environmental Research Development" (VP1-3.1-SMM-08-K-01-019)

REFERENCES

[1] OECD Health Data 2013, http://stats.oecd.org/index.aspx?DataSetCode=HEALTH_STAT

[2] Varoneckas A., Mackutė-Varoneckienė A., Martinkėnas A., Žilinskas A., Varoneckas G. Web-based tool for management of CAD patients after coronary bypass surgery//Computers in Cardiology 2005: 32nd Annual Conference on Computers in Cardiology, Lyon, September 25-28, 2005. vol. 32, p. 155-158.

[3] Varoneckas G., Martinkenas A., Podlipskytė A., Varoneckas A., Žilinskas A. A web-based data bank of

heart rate and stroke volume recordings during sleep // E-Health: proceedings of Med-e-Tel 2006: the international trade event and conference for eHealth, telemedicine and health ICT: April 5-7, 2006. p. 371-375.

[4] dbForge Studio for MySQL// https://www.devart.com/dbforge/mysql/studio/

[5] Zigmond A.S., Snaith R.P. The Hospital anxiety and depression scale. Acta Psychiatrica Scandinavica 1983, 67: 361-70.

[6] Beck A.T, Steer R.A, Brown G.K. 1996. Beck Depression Inventory. Second Edition manual. San Antonio, TX: The Psychological Corporation.

[7] Varoneckas G. The subjective assessment of sleep by the Pittsburgh Sleep Quality Index (Lith.). Nervų ir psichikos ligos [Nervous and Mental Illness]. 2003, 4 (12), 31-33.

[8] Buysse D.J., Reynolds III C.F., Monk T.H. et al. The Pittsburgh sleep quality index: A new instrument for psychiatric practice and research. Psychiatry Research 1988, 28: 193-213.

[9] Murray W. Johns. A New Method for Measuring Daytime Sleepiness: The Epworth Sleepiness Scale. Sleep 1991, 14(6): 540-545.

[10] Jenkinson C., Layte R., Wright L., Coulter A. The U.K. SF-36: an analysis and interpretation manual. A guide to health status measurement with particular reference to the Short Form 36 health survey. University of Oxford. - 1996. 65 p.

[11] Puzaras, P., Ančerytė, D., Martinkėnas, A., Varoneckas, G. Catholic faith and life quality (Lith). Sveikatos mokslai. – 2000, 2: 28-35.

[12] Stankus A. Multidimensional Fatigue Inventory (Lith). Biologinė psichiatrija ir psichofarmakologija [Psychopharmacology and Biological Psychiatry] 2007, 9 (2): 86-87.

[13] Tiesinga L.J, Dassen T.W.N, Halfens R.J.G. et al. 1997. Measuring fatigue with the Dutch Fatigue Scale (DUFS) and measuring exertion fatigue with the Dutch Exertion Fatigue Scale (DEFS): Manual. University of Groningen: Department of Health Sciences.

[14] Bringsén Å, Andersson HI, Ejlertsson G. Development and quality analysis of the Salutogenic Health Indicator Scale (SHIS). Scand J Publ Health 2009, 37: 13-9.

[15] Nilsson P, Bringsén Å, Andersson HI, Ejlertsson G. Development and Quality analysis of the Work Experience Measurement Scale (WEMS). WORK 2010, 35: 153-161.

[16] Björkqvist, Österman, 1992. Work Harassment Scale. Vaasa, Finland: Department of Psychology, Abo Academy University.

[17] M Malik, A.J. Camm, Heart rate variability. Clinical Cardiology, 13 (1990), pp. 570–576

[18] Pincus S. Approximate entropy as a measure of system complexity, Proc. Nat. Acad. Sci. USA, 88, (1991), 2297-2301.

[19] Chen Z, et all. Effects of nonlinearities on detrended fluctuation analysis, Phys. Rev. E, 65, (2002), p. 1-15.

[20] Goldberger AL, et all. Fractal dynamics in physiology: alterations with disease and aging. Proceedings of National Academy of Sciences, v.99, (2002), 2466-2472.

Implementation of CSR Aspects in Human Resources Management (HRM) Strategies of Maritime Supply Chain's Main Involved Parties

T. Pawlik
Bremen University of Applied Sciences, Centre of Maritime Studies, Bremen, Germany

S. Neumann
University of Vechta, Germany

ABSTRACT: As stated by Efthimios Mitropoulos, the shipping industry needs a sense of CSR as much as for advances in the field of technology and safety. He argues that it is time for shipping to adopt a formal, standardized approach to CSR (Lloyd's List, 2010). Many carriers, terminal operators and forwarding agents are already committed to CSR topics such as environmental friendliness. However, CSR programs have to be fully integrated throughout all company's operations to be effective. Employee involvement is a critical success factor for CSR performance. Consequently, when striving for becoming a high performing CSR organization, it is necessary for companies to understand that HR professionals play a key role in achieving CSR objectives. This paper contributes to identifying links between CSR and HRM. Furthermore, it delivers HR related CSR examples implemented by terminal operators as one of the maritime supply chain's main involved parties. Discussing these examples can support other actors within the maritime supply chain in the evaluation of existing and the development of additional CSR components within HRM.

1 INTRODUCTION

1.1 *Corporate Social Responsibility*

Starting in the 1980s consumers' and wider society's awareness of social and environmental concerns have increased and resulted in a trend towards socially responsible consumerism and investing (Jonker et al., 2011 and Fafaliou et al., 2005). This trend affected business policies of companies and made them invest in activities that aim for more than just profit maximization. According to the European Commission's definition of 2006 Corporate Social Responsibility (CSR) means that companies voluntarily integrate social and environmental concerns in their business operations and their interaction with stakeholders. Likewise, stakeholders expect CSR actions being taken along the whole supply chain, including not only the production but also the shipment and transhipment of goods (Pawlik et al., 2012): "Many users of shipping services want to ensure their goods are being shipped in a 'socially responsible' way" (Matthews, 2010). Terminal operators play an important role within the maritime supply chain as a service provider in ports where goods begin, continue or end their sea journey. The focus of this paper will therefore be on terminal operators and their CSR concepts.

It seems to be widely accepted that a successful CSR concept has to be integrated on all levels of a company – top management as well as all employees need to be sensitized to the CSR objectives (Pawlik et al., 2012). According to Strandberg (2009) human resource managers have the tools and opportunity to leverage employee commitment to, and engagement in the company's CSR strategy. Consequently, proper Human Resource Management is an essential component of an authentic and successful CSR concept.

1.2 *Human Resource Management*

Nowadays HR professionals could be regarded as strategic business partners who are involved in corporate decision-making and policy formulation (Inyang et al., 2011). Inyang et al. (2011) stress that HR policies and strategies form the framework for culture in the organization and that human resource professionals are change agents who can improve an organization's capacity for a change by shaping processes and culture. The HR function is the responsible party to focus on actions that build employee competencies and motivation to engage in a company's strategy (Friedman, 2009). Thus, HR professionals can help the organization to achieve its CSR goals and by doing so, to deliver greater benefits for the business, for employees, for society, and for the environment.

1.3 The nexus between Corporate Social Responsibility and Human Resource Management

Following the discussion above it seems quite obvious that there is a strong nexus between Corporate Social Responsibility and Human Resource Management: "There is no doubt that CSR is a strategic issue that permeates departmental boundaries and influences the way the organization does its business and relates with its stakeholders, both internally and externally. The HRM function is equally a pervasive responsibility which affects all units and departments in the organization" (Inyang et al., 2011). If employees throughout the whole company are engaged in CSR, this might have a positive impact on its image and economic performance of the company (Strautmanis, 2008). On the contrary, if employees are not engaged, CSR becomes only an exercise in public relations and in respect thereof the company's credibility will become damaged when it becomes evident that the organization is not "working the talk" (Mees and Bonham, 2004). The HRM can provide action plans on how to strategically and successfully implement a CSR program, lead and educate organizational members on the value of CSR, and motivate them to participate in CSR actions. In the following, HRM activities of terminal operators on implementing a CSR program will be outlined.

2 CONCEPTUAL FRAMEWORK AND SAMPLE

2.1 The Human Resource Cycle

As mentioned above, in today's management thinking it is widely accepted that human resource management has to be an integral part of strategic management. Since the early work of Galbraith and Nathanson (1978), who added HRM to the arena of strategic management, several HRM concepts have been proposed, including among others the Harvard-concept and the Michigan-concept (Felger et al., 2004). In this paper, the Michigan-concept will be used as the guiding conceptual framework. This concept, developed by Tichy, Fombrun and Devanna (1982), still fascinates by its distinct focus on four generic human resource management functions – selection, appraisal, rewards and development, which can be represented as a human resource cycle (HR-cycle, see fig. 1). The dependent variable in this cycle is performance as "a function of all the human resource components: selecting people who are best able to perform the jobs defined by the structure; motivating employees by rewarding them

judiciously; training and developing employees for future performance; and appraising employees in order to justify the rewards. In addition, performance is a function of the organizational context and resources surrounding the individual. Thus, strategy and structure also impact performance (…)." (Tichy et al., 1982).

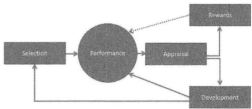

Figure 1. The Human Resource Cycle (adapted from Tichy et al., 1982)

Enhancing the four generic functions by CSR-aspects leads to the following questions concerning the sample:
1 Are there examples for links between CSR and the recruiting process (selection)?
2 Does the appraisal process involve CSR matters?
3 Is CSR linked to the rewards function?
4 Does personnel development include CSR elements?

In order to approach these questions, we analyzed the websites of our sample in the period between November 2014 and January 2015. All websites were finally verified before transmitting this paper to the editors. The results of this analysis shall deliver a starting point for future research. At a later stage, studies that are more comprehensive have to deploy research methods such as questionnaires and/or expert interviews to enhance our initial findings. Of course, the analysis of websites can only identify those aspects of CSR-related HRM issues that the companies communicate explicitly. On the one hand, this may sound like a limitation but on the other hand, analyzing HR topics communicated on an organization's website allows interpretations in the light of the signaling theory. Signaling theory belongs to the research agenda of economics of information. Applying this concept to the context of HRM roots back to Spence (1973), who used signaling theory inter alia to explain that employers have to interpret certain signals transmitted by job applicants since various attributes of potential employees are not observable. However, signaling theory can also be used to examine the perspective of potential employees (Suazo et al., 2009): "Signaling theory suggests that information is important because job seekers facing uncertainty and incomplete information use the information available as signals about job and organizational attributes" (Allen et al., 2007; Rynes, 1989; Spence, 1973). As stated above, information on CSR activities and practices form an essential part of those signals that companies send to potential employees (Gond et al., 2010). Nowadays, companies' web sites are in many cases the initial point of contact between job seekers and their potential employers. Thus, all human resources related information delivered on organizational web sites, including CSR matters, are crucial for a job seeker's perception of the employer-image of a company (Allen et al., 2007).

2.2 Sample

There are a number of different players in the container terminal operators industry. Notteboom & Rodrigue (2012) for instance suggest a distinction between stevedores, maritime shipping companies and financial corporations. In our analysis, we focus on the four largest global players: PSA International (PSA), Hutchison Ports Holdings (HPH), APM Terminals (APM) and DP World. PSA and HPH belong to the category of stevedores, using horizontal integration as their business model. APM's business model is vertical integration as the terminal activities are part of the Maersk Group. Dubai World has a major share in DP World, thus this terminal operator holding is strongly linked with the government of Dubai. In 2012, the four largest global container terminal operators had in total – measured by equity based throughput – a market share of ca. 26 percent (see table 1) which remained on the same level in 2013 (Drewry, 2015).

Table 1. Global Terminal Operators (adapted from Drewry, 2013)

Ranking	Operator	Mio. TEU	Share of World Throughput
1	PSA International	50.9	8.2 %
2	Hutchinson Port Holdings	44.8	7.2 %
3	APM Terminals	33.7	5.4 %
4	DP World	33.4	5.4 %

It has to be noted that the following survey outcomes only result from the analysis of the global websites of the corporations of our sample. A more detailed analysis of e.g. regional websites should be has to be conducted in the future to verify below findings.

3 SURVEY RESULTS

3.1 Selection

The analysis of the terminal operators' websites shows that most CSR/HR-signals linking to the HR-cycle segment of "selection" relate to the field of diversity. According to Williams & Bauer (1994) such diversity signals are an important element in the recruitment process and can have a positive impact on a company's attractiveness to potential employees. It seems to be quite common that companies interpret their diversity practices as an integral part of their CSR framework, i.e. understanding diversity as a sub-domain of CSR (Subeliani & Tsogas, 2007).

APM is aware of the fact that they are dependent "on a strong talent pipeline to deliver on its growth ambitions, especially in growth markets where APM Terminals is increasing its presence" (APM, 2014). They see themselves as being "committed to providing a positive workplace for [their] people where opportunities are equal and differences are valued, and to exploring the largest possible talent pool when hiring. [APM wants] to attract the best and brightest people from the broadest pool possible, and ensure that [they] select the right candidates, based on merit, skill and personality" (ibid.) APM also addresses shortcomings in their diversity activities: "While there is a constant influx of emerging market talent, the representation of women in senior leadership positions (director and general manager level) has not grown at the same pace, achieving a growth of 1% compared to last year. APM Terminals is committed to drive change in this area, and we measure our employees' views on our commitment to Diversity & Inclusion in the annual Global Employee Engagement survey" (ibid.).

Under the heading, "People and Safety" diversity messages can also be found on DP World's website: "We recognise that our success is enhanced by the diversity of our people. We have active equal opportunities polices that support our drive towards a culture of diversity, inclusion and equality. We believe it is important to help unleash the talent and potential of everyone working at DP World and the rich and diverse mix of backgrounds, beliefs, cultures, skills and knowledge is a major contributor towards our continual business success and performance" (DP World, 2015).

PSA (2015) does not directly link CSR and diversity management and on its CSR webpage, there are no direct hints towards personnel selection aspects. However, on the webpage "Our people" PSA states to be "guided by [.] core beliefs in Diversity, Equal Opportunity, Leadership Grooming and Learning & Development." PSA believes "in a work environment that attracts, retains and fully engages diverse talents, leading to enhanced innovation and creativity in our services." Furthermore, PSA proclaims: "We continue to build a work culture where all employees feel welcome, valued, and able to express their ideas and beliefs freely. By doing so, we are better able to encourage diversity of thought and approach; and more likely to discover innovative ways to enhance our services."

In the case of HPH, it is difficult to identify a clear signal that links CSR and their process of personnel selection. Although, HPH carries out many activities in the field of CSR and reports on them on their webpage "Corporate Social Responsibility", there are no specific messages recognizable on the "Career Centre" webpage. The only statement relating to diversity is "The 30,000-plus people at HPH represent a wide variety of nationalities, cultures and backgrounds, all working together in a supportive and cooperative environment" (HPH, 2015).

In addition to "diversity management", CSR/HR-signals related to the HR-cycle segment of "selection" can also embrace the creation of local employment opportunities. DP World as well as APM provide respective statements on their websites, whereas such explicit statements could not be found on the global websites of HPH and PSA.

3.2 *Appraisal*

According to Devi (2009) the "designing of [a] Performance Management System should be done in such a manner that it measures the socially responsible initiatives taken by employees. This becomes important as the internalization of CSR in an organizational culture requires that appropriate behaviours get appraised, appreciated as well as rewarded. Otherwise, the organization might fail to inculcate it amongst all employees due to lack of positive reinforcement." Also, Lame & Care (2011) argue that "CSR related goals should be jointly set between the supervisor and employee so that both feel a sense of commitment to the goals and accountability for their achievement. On-going performance feedback, reasonable resources, and other support should be provided to employees to assist them in realizing their goals. (…) Rewards for CSR can then be linked to the achievements of either group or individual level goals."

Apparently, the websites of the companies in our sample do not give any direct signals of CSR related appraisal systems. However, DP World's sustainability section of its annual report indicates such signals: "We regularly measure our progress against our corporate responsibility strategy. (…) We also believe that it is important to measure changes in behaviour and embed such change through individual performance objectives. Significantly, an outcome of the Corporate Responsibility Advisory Committee in 2013 was the decision to incorporate corporate responsibility focused objectives for senior management with ongoing monitoring to be implemented in 2014 to chart progress" (DPW, 2014).

Similar to this, an extended web research identified a link between CSR and HR appraisal in the case of HHLA, the Hamburg based terminal operator. In its publicly available remuneration report, HHLA describes the assessment process for their executive board members that explicitly considers ecological as well as societal CSR components in addition to financial targets (HHLA, 2014).

3.3 *Rewards*

When CSR related goals – subject to formal appraisal – are being achieved by employees, their performance should be rewarded in order to sustain the level of motivation and commitment towards those goals. The list of possible rewards that exist in organizational settings is remarkably long – ranging from pay in its various forms (e.g. bonuses or stock options) to management praise, respect from colleagues or extended responsibilities (Tichy et al., 1982). Even if it appears at first sight that the companies of our sample do not give any direct signals of CSR related rewards, there are some initiatives that could be interpreted as rewards for CSR related goal achievement.

As mentioned in section 3.1 PSA is guided by their core beliefs in "Diversity, Equal Opportunity, Leadership Grooming and Learning & Development". Accordingly, people who are role models of the organization's beliefs and vision will be rewarded with the PSA Global Champions Award: "To celebrate the people and spirit of PSA, we launched the inaugural PSA Global Champions Award. The annual Award recognises our 'heroes' – role models of our shared vision for employees around the world to emulate as we rally together to shape our future." (PSA, 2012). Another example of how PSA rewards employees' CSR engagement is presented in the Annual Report 2012: "In 2012, selected staff from PSA Antwerp and PSA Zeebrugge were given the opportunity to attend the London Paralympics, as a token of gratitude for their volunteer work rendered to the Belgian Paralympics Committee." (PSA, 2012).

DP World (2014) mentions in its Annual Report 2013 that employee engagement in CSR actions, such as corporate volunteering is important. But it is not explained if employees will be rewarded for their engagement.

APM set a variety of goals regarding CSR (such as the reduction in CO_2 per TEU) and it is mentioned that goal achievement is regularly assessed. Generally, goal achievement will be rewarded. But from the available information it is not judgeable what kind of rewards employees get for goal achievement (APM, 2014): "Since talented employees are the foundation of our company culture and success, your performance will be rewarded in a way that promotes continued achievement. APM Terminals offers employees fair, competitive compensation and rewards which include continuous development opportunities".

HPH also set CSR goals (such as waste recycling) but it cannot be seen if and what kind of reward employees get if these goals are achieved (HPH, 2014).

3.4 *Development*

In their Annual Report 2013 of PSA argues that personal development is taken seriously in all areas (PSA, 2013). In this context no information is provided as to what extent the company's learning and development offerings include activities designed to ensure that employees are properly equipped with skills and knowledge to carry out CSR related actions and achieve set goals.

In DP World's Annual Report 2013 "creating a learning and growth environment" is presented as one of the four strategic priorities which are meant to support the idea of creating value for all stakeholders (DP World, 2014). Offerings concerning CSR learning opportunities are explicitly mentioned: "We also launched an online e-learning module for our people to improve their understanding of what corporate responsibility is and what it means to our Company." (DP World, 2014).

APM Terminals also presents e-learning opportunities (e.g. anti-corruption e-learning course) for employee development in order to sensitize them for the company's CSR policy (APM, 2014). Furthermore, APM invested in other training and networking opportunities: "APM Terminals' encourages you to assume ownership of your development with the full support of the organisation. To enable you, we offer some of the best learning tools and training programs within the industry. We partner with our leading professionals, run top talent programs and develop leaders to be role models".

HPH displays the prominent meaning of employee development and training: "We are committed to investing in the development and professional growth of our employees. To this end, we regularly arrange forums such as training courses, seminars, sharing sessions, web-based conferences, or in collaboration with external training institutes" (HPH, 2014). From an outside perspective it is not detectable if these trainings are offered to provide employees with necessary skills to achieve CSR goals.

4 CONCLUDING DISCUSSION AND RECOMMENDATIONS

The review of the companies' websites above indicates that a wide-ranging variety of approaches for stimulating employees' CSR involvement is already in place. Using the idea of signaling theory,

we assume the perspective of a potential employee, thus we were only able to identify explicit CSR statements. Even if the analysis of websites can only identify those aspects of CSR-related HRM issues that the companies communicate explicitly - what may seem as a limitation – the idea behind this approach is to hold up a mirror to the companies and show them that there is room for more communication. It is most likely that the companies of our sample consider many additional CSR measures in their HRM system than those aspects which we could identify from an analysis of their websites. Further research could be conducted to assess whether the non-exhaustive communication of CSR issues is intentional or due to inattentiveness.

Against the background of potential employees' preferences, such non-exhaustive communication could be regarded as a wasted chance to attract as many good applicants as possible. All human resources related information delivered on organizational web sites, including CSR matters, are crucial for a job seeker's perception of the employer-image of a company (Allen et al., 2007). Young professionals pay attention to CSR actions taken by companies. According to Zappala (2004) an organization's image as a good corporate citizen attracts potential employees and PricewaterhouseCoopers (2008) revealed in a study that "corporate responsibility is critical – 88% of millennials said they will choose employers who have CSR values that reflect their own". Neumann et al. (2012) confirmed these findings in a more recent study: "young people expect companies to act not only profit-oriented, but recognize their overall responsibility for society and the environment and act accordingly".

Hence, we suggest that companies strategically implement CSR into HRM and communicate this thoroughly. There are various recommendations for integrating CSR into HRM available, e.g. Industry Canada (2009) proposes the following ten steps:
1 Vision, mission, values and CSR strategy development
2 Employee codes of conduct
3 Workforce planning and recruitment
4 Orientation, training and competency development
5 Compensation and performance management
6 Change management and corporate culture
7 Employee involvement and participation
8 CSR Policy and Program Development
9 Employee Communications
10 Measurement, Reporting – and celebrating successes along the way!

This paper represents the starting point of a larger research project in which other actors of the maritime supply chain such as logistics providers as well as container shipping lines will be considered in the same way. The CSR-HRM scope leaves room for a multitude of further research approaches and we look forward to elaborate in more detail on this ourselves and to follow the appreciated work of other researchers.

REFERENCES

Allen. 2007. Web-Based Recruitment: Effects of Information, Organizational Brand, and Attitudes Toward a Web Site on Applicant Attraction. *Journal of Applied Psychology*, 92: pp. 16961696-17081696-1708.

APM Terminals, 2014. *Sustainability Report* 2013.

Devi, A. 2009. Corporate Social Responsibility: The Key Role of Human Resource Management. *Business Intelligence Journal*, 2(1).

DP World. 2014. *Connecting global markets* (Annual Report). Available: http://web.dpworld.com/wp-content/uploads/2014/05/2013DPWorldAnnualReportAccounts.pdf [2/7/2015].

DP World. 2015. *People and Safety*. Available: http://web.dpworld.com/sustainability/people-and-safety/#zvAAWkroIfo4QhR1.99 [1/30/2015, 2015].

Drewry Shipping Shipping Consultants Ltd. 2014. *The Independent Maritime Adviser 2014-last update*. Available: http://www.drewry.co.uk/news.php?id=293 [2/3/2015, 2015].

Drewry Shipping Consultants Ltd. 2013. *The Independent Maritime Adviser 2013-last update*. Available: http://www.drewry.co.uk/news.php?id=293 [2/3/2015, 2015].

Friedman, B. 2009. Human resource management role implications for corporate reputation. *Corporate Reputation Review*, 12 (3): pp. 229-244.

Fafaliou, I., Lekakou, M. and Theotokas, I. 2006. Is the European shipping industry aware of corporate social responsibility? The case of the Greek-owned short sea shipping companies. *Marine Policy*, 30(4): pp. 412-419.

Felger, S., Paul-Kohlhoff, A. and Marginean, A. 2004. *Human Resource Management Konzepte, Praxis und Folgen für die Mitbestimmung*. Düsseldorf: Hans-Böckler-Stiftung.

Galbraith, J.R. and Nathanson, D.A. 1978. *Strategy implementation the role of structure and process*. St. Paul, Minn: West Pub. Co.

Gond, J., El-Akremi, A., Igalens, J. and Swaen, V. 2010. *Corporate social responsibility influence on employees*. ICCSR Research Paper Series—ISSN, pp. 1479-1512.

HHLA. 2014. *Annual Report 2013*. Hamburg.

Hutchison Port Holdings Limited. Available: http://www.hph.com/en/webpg-22.html [1/30/2015, 2015].

HPH Limited. 2014. *Annual Report*.

Industry Canada. 2015, *Human Resources for Sustainability - Corporate Social Responsibility*. Available: http://www.ic.gc.ca/eic/site/csr-rse.nsf/eng/h_rs00552.html [2/2/2015, 2015].

Inyang, B., Awa, H. and Enuoh, R. 2011. *International Journal of Business and Social Science*, 2(5) Special Issue March 2011: pp. 118-126.

Jonker, J., Stark, W. and Tewes, S. 2011. *Corporate Social Responsibility und nachhaltige Entwicklung: Einführung, Strategie und Glossar*. Berlin and Heidelberg: Springer.

Lam, H. and Khare, A., 2010. HR's crucial role for successful CSR. Journal of International Business Ethics, 3(2), pp. 3-15.

Lloyd's List, 2010. CSR and Shipping. 25 March 2010, *Lloyd's List*: p. 4.

Matthews, S. 2010. Shipping sees the broader benefits of acting responsibly. *Lloyd's List*, 25 May 2010, p. 4.

Mees, A., and Bonham, J. 2004. Corporate social responsibility belongs with HR. Canadian HR Reporter, April 2004, pp. 11-13.

Muirhead, S.A. 2002. *Corporate citizenship in the new century: Accountability, transparency, and global stakeholder engagement*, 2002, Conference Board.

Müller-Christ, G. and RKW GmbH (eds.) 2011. *Der Nachhaltigkeits-Check. Die Sicherung des langfristigen Unternehmensbestandes durch Corporate Social Responsibility*. Bremen: „initiative umwelt unternehmen". Eine Aktivität des Bremer Senators für Umwelt, Bau, Verkehr und Europa.

Neumann, S. and Pawlik, T. 2012. Corporate Social Responsibility and Employer Branding. In Lemper, B., Neumann, S. and Pawlik, T. (eds.), *The Human Element in Container Shipping*. Frankfurt a. Main: Peter Lang Verlag: pp. 39-54.

Notteboom, T. and Rodrigue, J., 2012. The corporate geography of global container terminal operators. *Maritime Policy & Management*, 39(3): pp. 249-279.

Pawlik, T., Gaffron, P., Drewes, P. 2012. Corporate Social Responsibility in Maritime Logistics. In Dong-Wook, S. and Panayides, P. (eds.), *Maritime Logistics: Contemporary Issues*. Bingley: Emerald Publishing Group: pp. 205-226.

Preuss, L., Haunschild, A. and Matten, D. 2009. The rise of CSR: implications for HRM and employee representation. *The International Journal of Human Resource Management*, 20(4): pp. 953-973.

PricewaterhouseCoopers. 2008. *Managing tomorrow's people – Millenials at work: Perspectives from a new generation*. Report. Available: https://www.pwc.de/de/prozessoptimierung/assets/millennials_at_work_report08.pdf [2/7/2015].

PSA International, 2012. *Annual Report 2012*. Available: http://www.internationalpsa.com/about/pdf/AR2012/AR2012.pdf [02/07/2015].

PSA International, 2013. *Annual Report 2013*. Available: http://www.internationalpsa.com/about/pdf/AR2013/AR2013.pdf [2/7/2015].

Rynes, S.L., 1989. Recruitment, job choice, and post-hire consequences: A call for new research directions. *CAHRS Working Paper Series*: pp. 398.

Strandberg, C. 2009. *The role of human resource management in corporate social responsibility issue brief and roadmap*. Report for Industry Canada. Burnaby, B.C: Strandberg Consulting.

Strautmanis, J. 2008. Employees' values orientation in the context of corporate social responsibility. *Baltic Journal of Management*, 3(3): pp. 346-358.

Suazo, M.M., Martinez, P.G. and Sandoval, R. 2009. Creating psychological and legal contracts through human resource practices: A signaling theory perspective. *Human Resource Management Review*, 19(2): pp. 154-166.

Subeliani, D. and Tsogas, G. 2005. Managing diversity in the Netherlands: a case study of Rabobank. *The International Journal of Human Resource Management*, 16(5): pp. 831-851.

Tichy, N.M., Fombrun, C.J. and Devanna, M.A. 1982. Strategic Human Resource Management. *Sloan management review*, 23(2): pp. 47.

Williams, M.L. and Bauer, T.N. 1994. The Effect of a Managing Diversity Policy on Organizational Attractiveness. *Group & Organization Management*, 19(3): pp. 295-308.

Zappala, G. 2004. Corporate Citizenship and Human Resource Management: A New Tool or a Missed Opportunity? *Asia Pacific Journal of Human Resources*, 42(2): p. 185-201.

Analysis of Factors Influencing Latvian Seafarers' Outflow Rate

R. Gailitis
Latvian Maritime Administration, Riga, Latvia

ABSTRACT: In general the seafarers' pool is stock model where the future supply of seafarers depend on present stock and difference between the inflow of new seafarers and outflow from seafarers' pool due to retirement or work possibilities in other industries ashore. The outflow can be viewed as probability that decision to leave employment of board is taken and therefore outflow depend on employment conditions on board and attraction of other factors ashore. Even the seafarers' employment can be viewed as global, often the residence of seafarers remains linked with particular country therefore interaction exist between global and local employment conditions. The aim of this article is to analyse and describe factors which can be used to determine and analyse Latvian seafarers' outflow.

1 INTRODUCTION

One of maritime industry's problems linked with growth of world fleet is concerns related to shortage of manpower. International Maritime Organization emphasises that already global shortage of seafarers, especially officers, has already reached serious proportions, threatening the very future of the international shipping industry, which is the lifeblood of world trade. (IMO) In view of maritime industry seafarer's profession is becoming less attractive which results in less entrants in the labour market. One of the counteractions of IMO is campaign "Go to sea" to attract entrants to the shipping industry (IMO 2008). However number of available seafarers depend not only on inflow of new entrants in seafaring profession but also from the outflow of existing pool.

This topic is not widely researched by applying statistical methods as reliable data regarding seafarers is scare and often not available (European Commission, 2011). Therefore, this research was carried out by analysing information from the Certification database of The Seamen Registry of Maritime Administration of Latvia.

1.1 Data gathering

Reliable data is the basis for any quantitative analysis. Data stored in the Certification database of The Seamen Registry of Maritime Administration of Latvia gives possibility to research seafarer's employment from different viewpoints (Gailitis R., 2014). Database collects information about seafarers' employment from Latvian crewing companies and additionally from seafarers' themselves. Due to fact that part of Latvian seafarers is employed directly from shipowners or work through foreign crewing companies data about their employment regarding last years is fragmented as data about their employment is corrected only when such person submit documents for certification. To decrease impact from the lack of employment data as the time period for analysis was chosen 10 years (from 2003 till 2012). Increase of outflow rates in the last years can be partly explained by insufficient information about all employed persons.

1.2 Context of seafarer in this research

The definition of a seafarers as any person working on board is too broad due to internal differences among the employed seafarers on board. Employment on board differs in terms of responsibility level, qualification requirements and department were a person works. As mentioned by David Glen it is relatively easy to enter and leave employment on board at lower skill levels (Glen, 2008), therefore outflow rates for different qualification levels should be analysed to determine differences between qualification levels.

David Glen refers to criteria developed by Li and Wonham (Li & Wonham, 1999) regarding to measuring seafarers' number. One of the criteria which characterises seafarer is seafarer's qualifications therefore this criteria was used in analysis to differentiate seafarers. Internationally STCW convention sets standards for seafarers' qualifications, therefore analysis of this research were limited to seafarers for which position on board is mainly based on STCW requirements covering ratings and officers from the deck and engine department. See Table 1. In this research ratings in engine department also include equivalent and similar qualifications such as fitters, oilers, wipers, greasers etc. even STCW convention covers only standards for ratings forming part of an engine room watch and able seafarer engine.

Table 1. Seafarers' qualifications, department and positions

Qualification Level		Department	
		Deck	Engine
Ratings	Support	Deck ratings	Engine ratings
Officers	Operational	3rd Off, 2nd Off	4th Eng, 3rd Eng
	Management	Ch. Off, Master	2nd Eng, Ch. Eng

1.3 Employed seafarers

In this research as employed seafarer is considered person who worked on board ship at least once in particular year according available data from the seafarers' database. Table 2 shows number of employed Latvian seafarers according to the year and qualification.

Table 2. Number of analysed Latvian seafarers (2003 – 2012)

Year	Deck ratings	Engine ratings	Deck officers	Engine officers
2003	2654	1690	1599	1715
2004	2705	1720	1647	1751
2005	2611	1668	1745	1793
2006	2478	1552	1807	1874
2007	2096	1273	1863	1890
2008	1960	1194	1910	1934
2009	2174	1276	1992	1983
2010	2204	1277	2015	2028
2011	2155	1271	2037	2003
2012	2054	1247	1979	1938
Average	2309	1417	1859	1891
δ	275	213	156	108

The data shows that there were employed equal number of deck and engine officers, while number of deck ratings exceeds employed engine ratings on average.

Table 3 refers to the qualifications and average frequency of employment regarding ship's type. Ships are classified according applied methodology by EQUASIS (EQUASIS, 2014). It can be seen that main ship's types on which Latvian seafarers were employed in time period 2003 – 2012 are tankers and general cargo ships.

Table 3. Employment of Latvian seafarers (2003 - 2012) (MAL 2015)

Ship's type	Deck rating	Engine rating	Deck officers	Engine officers
Oil and Chemical Tankers	40%	52%	50%	50%
General Cargo ships	19%	17%	17%	15%
Gas Tankers	6%	7%	13%	13%
Offshore Vessels	6%	2%	3%	4%
Heavy load carriers	5%	5%	3%	3%
Passenger ships	6%	4%	2%	3%
Ro-Ro Cargo Ships	4%	3%	3%	3%
Other type	14%	10%	9%	10%
Average empl. seafarers	2 309	1 417	1 859	1 891

Figure 1 reflects average age of employed seafarers as well as +/- standard deviation from average age for seafarers employed in the given position. It can be seen that they are similar between parallel positions at deck and engine departments comparing average ages and standard deviation. For example average age for 2nd officers and 3rd engineer is 35 years, while standard deviations are 10 and 9 years accordingly.

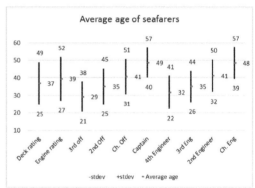

Figure 1. Average age of employed seafarers (2003 - 2012) (MAL 2015)

1.4 The impact of the economic crisis on Latvian economy

World economy faced one of the most severe crisis in 2008 since the Great Depression of the 1930s (United Nations, 2009). The economic crisis had considerable impact on the Latvian economy, which also reflected in decrease of gross domestic product of Latvia and sharp increase in unemployment in 2009 (Fig. 2).

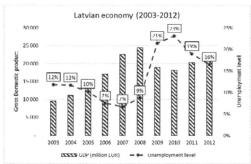

Figure 2. Changes of Latvian GDP and unemployment (2003 - 2012) (CSB 2015)

Due to the changes in national economy it can be assumed that crisis had impact on changes in outflow rates for seafarers as well as employed seafarers were willing to keep their employments positions on board and stable source of income.

2 OUTFLOW OF SEAFARERS

Waal describing maritime officers' model indicates that outflow from the nautical sector comprises people who retire and who stop working on board of vessels for other reasons like another job inside or outside the sector or get incapacitated or unfit to work (Waals & Veenstra, 2002). The number of people who retire can be estimated relatively easy by researching age profile of seafarers and making conclusion about the age above which persons retires and leaves the nautical sector. The main difficulties lies with determination of number or percentage of persons who leave nautical sector before the age of retirement as usually no figures are available. According to Waal, outflow rate is not a fixed figure, but follows a certain development over time. He assumes that at the start of the career of a seamen, the outflow is quite high due to the fact that these are easily persuaded to find more interesting job. This process of outflow is assumed to be decreasing with age and reaches a minimum at age of 40-44. From then on, the wastage rate increases again because more and more people become incapacitated and reaches a maximum just before the retirement age. According to Waal, the main factor influencing seafarers' outflow is age. According to David Glen, for seafarers' working at lower qualification levels it is relatively easy to enter and exit active seafarers' pool. Therefore it can be assumed that the outflow rate for ratings will be higher than for officers. By analysing links between development of economy and supply of seafarers, Glen concludes that national supply will reduce as real incomes in country rise, as the 'opportunity cost' of a seafaring career (time spent away at sea) rises. (Glen, 2008). Therefore an assumption can be

made that changes in economy influences outflow rate as well. The opportunity costs are linked both with the employment conditions on board (earned salary, working far away from home and families, length of contract etc.) and with possibility to get job ashore which will satisfy a person.

For outflow analysis the real outflow rates from employed seafarers' pool were compared in relation to different factors such as qualification, age profile and ship's type.

2.1 *Outflow and qualification*

The data about outflow rates were analysed between officers and ratings from deck and engine departments (Fig. 3). It can be assumed that changes in GDP (Fig. 2) had influence to outflow rate for ratings. Increase of outflow from 2010 can be explained by data error described earlier and also recovering of Latvian economy after sharp fall.

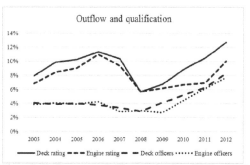

Figure 3. Outflow rates of officers and ratings (2003 – 2012) (MAL 2015)

It can be easily concluded that outflow for ratings differs from outflow for officers and outflow rate for similar qualifications such as engine ratings and deck ratings or engine officers and deck officers tend to be similar.

Table 4. Results of t-test comparing outflow rates for officers and ratings (Author's calculations)

Qualification	Ratings	Officers
Department	Deck/ Engine	Deck/ Engine
Mean	9.41% / 8.01%	4.59% /4.27%
Variance	0.046% / 0.032%	0.026% / 0.024%
Observations	10	10
Pearson Correlation	0.859	0.938
Hypothesized Mean Difference	0	0
df	9	9
t Stat	4.05	1.79
P(T<=t) one-tail	0.00	0.05
t Critical one-tail	1.83	1.83

Differences in outflow rates were checked by using paired t-test with significance level α = 0,05 to decide whether there is a significant mean difference between results of outflow for deck and engine departments. In case of ratings t_{stat} > $t_{critical}$, so it can be concluded that even the outflow rates varies similar as Pearson correlation coefficient is 0,859, the average outflow rate for deck and engine ratings can be regarded statistically different. In case of deck and engine officers the outflow rates tend to have the same variation pattern as there is strong correlation between the outflow rates (R=0,938). Hypothesis that average outflow rate for deck and engine officers is similar cannot be rejected with probability of 95% as t_{stat} < $t_{critical}$, however actual p value of test results is 0.0536 therefore hypothesis can be rejected with maximum probability of 94%. The outflow rates in time period of 2003 – 2008 for deck and engine officers tend to be almost similar as p value for this time period is 0.433 which shows that maximum probability to reject hypothesis, that the average outflow rates for deck and engine officers are different, is 57%.

It can be concluded that average outflow rates for deck ratings and also officers is higher than for engine ratings and officers, which can be explained due to the higher supply level of deck ratings and officers in the last years.

Anova two factor analysis without replication were used to check at which extent position on board for officers influences outflow rate. Results shows that factors such as the position and year influence the outflow rate for deck officers with probability higher than 95%. For engine officers the position as factor influences outflow with probability of 70% and year influences the outflow with probability of 90%.

2.2 Analysis of ratings' outflow rates

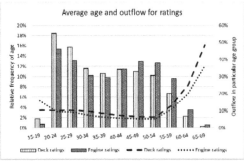

Figure 4. Average age structure of employed ratings (2003 - 2012) and responding outflow rates (MAL 2015)

Figure 4 shows average age structure of employed deck and engine ratings and the corresponding outflow rate for the age group. This graph clearly shows retirement age, which can be considered 60 years as only 2% of deck ratings and 4% of engine ratings were employed at the age group 60-64.

Table 5. Z test values for ratings' outflow rates in particular age group (Author's calculations)

Age group	Deck ratings	Engine ratings	Z test value
20-24	10%	10%	0.36
25-29	10%	9%	0.85
30-34	10%	7%	0.79
35-39	8%	6%	0.66
40-44	7%	6%	0.49
45-49	6%	5%	0.42
50-54	6%	5%	0.62
55-59	13%	11%	1.28
20-59	9%	7%	0.30

Z test for two population proportions were carried out with significance level of 95% (α = 0,05), to determine, if average outflow differ significantly comparing deck and engine ratings,

The results of Z test show that Z_{test} < $Z_{critical}$ (1.96) for all compared age groups. See Table 5. Therefore it can be concluded that the average outflow with regard to the age groups doesn't differ significantly, however according to the Figure 3 there can be statistical differences in particular years. Z_{test} values for two population proportions comparing proportion of outflow for deck and engine ratings in 2010 is 2.17, in 2011 is 3.44 in 2012 is 2.7 which exceed $Z_{critical}$ and supports this conclusion.

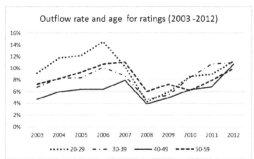

Figure 5. Outflow rates of ratings comparing the age groups (2003-2012) (MAL 2015)

Figure 5 shows average outflow rate for ratings taking into account the age group and year. It can be seen that there are noticeable differences comparing the age groups from 2003 till 2006. Outflow for younger ratings at age 20-29 were the highest, while for ratings at age 50-59 outflow was the lowest. The outflow for ratings at the age of 30 - 39 and 40-49 was similar in this time period. Starting from 2007, the differences are not so considerable and fluctuate around the mean values.

2.3 Analysis of officers outflows rate

Figure 6 shows the age structure of employed deck and engine officers and corresponding outflow rate for age group. This graph shows that seafarers retire at age 60 – 64 years as only 3% of deck officers and 3.6% of engine officers were employed at this age group.

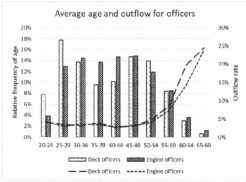

Figure 6. Average age structure of employed officers (2003 - 2012) and responding outflow rates (MAL 2015)

The average outflow for the deck and engine officers does not differ significantly as Z_{test} value comparing with the average outflow's proportion of employed deck and engine officers within age from 20 – 64 years is 0.71, which is less than $Z_{critical}$ value 1,96.

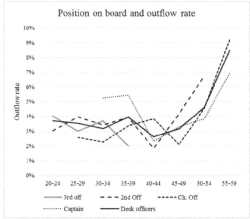

Figure 7. Average outflow rate for employed deck officers (2003 - 2012) regarding position on board (MAL 2015)

Figure 7 shows average outflow rates comparing employed deck officers at the age from 20 – 60 years and positions on board. Only age groups where position exceeded 10% from all deck officers employed in particular age group were included in analysis. For example 3rd officers were analysed in age groups from 20 - 39, 2nd officers were analysed in age groups from 20 - 55 etc. One sample z-test ($Z_{critical}$ = 1.96) for proportions were used for analysis regarding the outflow rate of all deck officers and

the outflow rate of deck officers in particular position. Results of analysis is given in the Table 6.

Table 6. Z test results for outflow rates regarding deck officers position on board and age group (Author's calculations)

Z test	3rd off	2nd Off	Ch. Off	Captain
20-24	0.63	1.34		
25-29	1.70	1.31	**2.97**	
30-34	1.47	0.73	**2.61**	**5.97**
35-39	**4.27**	0.09	1.29	**3.21**
40-44		**2.14**	**3.30**	0.88
45-49		**3.01**	**3.19**	0.31
50-54		**5.25**	0.11	1.86
55-59			1.00	**2.22**

According to the Table 6, outflow rate for chief officers statistically differs from the average deck officers outflow rate in age groups 25 – 34 and 40 - 49. For example, outflow rate for 3rd officers at age group 35-39 is lower as shown in Figure 6 while result for z test is 4.27 which considerably exceeds z critical value.

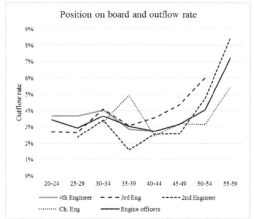

Figure 8. Average outflow rate for the employed engine officers (2003 - 2012) regarding position on board (Author's calculations)

The analysis for outflow rates regarding position on board and age group were carried out for the engine officers similar to the deck officers. Results are shown in Figure 8 and corresponding Z values are given in Table 7. For example, at age group 35-39 chief engineers and 2nd engineers outflow rates are statistically different from the average outflow rate for engine officers. Graph clearly shows that in case of chief engineers, outflow rate is higher and for 2nd engineers outflow rate is lower.

The results should be interpreted with care as test results gives only indication that on average proportion of employed captains or chief engineers which leave employment on board at age 35 – 39 is larger than average proportion of officers leaving employment on board at age 35-39.

Table 7. Z test results for outflow rates regarding engine officers position on board and age group (Author's calculations)

Z test	4th Eng	3rd Eng	2nd Eng	Ch. Eng
20-24	0.39	1.09		
25-29	**2.15**	0.86	1.65	
30-34	1.03	1.20	0.72	1.00
35-39	0.53	0.08	**4.37**	**5.71**
40-44	0.04	**2.62**	0.54	0.91
45-49		**3.62**	1.74	0.11
50-54		**4.76**	1.66	**2.17**
55-59			1.92	**2.74**

Table 8 gives an overview of the average outflow for deck or engine officers position' and age groups from 20 - 54 as the sharp increase regarding outflow rate starts with the age group of 55 - 59. It can be seen in the Table 8 that outflow rates tend to be similar independently on position or age group.

Table 8. Average outflow regarding officers' position and age group (Author's calculations)

	Age group						
	20-24	25-29	30-34	35-39	40-44	45-49	50-54
3rd Off	4%	3%	4%	2%			
4th Eng.		4%	4%	4%	3%	3%	
2nd Off	3%	4%	3%	4%	2%	4%	7%
3rd Eng.	3%	3%	4%	3%	4%	4%	6%
Ch. Off		3%	2%	3%	4%	2%	5%
2nd Eng.		2%	3%	2%	3%	3%	5%
Captain			5%	5%	2%	3%	4%
Ch. Eng.			3%	5%	2%	3%	3%

Anova two factor analysis without replication were applied with significance level of α = 0,05 to test hypothesis that factors such as age and position on board has limited impact on the average outflow rates. Results of the anova analysis is shown in Table 9.

Table 9. Results of Anova two factor analysis for officers (Author's calculations)

Factors	F	P-value	F crit
Group 1: 3rd Off, 4th Eng, 2nd Off, 3rd Eng; Age (20-39)			
Position	0.62	0.62	3.86
Age	1.18	0.37	3.86
Group 2: 4th Eng, 2nd Off, 3rd Eng, Ch.Off, 2nd Eng Age (25-44)			
Position	0.84	0.53	3.26
Age	0.46	0.72	3.49
Group 3: 2nd Off, 3rd Eng, Ch.Off, 2nd Eng, Captain, Ch.Eng Age (30-49)			
Position	1.26	0.33	2.90
Age	1.12	0.37	3.29

F values are less than F critical values for all analysed groups and age ranges which shows that hypothesis that age and position factors does not influence outflow cannot be rejected with probability of 95% . Comparing p values of anova test it can be concluded that position influences outflow with significance (1-p) 38% for the first analysed group, 47% for the second group and 67% for the third group. Age influences outflow with significance 63% for first group, 28% for second group and 63% for third group.

The average outflow rates for officers where compared with regard to age group and year as shown in Figure 9.

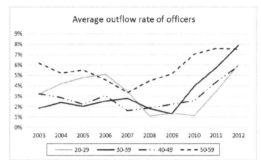

Figure 9. Outflow rates of officers comparing age groups (2003 -2012) (MAL 2015)

Graph shows that the outflow rates for the officers employed at age groups 30-39 and 40-49 tend to be similar while outflow rates for younger officers at the age group 20 - 29 and older officers have more variety. Careful conclusion can be drawn that balance between working conditions on board and local economic development had more impact particularly in those age groups.

3 OUTFLOW AND SHIPS' TYPE

Taking into account different employment conditions like salary levels, length of contracts, availability of Internet, it can be assumed that different outflow rates are on different types of ships (Fig. 10). Figure 10 shows average outflow rates on board ships for officers and ratings. Only the seafarers within age 20 - 50 years were included in analysis to exclude age factor. According to the available data, the highest outflow rates are on offshore vessels, however this assumption should be treated with care as Latvian seafarers are often employed on offshore vessels directly not through Latvian crewing agencies therefore data about their employment is regarded as fragmented as it is only entered in the database when seafarers came to renew their certificates. Therefore additional qualitative analysis should be made to validate seafarers' outflow on the offshore vessels. The differences can be explained by different salary levels, different possibilities after employment on board to move ashore etc.

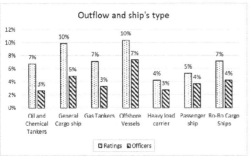

Figure 10. Average outflow rates for employed ratings and officers (2003 - 2012) on different types of vessels (MAL 2015)

In addition the outflow rates were analysed regarding main ships' types on which Latvian seafarers are employed according data in Table 3.

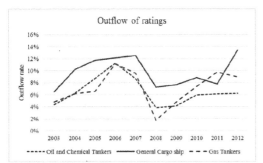

Figure 11. Ratings' outflow rates on tankers and general cargo ships (2003 – 2012) (MAL 2015)

Figure 11 shows the outflow rates for the ratings on tankers and general cargo ships. It can be concluded that difference in the outflow rates between tankers and general cargo ship's remains stable and fluctuations are mainly caused by changes in local economic conditions.

Figure 12 (in the same way) shows the outflow rates for employed officers on board tankers and general cargo ships'.

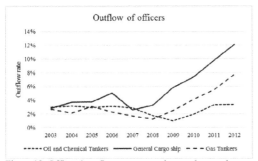

Figure 12. Officers' outflow rates on tankers and general cargo ships (2003 – 2012) (MAL 2015)

It can be seen that till 2008 differences in outflow rates are not so considerable as from 2008 – 2012, when clearly can be seen difference between the outflow rates on general cargo ships and tankers.

4 INFLUENCE OF ECONOMIC FACTORS ON THE SEAFARERS' OUTFLOW RATES

Several economic parameters were chosen for analysis to determine magnitude of economic factors influencing seafarers' outflow rates. Employment and relative outflow were represented by employed officers and ratings on tankers as tankers are main ships' type on which Latvian seafarers are employed. Correlation between the outflow rates in particular age group and parameters such as salary on board, relative differences between salary on board and salary ashore, gross domestic product, growth of GDP, average net salary ashore, unemployment and changes in unemployment were analysed from 2003 till 2012.

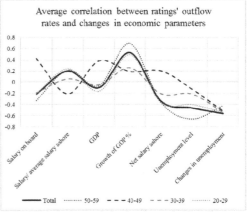

Figure 13. Average correlation between ratings' outflow rates and economic parameters (Author's calculations)

According to the results for ratings, it can be seen that there is not a strong correlation between particular economic parameter and the outflow rate for ratings and only medium positive correlation exists between growth of GDP and medium negative correlation exists between the outflow rates and changes in the unemployment level. It can be assumed that more different economic factors influences person's decision to leave employment on board due to the lower difference between salary on board and average salary ashore.

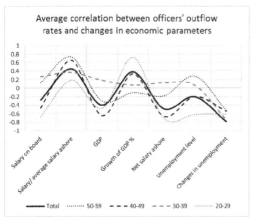

Figure 14. Average correlation between officers outflow rates and economic parameters (Author's calculations)

Figure 14 shows the results of analysis for correlation between the economic indicators and average outflow rates for officers. It can be seen that for different age groups significance of correlation differs. Limited significance is for officers employed at age 30-39 while for other groups as main factors influencing outflow rate can be assumed difference between salary on board and salary ashore, growth of GDP and changes in unemployment level ashore.

5 CONCLUSIONS

The outflow analysis taking into consideration age groups supports Waals assumption that outflow rate follows certain development over time. However outflow rates for officers in the age from 20 - 49 differ relatively little. Outflow analysis also approve David Glen's assumption that for the seafarers with lower qualification it is easier to enter and exit active seafarers' pool as the outflow rates for ratings is higher than for officers comparing the age groups and ships' type were seafarers were employed.

By analysing correlation of outflow rates and economic parameters it can be concluded that economic factors with highest influence on seafarers' decision to leave are changes in well–being which can be viewed through changes in GDP and changes in opportunities for job ashore which are reflected through changes in the unemployment level.

Even the outflow rates for parallel positions in deck and engine departments tend to be similar, average outflow rates for deck ratings and also officers is higher than for engine ratings and officers, which can be explained due to higher

supply level of deck ratings and officers in last years and also differences in employment on different types of ships as ship's type also influences outflow rate.

These analysis did not consider the level of maritime education programme graduated by persons before starting career on board. However, according assessment of contributions of maritime education institutions in Latvian seafarers pool (Gailitis, 2013) the outflow rates for officers graduating from ISCED level 6 (Bachelor's or equivalent level) programmes are higher in first years in comparison with outflow rates for officers graduated from ISCED level 5 (Short-cycle tertiary education) programmes.

Analysis of the outflow rates gives possibility to include those results in seafarers' supply model which will give possibility to predict future supply of seafarers better.

REFERENCES

Central Statistical Bureau. Statistics Database. Retrieved Janurary 15, 2015, fromhttp://www.csb.gov.lv/en/dati/statistics-database-30501.html

EQUASIS. (2014). The world merchant fleet in 2013. European Quality Shipping Information System. Retrieved Janurary 15, 2015, from http://www.equasis.org/

European Commission. (2011). STUDY ON EU SEAFARERS EMPLOYMENT. Brussels: European Commission.

Gailitis R., Fjodorova A. (2014). Determination of Parameters to Model Seafarers' Supply in Latvia. TransNav, the International Journal on Marine Navigation and Safety of Sea Transportation, 8, 245-252.

Gailitis, R. (2013). Assesment of Contribution of Maritime Education Institutions in Latvian Seafarers Pool. Journal of Maritime Transport and Engineering, 2(1), 4-12.

Glen, D. (2008). What do we know about the labour market for seafarers? Marine Policy, 845-855.

IMO. Studies and Careers in Shipping. Retrieved from International Maritime Organization: http://www.imo.org/KnowledgeCentre/ShipsAndShippingFactsAndFigures/StudiesandCareersinShipping/Pages/Default.aspx

Li, K., & Wonham, J. (1999). A method for estimating world maritime employment. Transportation Research, 183-189.

United Nations. (2009). The World Financial and Economic Crisis and its Impact on Development. New York: General Assembly. Retrieved January 10, 2015

Waals, F. A., & Veenstra, A. W. (2002). A Forecast Model and Benchmarking of the Supplu and Demand of Maritime Officers. International Annual Confference of International Association of Maritime Economists 2002 Conference Proceedings. Panama. Retrieved January 10, 2015, from www.cepal.org/transporte/perfil/iame.../Waals_et_al.doc

Wu, B., & Winchester, N. (2005). Crew study of seafarers: a methodological approach to the global labour market for seafarers. Marine Policy 29, 323–330.

Maritime Education and Training (MET)

Maritime Education and Training (MET)
Safety of Marine Transport – Marine Navigation and Safety of Sea Transportation – A. Weintrit & T. Neumann (eds.)

Investigation of Sea Training Conditions of Deck Cadets: a Case Study in Turkey

S. Yıldız, Ö. Uğurlu & E. Yüksekyıldız

Maritime Transportation and Management Engineering Department, Karadeniz Technical University, Trabzon, Turkey

ABSTRACT: Maritime is a difficult and arduous profession. Seafarers were faced with many negative factors in this profession. Challenges of the profession, makes it difficult to be preferred. Fatigue is one of the factors that affect continuity in the profession of seafarer and also it affect the formation of ship accidents occurring. Cadet is one of the seafarers who will form the future key navigation officers, has a different significance. Purpose of this study was to examine deck cadets' responsibilities, working load and occupational difficulties they encountered for identification of these factors and to determine fatigue connected with these factors. In order to identify the difficulties encountered by deck cadets during their sea trainings, render their lives during trainings easier, and establish their expectations, questionnaire and interviews were conducted with 618 deck cadets, 3 trainers and 10 maritime companies that have substantial shares in Turkey's maritime trade. As a result of this study identified factors that increase fatigue mostly, reduction measures have been exposed and have tried to offer solutions. This study is an advisory to carry out for improve the continuity of trainees in the profession.

1 INTRODUCTION

Vocational education and training is about developing special skills and talents (Gekara et al. 2011). In the twenty first century, vocational education and training has a significant role in preparing the individuals for life and business (Saunders 2012, Rauner et al. 2008). Vocational education and training is highly important for the maritime industry.

Over 90% of the world's trading operations are performed by sea and via maritime transportation sector (IMO 2011). The shipping industry worldwide has been significantly transformed by the extensive globalization, of both labour and capital, which has characterized the maritime sector since the 1960s (Alderton et al. 2004, Kahveci & Nichols 2006, Paixao & Marlow 2001). In maritime transportation sector, seafarers highly contribute to the maritime operations. Especially watchkeeping officer has a significant role among all seafarers. Global maritime sector have a need for trained and qualified officers to feed their global commercial fleets (Gekara 2009). However, due to the low occupational continuity, one of the sectoral problem is to find qualified and well trained seafarers. IMO (International Maritime Organisation) is aware of this issue. STCW 2010 Manila Conference's 12th decision has been referred to the subject of occupational continuity and made recommendations such as; providing reasonable working conditions, stable working hours and improved social opportunities in order to increase continuity (IMO 2011). However, previous studies confirmed that watchkeeping officers are subject to unreasonable work hours and conditions (Cole-Davies 2001, Jones et al. 2005, McNamara et al. 2000, Reyner & Baulk 1998, Uğurlu et al. 2012). Seafaring is a dangerous occupation because of these working conditions (Uğurlu et al. 2012). Improving such conditions and offering alternative that will encourage officer candidates to work permanently on sea are necessary for ensuring permanency of watchkeeping officers onboard.

STCW is an attempt to improve the quality of seaman as well as produce a uniform product (Lovett 1996). STCW Code regulates the necessary qualification standards governing training and education of all seafarers, their experience and skill requirements as well as performance of duties in a manner ensuring protection of security and safety of life and property on sea and protection of sea environment (IMO 2011, Kostylev & Loginovsky 2007). Training, education certifications and principles of watchkeeping onboard of cadets are according to this code. As per this code, cadets should complete training meeting STCW Code A-II/1 requirements in order to achieve certificate of competency. A year of this training is called training onboard (IMO 2011). Training onboard is a significant part of training process where the students could practice the knowledge they learned during their education life and start the professional life. Thus, this study covers the training periods of cadets. Accordingly, this study discusses working conditions and working hours of deck cadets during onboard training.

Method

In this study was made a survey with 618 deck cadets who received undergraduate level maritime education in Turkey and completed training onboard, as required by STCW Convention. The age range of deck cadets are varies from 18 to 24. This study has 4 sections. The first section covers the details of ship, voyage where deck cadets completed their training and attitudes, behaviors of the crew whereas the second section is about training conditions and social opportunities, third section is about their expectations in order to work permanently onboard and the last section is about 10 major shipping companies questionnaire results concerning their training and management policies about deck cadets.

In this study SPSS 22.00.00 analysis program was used to evaluate the answers and to check the reliability of the questionnaire's scale. The goal of this study is to determine deck cadets' training conditions, their work load and their perspective about occupation. Moreover, in light of this information to make recommendations in order to increase occupational continuity.

2 SURVEY DATA

2.1 Step I

According to the STCW code A II/2 deck cadets have to carry out training period on vessels over 500 grt for 12 month. So training period, type of ship trained on, navigation zone and attitudes, behaviors of the crew were examined at this step (IMO 2011). The review of ship types chosen for training confirmed that deck cadets generally preferred dry cargo and tanker type vessels as a training ship and most of ships preferred have a gross tonnage (grt) of 0 to 30000. As for the training periods of deck cadet, 79 % of the trainings onboard are 0 to 3 months whereas remaining 21 % is over 3 months (Table 1).

A criteria was recommend to achieve this, there should be a kind approach to trainees (IMO 2011). Survey data showed %73 of masters, chief officers and other crew members have "Good" attitudes to

deck cadets. Seventeen percent of masters, chief officers and other crew members have "Moderate" attitudes and %10 of superiors have "bad or very bad" attitudes to cadets. The review on attitudes and behaviors of masters, chief officers and crews towards the deck cadet onboard confirmed that the level of satisfaction is good. The other issue was training area because when those cadets graduated they have got an unlimited license and they were capable to navigate all around the world. According to survey data, deck cadets completed their training period navigated in the Black Sea (29%), Mediterranean Sea (25%), coastal navigation 22% and the remaining preferred other regions.

Table 1. Features of the vessels and training periods preferred by deck cadets (n=618)

Training Periods	n	Ship Type	n	Ship Gross Tonnage	n
0-1 Month	20	Tanker	147	0-3000	173
1-2 Months	161	Dry Cargo	294	3000-10000	248
2-3 Months	309	RoRo	108	10000-30000	127
4-6 Months	18	Passanger	17	30000-50000	37
> 6 Months	110	Container	49	>50000	33
		Other ship types	3		

n= number of deck cadets

2.2 Step II

The second section is about training conditions and social opportunities. According to MLC and STCW Conventions seafarers' working and resting periods arranged in two different systems, one of them was maximum working hours and other one minimum resting hour. In Turkey generally applying minimum resting hours which means provided with a rest period not less than minimum 10 hours in any 24 hours period and 77 hours in any 7 days period. Also daily 10 hours of rest can divided into 2 parts maximum and one of these two parts must be at least 6 hours (IMO 2011, ILO 2006). According to survey results, 68% of deck cadets stated that they worked 0 to 13 hours during their training onboard whereas 32% worked over 13 hours (Figure 1).

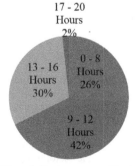

Figure 1. Total average working hours of deck cadets per day

According to the 2010 STCW Manila Conference decision 13, suitable living conditions must be provided for trainees' adequate training (IMO 2011). Twenty eight percent of the deck cadets did not have their own cabins at the ships where they were trained, they stayed at other cabins and had toilet, refrigerator, TV and internet problems at such cabins (Table 2). Seventy two percent of them stay their own cabin, 23% stay 2 persons in one cabin, %1 stay 3 persons, %4 stay 4 persons and %1 stay 5 persons in one cabin.

Table 2. Deck cadet cabin conditions (n= 618)

	Cadet Cabin (n)	WC (n)	Refrigerator (n)	TV (n)	Internet (n)
Yes	316	344	80	134	32
No	302	274	583	484	586

n= number of deck cadets

The review made on attitudes adopted towards trainees by the companies which hired deck cadets confirmed that 62% of those do not have a trainee policy and 38% of the ones having a policy tried to implement trainee policy onboard. One of the trainings that should be received by the seamen before going onboard training is orientation training offered by the companies.

Table 3. Assessment of Companies hired deck cadets for training (n=618)

Company Assessment	Yes (%)	No (%)
Does the company have a trainee policy?	38	62
Does the company implement its trainee policy onboard?	30	70
Did you have difficulties while disembarking from ships?	20	80
Did you receive orientation training at the company before the training onboard?	22	78
Did you visit the company when the training period is over?	45	55
Did the company make an assessment at the end of the training?	23	77

It was confirmed that 77% of the deck cadets going onboard training did not receive such training and training assessment of 77 % of the deck cadets were not performed by the company after the training. Besides 20% of the deck cadets had difficulties while disembarking from ship after their training period (Table 3).

2.3 *Step III*

The third section covers alternatives to be offered for continuing the sea life, period of working onboard, job preferences after the life at sea and overall review of traineeship and live as an officer. 327 of the total number of candidates participating to the survey plans to work 5 to 10 years at sea whereas 183 of them plan to work over 10 years and 108 of them plan to work between 0 to 5 years.

Considering the difficulties of life at sea, the alternatives that should be presented for improving the duration of working at sea should be satisfactory salaries, shorter contract periods, improved social opportunities, easy internet access and qualified crew etc. Following Table 4 shows demands of deck cadets to improve their occupational continuity when they are watchkeeping officers.

Table 4. Alternatives to be offered for sustaining life at sea

Measures to Increase Occupational Continuity	Number of deck cadets
Easy communication and internet access	22%
Improved social opportunities (TV, gym, sauna, etc.)	19%
Satisfactory salary	18%
Shorter contract periods	13%
Better working hours and conditions	6%
Good provision	6%
Working with qualified crew	6%
Working with family	4%
Working on close navigation zone	3%
Working with professional company	2%
Sign on and sign off as planned	1%

Total number of deck cadets= 618

Another issue examined here was difficulty of being a trainee and a watchkeeping officer. During their yearly sea training period, cadets were to have prior knowledge about the profession and to recognize the difficulties of the profession. According to data obtained from questionnaire; being unqualified, unreasonable working hours and working conditions could be listed as the common difficulties of being a trainee, whereas homesickness, being isolated from social life and unreasonable working hours were the difficulties of the profession (Table 5).

Table 5. Difficulties of the profession and traineeship

	Difficulties of traineeship (n)	Difficulties of the profession (n)
Being unqualified (there is no task definition, insufficient training and experience)	238	-
Unreasonable working hours and conditions	233	136
Homesickness	-	196
Being isolated from social life	6	147
Crew management and responsibility	12	48
Uneducated and insufficient crew, bad company	22	51
Dangerous and difficult profession	14	25
Insufficient communication with the land	2	76
Long contract periods	-	9
Bad attitudes of superiors	83	8
Being female	4	11
Difficulty of finding trainee vacancies	10	-
Being unauthorized	40	-
No answer	24	30

n= number of deck cadets. One cadet may choose more than one difficulty.

One of the factors that affecting sustainability was the occupational perspective of deck cadets. The majority of cadets define maritime profession as a stepping stone to another career dream. Twenty five percent of the deck cadets plan to work for a shipping company after the life at sea whereas 25% were planned to do trading, 21% were intended to work for government offices and 14% were planned to work in private sector.

2.4 Step IV

Examined companies have a great, professional background in Turkish and World Maritime Trade. Companies have different types of vessels such as dry cargo vessels, container vessels, tanker vessels etc. which working in international countries and ports. They have totally 114 pieces of vessels with different sizes and different nationality flags, total shipping capacity of these vessels 4372655 deadweight (DWT).

Ten major shipping companies' questionnaire answers were examined and presented at this step. First of all 50% (5/10) companies define deck cadets as a candidate officer "In addition to learning they must take responsibility about their job." 30% (3/10) of companies define deck cadets as a student "They are only being there for training." and 20% (2/10) of companies define deck cadets as an officer "He/she always be a step forward.".

Second important implication is companies training strategy about deck cadets, 9 companies apply occupational training program before embarkation. Eight companies apply planned on board training program under control of chief officer. Also 2 companies guaranteed to provide a personal cabin for their cadets; others may stay with another cadet(s) in one cabin. All examined companies give salary to cadets. Expectations of shipping companies from deck cadets are the other important inference. Deck cadets, should be aware of these expectations, and must work in order to meet them.

Expectations could be listed as follows;
1 Cadets should be eager to learn, ambitious and hardworking.
2 Cadets should be able to work as an officer and take responsibility about their job.
3 Cadets should aim persistence in the company and occupation.
4 Cadets should have sufficient English

Statistical test carried out to determine reliability of the question scales. First reliability test applied between 4 questions "Master's attitudes to cadets, chief mate's attitudes to deck cadets, crew's attitudes to deck cadets and how it feels to be a deck cadet". These questions have same systematic scale which was called Likert-type scale "Very good, good, moderate, bad and very bad". Cronbach's alpha test applied to these questions because that is one of a common test in Likert-type scale. Alpha value should be between 0-1; 1 means highest reliability (100%) and 0 means lowest reliability (0%). The alpha value was 0,782 (78%). That means the questionnaire scale is reliable. Another alpha test applied between "Trainee policy existence question and trainee policy implementation question". The alpha value was 0,921 (92%) which mean high reliability.

3 RESULTS AND DISCUSSION

Maritime is a difficult and tedious profession by definition. Seafarers suffer from heavy weather conditions, hard working conditions, unreasonable working hours (Bloor et al. 2004), being isolated from family and social life which could have negative impact on continuity of working in this profession. The first job practice of a deck cadet starts with training onboard. Thus such training is an important (Kaida et al. 2006, Uğurlu et al. 2012) part of a deck cadet's professional life. Their occupational perspective taking shape according to their working and living conditions on board. If the conditions are close to their expectations, their occupational continuity is getting higher.

The survey confirmed that majority of deck cadets prefer dry cargo and tanker vessels for training onboard. This might be due to the fact that dry cargo and tanker type ships make up majority of the Turkish ship market. According to 2013 official statistics, 53% of Turkish merchant marine is dry cargo vessels whereas 15% is tanker vessels and 8% is container vessels (Shipping 2014). In terms of navigation zones and ship tonnages, this could be seen deck cadets generally prefer completing their training on short sea carriers (ships navigating on the Mediterranean sea, Black Sea and within the coastal navigation). Although navigation period is short and work load is heavy on short sea carrier ships, the review of survey results confirmed that generally the shipmaster, first officer and crew have positive attitudes and behaviors towards deck cadets and we might say that this will not have negative impact on deck cadets' professional continuity.

The review of working hours experienced by deck cadets during training confirmed that 68% worked approximately 13 hours and less whereas 32 % worked for 14 hours and more per day. According to STCW and Maritime Labour Convention 2006 (MLC); all persons who are assigned duty as officer in charge of a watch and those whose duties involve safety, prevention of pollution and security duties shall be provided with a rest period of not less than minimum of 10 hours of rest in any 24 hour period and 77 hours in any 7 day period (IMO 2011). Thus 32 % of deck cadets work under conditions violating

STCW convention during their training. The review of cabin and social opportunities on ships stayed during training revealed that more than half of deck cadets did not have their own cabins, the ones having a cabin had limited facilities such as bath, toilet, television and internet access as well as limited social life. These are basic human needs so cabin and social life restrictions can be considered as factors having negative impact on continuity to work on sea. The review of companies' approach towards deck cadets confirmed that most of the shipping companies do not have a certain trainee policy or deck cadets are not informed about such policies and they feels like not as valuable as an officer. The companies having a trainee policy have difficulty in implementing such policies and the relationship between the companies and deck cadets lack efficiency. For example, most of the deck cadets did not have orientation training before going onboard. This fact can be considered as one of elements threatening deck cadet's safety at sea. It is without doubt that cadets working under all these negative conditions will constitute a great risk in terms of safety of life and goods at sea.

The survey data at hand revealed that 435 cadets considered working at sea for a period shorter than 10 years. Thus, that can conclude, deck cadets do not consider continuing to work for this profession. Unsatisfactory salaries, being isolated from family and social life, long contract terms, unreasonable working hours and conditions, limited social life, difficulty of internet and phone access can be listed as the most important factors impairing continuity in the profession. These issues should be improved in order to ensure continuity in this profession. Being unqualified is the most difficult aspect of a trainee's life. This stems from the attitude that a trainee would do any task onboard or is a joker crew who should adapt to do any task.

According to the companies' interviews 50% (5/10) of companies describe deck cadets as "Cadet is a candidate officer, in addition to learning he/she must take responsibility in some subjects.", 20% (2/10) of companies describe deck cadets as "Cadet have to work as an officer, he/she always be a step forward.". So companies expect some additional vocational skills from deck cadets. Trainees should be able to feel responsible, to meet those expectations. For this purpose, necessary to provide equivalence between authority and responsibility, it is one of the basic principles of management. Companies' training and personnel departments should prepare a trainee policy together and to provide implementation of this policy. Thus trainees working conditions, duties and responsibilities will be clearly certain, so they feel self confident, responsible and they will gain professional stance.

4 CONCLUSION

The findings and results presented in this study should not be considered as national issues only. Unfortunately, these issues are common problem of the entire maritime community. The solutions of these issues have a great importance to improve occupational continuity of deck officers. Training period is the first stage of the deck officers met their profession. Improvement of social opportunities and working conditions are very important for encouraging deck officer candidates. Because, increasing the quality of professional seafarers will be able to ensure continuity.

As a result, recommendations to increase occupational continuity can be listed as follows.

1 To provide easy and cheap communication with land and internet access,
2 Social opportunities should be improved; TV, video game room, gym, sauna, pool etc.
3 Contract periods should be shortened, obey the STCW and International Labour Organization (ILO) regulations about working and resting hours. Also inspectors should examine working hours more carefully, in order to increase applicability of regulations.
4 Companies should have a standard trainee policy, these policies should be implemented on ships and practices should be monitored by official authorities,
5 Preparing a job description for the trainee, specifying daily working hours of the trainee on deck and bridge and respecting such job description,
6 Implementing policies that will eliminate the views considering a trainee as manpower on the ship,
7 Cadets should be seen as a crew member, and to provide to use their occupational rights.

REFERENCES

Alderton, T. , Blur, M. , Kahveci, E. , Lane, T. , Sampson, H. , Thomas, M. , Winchester, N. , Wu, B. & Zhao, M. 2004. *The Global Seafarer: Living and Working Conditions in a Globalized Industry*, International Labour Organization.
Bloor, M. , Pentsov, D. , Levi, M. & Horlick-Jones, T. 2004. Problems of Global Governance of Seafarers' Health and Safety. Cardiff University Seafarers International Research Centre (SIRC).
Cole-Davies, V. 2001. Fatigue, health and injury offshore: A survey. *Contemporary Ergonomics*, 485-492.
Gekara, V. 2009. Understanding attrition in UK maritime education and training. *Globalisation, Societies and Education*, 7, 217-232.
Gekara, V. O. , Bloor, M. & Sampson, H. 2011. Computer-based assessment in safety-critical industries: the case of shipping. *Journal of Vocational Education & Training*, 63, 87-100.
Ilo 2006. Maritime Labour Convention. *In:* OFFICE, I. L. (ed.). Geneva.

Imo 2011. STCW Including 2010 Manila Amendments STCW Conventions and STCW Code.

Jones, C. B. , Dorrian, J. , Rajaratnam, S. M. & Dawson, D. 2005. Working hours regulations and fatigue in transportation: A comparative analysis. *Safety science,* 43, 225-252.

Kahveci, E. & Nichols, T. 2006. *The Other Car Workers: Work, organisation and technology in the maritime car carrier industry,* Palgrave Macmillan.

Kaida, K. , Takahashi, M. , Åkerstedt, T. , Nakata, A. , Otsuka, Y. , Haratani, T. & Fukasawa, K. 2006. Validation of the Karolinska sleepiness scale against performance and EEG variables. *Clinical Neurophysiology,* 117, 1574-1581.

Kostylev, I. & Loginovsky, V. 2007. Comprehensive review of the STCW 78 Convention and Code: Some concepts and trends. *8th IAMU Annual General Assembly.* Odessa: IAMU.

Lovett, W. A. 1996. *United States shipping policies and the world market,* Greenwood Publishing Group.

Mcnamara, R. , Collins, A. & Mathews, V. 2000. A review of research into fatigue in offshore shipping. *Maritime review,* 118-122.

Paixao, A. & Marlow, P. 2001. A review of the European Union shipping policy. *Maritime Policy & Management,* 28, 187-198.

Rauner, F. , Maclean, R. & Boreham, N. C. 2008. *Handbook of technical and vocational education and training research,* Springer.

Reyner, L. & Baulk, S. 1998. *Fatigue in Ferry Crews: a Pilot Study,* Cardiff, Seafarers International Research Centre (SIRC).

Saunders, R. 2012. Assessment of professional development for teachers in the vocational education and training sector: An examination of the concerns based adoption model. *Australian Journal of Education,* 56, 182-204.

Shipping, T. C. O. 2014. Turkish Shipping Sector Report 2013.

Uğurlu, Ö. , Köse, E. , Başar, E. , Yüksekyıldız, E. & Yıldırım, U. Investigation of Working Hours of Watchkeeping Officers on Short Sea Shipping: A Case Study in an Oil Tanker. The 2012 International Association of Maritime Economists Conference, 2012.

Maritime Education and Training (MET)
Safety of Marine Transport – Marine Navigation and Safety of Sea Transportation – A. Weintrit & T. Neumann (eds.)

Sleep Quality, Anxiety and Depression Among Maritime Students in Lithuania: Cross-sectional Questionnaire Study

J. Andruskiene
Klaipeda State University of Applied Sciences, Klaipeda, Lithuania
Marine Science and Technology Centre, Klaipeda University, Klaipeda, Lithuania

S. Barseviciene
Klaipeda State University of Applied Sciences, Klaipeda, Lithuania

G. Varoneckas
Marine Science and Technology Centre, Klaipeda University, Klaipeda, Lithuania

ABSTRACT: Background. The research in the area of marine students' sleep quality and mental health is lacking in Lithuania, as well as other European countries. The aim was to assess the frequency of poor sleep and the relations among poor sleep, anxiety and depression in the sample of maritime students. Methods and Contingent. Questionnaire survey was conducted in 2014 at the Lithuanian Maritime Academy, 393 (78.9% of them males) students participated. Sleep quality was evaluated by Pittsburgh Sleep Quality Index. Anxiety and depression were assessed by Hospital Anxiety and Depression Scale. Sociodemographic questions were used. The Chi-square test or Fisher exact test was used to estimate association between categorical variables. P-values less than 0.05 were interpreted as statistically significant. Results. Poor sleep was found in 45.0% of the students. Mild depression was established in 6.9%, moderate in 2.3%, severe in 0.8% of the students. Mild anxiety was found in 19.1%, moderate in 14.8% and severe in 7.9% of the students. Depression (score ≥ 8) was significantly more frequent among third (fourth) year students (22.2%) with poor sleep, as compared to the students demonstrating good sleep (2.7%). Marine Engineering programme students whose sleep was poor more often had depression (22.0%), as compared to the students whose sleep was good (5.7%). Conclusions. Maritime students had poor sleep more often than anxiety or depression. Anxiety and depression were more common among the students demonstrating poor sleep rather than good sleep. Key words: maritime students, sleep quality, anxiety, depression.

1 INTRODUCTION

More than 60% of the USA students studying in the different areas were categorized as poor sleepers, were using medications and reported more physical and psychological health problems, as compared to good sleepers (Lund et al., 2010). More than one third (31%) of the medical students at the University of Tartu, evaluated the sleep quality as satisfactory, poor or very poor (Veldi et al., 2005). The prevalence of insomnia among nursing students in Italy increased significantly from 10.3% (<20 years) to 45.5% among older students, and was predicted by the severe depression, headache and poor quality of life (Angelone et al., 2011). The social activities were less frequent and not regular among the college students aged 18-39 years demonstrating poor sleep, as compared to the students having good sleep (Carney et al., 2006). Scientific studies, investigating psychoemotional problems among the students, have proved that disturbed sleep was significantly related with higher risk of depression

and vice versa (Regestein et al., 2010; Eisenberg et al., 2007; Geisner et al., 2012).

Poor sleep among the first year students could be related with changed living environment, especially noise and light in residential halls (Sexton-Radek & Hartley, 2013). However alcohol consumption was the significant predictor of worsened sleep quality and poorer academic performance in the randomly selected sample of 236 students (124 women) of an arts college (Singleton & Wolfson, 2009).

College students are vulnerable to a variety of sleep disorders, which can lead to sleep deprivation and impaired individual cognitive performance (Van Dongen et al., 2003). However the scientific literature critically reviewed the efficacy of relevant behavioural sleep medicine interventions and discussed special considerations for using them with college students who have unique sleep patterns and lifestyles (Kloss et al., 2011).

In summary, insufficient sleep and irregular sleep-wake patterns resulting chronic sleep debt and negative consequences are present at alarming levels among students. The research in the area of students' sleep quality and mood disorders is lacking in Lithuania, as well as other European countries, especially among marine students.

The aim of the study was to evaluate sleep quality and the state of mental health of maritime students in the institution of higher education in Lithuania and establish relations between sleep quality and mental health with regard to studying year and study programme.

2 MATERIAL AND METHODS

2.1 *Study sample*

The survey was conducted on March and April, 2014. The study sample consisted of 393 Lithuanian Maritime Academy students (78.9% were male), from 18 to 34 years of age. The first (34.9%), the second (28.0%), the third and fourth (37.1%) year students were involved in the study (Table 1). Mean age of the students was 20.71 (SD=1.971). Students were grouped according to the study programmes: Marine Navigation students' group (28.5%), Marine Engineering (37.4%) and the group of students, studying Port and Shipping Management or Finances of Port and Shipping Companies or Maritime Transport Logistics Technologies (34.1%). The grouping was performed in order to ensure the even distribution of the respondents as much as possible. The grouping was also performed according to the age: 18-19 years (22.7%), 20 (26.7%), 21 (25.2%) and $\geq$ 22 years (24.4%). Frequency of the subjectively perceived sleep quality and anxiety was compared in age groups, studying year groups and study programme groups.

The study was approved by Bioethics Committee of Klaipeda University.

2.2 *Questionnaires*

Pittsburgh Sleep Quality Index (PSQI) (Buysse et al., 1989) was used for subjective sleep quality evaluation. PSQI is a self-rated questionnaire which assesses sleep quality over a 1-month time interval. 19 individual items generate seven "component" scores: subjective sleep quality, sleep latency, sleep duration, habitual sleep efficiency, sleep disturbances, use of sleeping medication, and daytime dysfunction. The sum of scores for these seven components yielded one global score. Total PSQI score < 5 was evaluated as good sleep quality; > 5 – poor sleep.

Hospital Anxiety and Depression (HAD) scale (Zigmond & Snaith, 1983), a self-assessment scale was used to identify states of depression, anxiety and emotional distress among the students of Lithuanian Maritime Academy. The HAD scale has in total 14 items, with responses being scored on a scale of 0-3, with 3 indicating higher symptom

frequencies. Score for each subscale (anxiety and depression) ranged from 0-21 with scores categorized as follows: normal (0-7), mild (8-10), moderate (11-14), severe (15-21). Scores for the entire scale (emotional distress) ranged from 0-42, with higher scores indicating more distress. Prior to completing the scale respondents were asked to "fill it complete in order to reflect how they have been feeling during the past week" (Zigmond & Snaith, 1983).

Additional sociodemographic questions about respondents' age, gender, study programme and year of the studying were included in the questionnaire.

2.3 *Statistical Analysis*

The Chi-square test or Fisher exact test was used to estimate association between categorical variables. P-values less than 0.05 were interpreted as statistically significant.

3 RESULTS

3.1 *Subjective sleep quality and psycho-emotional status of the students*

Almost a half (45.0%) of the investigated students had poor sleep, according to PSQI. The third (fourth) year students more often had poor sleep (40.7%), as compared to the second year students (23.7%), p<0.001. The students of the programme Marine Engineering (40.7%) or other programmes (36.1%) (Table 1) more often had poor sleep, as compared to Marine Navigation programme students.

The students of the programme Marine Engineering had a significantly lower mean score (0.52) in the subscale of sleep efficiency, as compared to the students of the programmes Marine Navigation (0.85) and other programmes (0.82), p<0.05 (Table 2).

Global PSQI score was significantly higher among the students of Marine Navigation programme (6.8), as compared to Marine Engineering (5.09), p<0.05 (Table 3).

Mild depression was established in 6.9%, moderate in 2.3%, severe in 0.8% of the students. Mild anxiety was found in 19.1%, moderate in 14.8% and severe in 7.9% of the students.

Depression score, evaluated by HADS, was higher (4.43) among the students older than 22 years, as compared to the students aged 18-19 years (3.46), p<0.05. Anxiety score was significantly higher among the students older than 22 years (8.36), as compared to 18-19 years old students (6.42), p<0.05 (Table 3). Anxiety score was higher among the persons studying third or fourth year (8.5), as compared to the first year students (6.72), p<0.05 (Table 4).

Table 1. Sleep quality according to the PSQI, according to the age, year of studying and study programme

Variables	PSQI diagnosis				P
	Normal sleep (n=216)		Poor sleep (n=177)		
	n (%)	95 % CI	n (%)	95 % CI	
Age groups					
18-19	50 (23.2)	17.5 – 28.2	39 (22.0)	15.9 – 28.8	0.793
20	59 (27.3)	21.3 – 33.3	46 (26.0)	19.5 – 32.5	0.768
21	54 (25.0)	19.2 – 30.8	45 (25.4)	19.0 – 31.9	0.923
≥22	53 (24.5)	18.8 – 30.3	47 (26.6)	20.0 – 33.1	0.648
Year of studying					
First	74 (34.3)	27.9 – 40.6	63 (35.6)	28.5 – 42.7	0.783
Second	68 (31.4)	25.3 – 37.7	42 (23.7)	17.4 – 30.0	0.089
Third/Fourth	74 (34.3)	27.9 – 40.6	72 (40.7)**	33.4 – 48.0	0.191
**p<0.001, as compared to the second year					
Study programmes					
Marine Navigation	50 (23.2)	17.5 – 28.8	62 (35.0)	28.0 – 42.1	0.010
Marine Engineering	88 (40.7)**	34.2 – 47.3	59 (33.4)	26.3 – 40.3	0.132
Other[1]	78 (36.1)*	30.0 – 42.6	56 (31.6)	24.7 – 38.5	0.353

*p<0.05, **p<0.001, as compared to *Marine Navigation*
[1] – Programmes of Port and Shipping Management, Finances of Port and Shipping Companies, Maritime Transport Logistics Technologies

Table 2. Means of the PSQI subscales, according to the study programmes.

PSQI subscales	Study programmes			P
	Marine Navigation (n=112) Mean (SD)	Marine Engineering (n=147) Mean (SD)	Other[1] (n=134) Mean (SD)	
Sleep quality	1.09 (0.72)	0.92 (0.63)	0.93 (0.72)	0.290
Sleep latency	1.46 (0.89)	1.23 (0.92)	1.25 (0.87)	0.225
Sleep duration	0.39 (0.73)	0.27 (0.59)	0.34 (0.67)	0.819
Sleep efficiency	0.85 (1.15)	0.52 (0.89)	0.82 (1.14)	0.018
Sleep disturbance	1.24 (0.56)	1.21 (0.55)	1.32 (0.61)	0.081
Use of sleeping medication	0.1 (0.49)	0.12 (0.45)	0.22 (0.68)	0.328
Daytime dysfunction	0.99 (0.81)	0.82 (0.90)	0.87 (0.79)	0.092

[1] – Programmes of Port and Shipping Management, Finances of Port and Shipping Companies, Maritime Transport Logistics Technologies

Table 3. Global PSQI, depression and anxiety scores, according to the age, year of studying and study programme.

Variables	Global PSQI score Mean (SD)	Depression score Mean (SD)	Anxiety score Mean (SD)
Age groups			
18-19	5.24 (2.28)	3.46 (2.28)	6.42 (3.92)
20	5.57 (2.62)	3.75 (2.68)	7.02 (4.13)
21	5.87 (2.82)	4.06 (2.67)	7.93 (4.69)*
≥22	5.65 (2.89)	4.43 (3.07)*	8.36 (5.19)*
Year of studying			
First	5.55 (2.72)	3.84 (2.81)	6.72 (4.682)
Second	5.22 (2.44)	3.67 (2.31)	6.97 (3.62)
Third/Fourth	5.91 (2.77)	4.23 (2.89)	8.5 (4.91)*
Study programme			
Marine Navigation	6.8 (2.54)	4.26 (2.8)	7.54 (4.33)
Marine Engineering	5.09 (2.51)*	3.91 (2.83)	7.16 (4.85)
Other[1]	5.73 (2.86)	3.69 (2.5)	7.7 (4.44)

*p<0.05, as compared to the reference group (*italic*)
[1] – Programmes of Port and Shipping Management, Finances of Port and Shipping Companies, Maritime Transport Logistics Technologies

3.2 Relations among sleep quality, anxiety and depression

Depression (score ≥8) was more prevalent among third (fourth) year students who had poor sleep (22.2%), as compared with third (fourth) year students whose sleep was good (2.7%), p<0.001 (Table 4). First year, second and third (fourth) year students who had poor sleep, demonstrated anxiety (score ≥8) more often, as compared to the students who had good sleep, respectively 50.8% vs 21.6%, 52.3% vs 29.4%, 72.2% vs 29.7% (Table 4).

Depression (score ≥8) was more prevalent among the students with poor sleep studying Marine Engineering (22.0%), as compared to sleeping well students (5.7%) studying the same programme (Table 5).

Anxiety (score ≥8) was more prevalent among Marine Navigation students having poor sleep (56.5%) as compared to the students whose sleep was good (28.0%) (Table 5). Students of the Marine Engineering programme whose sleep was poor, demonstrated anxiety more often (57.6%), as compared to sleeping well students of the same study programme (25.0%). Persons, studying Port

79

and Shipping Management or Finances of Port and Shipping Companies or Maritime Transport Logistics Technologies which had poor sleep, reported anxiety (66.1%) more often, as compared to the students of the same programme which had good sleep (28.2%), p<0.001 (Table 5).

Table 4. Relationship among sleep quality, depression and anxiety according to the year of studying.

HADS	First year (n=137)		Second year (n=110)		Third and fourth year (n=146)	
	Good n=74	Poor n=63	Good n=68	Poor n=42	Good n=74	Poor n=72
D Score≥8	5 6.8%	8 12.7%	5 7.4%	4 9.5%	2 2.7%	16 22.2%**
A score ≥8	16 21.6%	32 50.8%	20 29.4%	22 52.3%	22 29.7%	52 72.2%**

*p < 0.05, as compared to good sleep
**p < 0.001, as compared to good sleep
D – Depression
A – Anxiety

Table 5. Relationship among sleep quality, depression and anxiety according to the study programme.

HADS	Marine Navigation (n=112)		Marine Engineering (n=147)		Other[1] (n=147)	
	Good n=50	Poor n=62	Good n=88	Poor n=59	Good n=78	Poor n=56
D score ≥8	5 10.0%	9 14.5%	5 5.7%	13 22.0%*	2 2.6%	6 10.7%
A score ≥8	14 28.0%	35 56.5%*	22 25.0%	34 57.6%**	22 28.2%	37 66.1%**

*p < 0.05, as compared to good sleep
**p < 0.001, as compared to good sleep
D – Depression
A – Anxiety
[1] – Programmes of Port and Shipping Management, Finances of Port and Shipping Companies, Maritime Transport Logistics Technologies

4 DISCUSSION

Our study results demonstrate that poor sleep and anxiety are common among maritime students, especially among third (fourth) year students, which experience increased work load during the last year of the studies and fully realize the specific of their future work, the stressors at the workplace and the possible outcomes. The results of our study are in line with other surveys, which demonstrated that students regarded their future profession as highly burdening and stressing already during the study process and at the beginning of the career. They realized that being a seaman required to collaborate with others, to perform complex mental tasks and to support co-workers. They also felt that their future job would involve elements of competition and problems related to interpersonal conflicts, and expected that their work would have to be performed under hard psychophysical conditions (Jezewska et al., 2006). Working on board was confirmed as a stressful workplace when the group of 1,578 Polish seafarers was examined and the level of experienced stress among seafarers was stated as an average (Jeżewska & Iversen, 2012). On the other hand our findings do not support the results of other studies, because in our study there are no significant sleep quality and anxiety prevalence differences between first and second year students. This difference could be due to the differences of the samples and other methodological issues.

Psychological problems which affected people working at sea were pointed out during the 12th International Symposium on Maritime Health held in France, June 6, 2013. The following psychological disorders were listed: suicides (auto-aggression), post-traumatic stress disorder, psychosis and depression, neurosis, personality disorders, addictions and behavioural disorders (Jeżewska & Iversen, 2013).

Our study results indicate close relations among sleep quality, anxiety and depression, especially among third (fourth) year students and studying in the programmes, where practices were held on ships. Our findings are in line with the study, carried out in the UK and Germany, during which the seafarers demonstrated poorer psychosocial health, as compared to the general population of Germany (Hinz et al., 2010). As for physical health, seafarers reported better health, as compared to the general population of Germany. This can be related with the "healthy worker effect", which is related with the requirement for seafarers to have medical check-ups before joining a ship, as a consequence ill persons may not be allowed to work on-board.

The results of our study demonstrate the necessity to monitor level of stress experienced while working at sea and teach seafarers how to cope with stressful situations in order to avoid psychological consequences. The results of our study could be used as a scientific basis for creation of postgraduate courses on stress control.

5 CONCLUSIONS

– Poor sleep was more prevalent among the maritime students than depression or anxiety.
– Sleep efficiency was significantly lower among the students of Marine Engineering programme, as compared with other study programmes.
– Poor sleep, depression or anxiety were more common among the third (fourth) year students, as compared to the second or first year students.

- Anxiety or depression was more prevalent among the students whose sleep was poor, as compared to the students reporting good sleep, independently of studying year or study programme.

ACKNOWLEDGEMENTS

We thank Inga Bartuseviciene, Deputy Director for Academic Affairs at Lithuanian Maritime Academy for help in organizing the questioning of students.
Presented research was carried out in Klaipeda University and funded by a European Social Fund Agency grant for national project "Lithuanian Maritime Sectors' Technologies and Environmental Research Development" (Nb.VP1-3.1-ŠMM-08-K-01-019).

REFERENCES

1. Angelone AM, Mattei A, Sbarbati M, Di Orio F. Prevalence and correlates for self-reported sleep problems among nursing students. J Prev Med Hyg. 2011;52(4):201-8.
2. Buysse DJ, Reynolds CF, Monk TH, Berman SR, Kupfer DJ. (1989). The Pittsburgh Sleep Quality Index (PSQI): A new instrument for psychiatric research and practice. Psychiatry Research. 1989;28(2):193-213.
3. Carney CE, Edinger JD, Meyer B, Lindman L, Istre T. Daily activities and sleep quality in college students. Chronobiol Int. 2006;23(3):623-37.
4. Eisenberg D, Gollust SE, Golberstein E, Hefner JL. Prevalence and correlates of depression, anxiety, and suicidality among university students. Am J Orthopsychiatry. 2007;77(4):534–542.
5. Geisner IM, Mallett K, Kilmer JR. An examination of depressive symptoms and drinking patterns in first year college students. Issues Ment Health Nurs. 2012;33(5):280–287.
6. Hinz A, Krauss O, Hauss JP. Anxiety and depression in cancer patients compared with the general population. Eur J Cancer Care. 2010;19:522–529.
7. Jezewska M, Leszczyńska I, Jaremin B. Work-related stress at sea self-estimation by maritime students and officers. Int Marit Health. 2006;57(1-4):66-75.
8. Jeżewska M, Iversen R. Stress and fatigue at sea versus quality of life. Int Marit Health 2012;63:106–115.
9. Jezewska M, Iversen RTB, Leszczyńska I. MENHOB — Mental Health on Board. 12th International Symposium on Maritime Health. Brest, France, June 6, 2013. Report of the MENHOB working group, workshop on mental health on board. Int Marit Health. 2013;64, 3:168–174.
10. Kloss JD, Nash CO, Horsey SE, Taylor DJ. The delivery of behavioral sleep medicine to college students. J Adolesc Health. 2011;48(6):553-61.
11. Regestein Q, Natarajan V, Pavlova M, Kawasaki S, Gleason R, Koff E. Sleep debt and depression in female college students. Psychiatry Res. 2010;176(1):34–39. doi: 10.1016/j.psychres.2008.11.006.
12. Lund HG, Reider BD, Whiting AB, Prichard JR. Sleep patterns and predictors of disturbed sleep in a large population of college students. J Adolesc Health. 2010;46(2):124-32.
13 Sexton-Radek K, Hartley A. College residential sleep environment. Psychol Rep. 2013;113(3):903-7.
14. Singleton RA Jr, Wolfson AR. Alcohol consumption, sleep, and academic performance among college students. J Stud Alcohol Drugs. 2009;70(3):355-63.
15. Van Dongen HPA, Maislin G, Mullington JM, Dinges DF. The cumulative cost of additional wakefulness: dose-response effects on neurobehavioral functions and sleep physiology from chronic sleep restriction and total sleep deprivation. Sleep. 2003;26:117-26.
16. Veldi M, Aluoja A, Vasar V. Sleep quality and more common sleep-related problems in medical students. Sleep Med. 2005;6(3):269-75.
17. Zigmond AS, Snaith RP. The Hospital Anxiety and Depression Scale. Acta Psychiatr Scand. 1983; 67:361-370.

Maritime Education and Training (MET)
Safety of Marine Transport – Marine Navigation and Safety of Sea Transportation – A. Weintrit & T. Neumann (eds.)

The Use of the Portuguese Naval Academy Navigation Simulator in Developing Team Leadership Skills

I.M.G. Bué, C. Lopes & Á. Semedo
Naval Academy – CINAV, Lisbon, Portugal

ABSTRACT: Teams play a key role in the current organizational contexts, being that the leadership is a conditioning process of team effectiveness. The objectives of this paper were, on the one hand to identify the leadership skills that are developed by the functional leadership model used in the Portuguese Naval Academy, and on the other hand to evaluate the potential of using the navigation simulator in the development of these skills, through the execution of non-structured tasks. Based on theoretical models of team effectiveness and functional leadership, skills such as clarification of the situation, clarification of the strategy, coordination, and facilitate team learning, have been identified as the leadership skills that students must develop by performing practical tasks of leadership. The use of the navigation simulator to perform this type of tasks, besides being considered one of the best learning tools for students, has the advantage to simulate real-life situations and events that future officers will be faced with in their professional life.

1 INTRODUCTION

In a permanently changing world, organizations of different sectors of economic activity, and in particular the education sector, constantly feel the need to adapt to the new challenges they face with. Considering this scenario, in the context of education, Europe saw the need to create an "European area of higher education" according to quality standards that would make it internationally competitive. The Bologna process has emerged as the new paradigm of higher education in Europe.

Towards the implementation of the Bologna process in Portugal, starting 2005 the Portuguese government approved several of structuring national higher education laws, also applicable to the military higher education system. This meant that the educational institutions should proceed with reforms in accordance with the changes produced in the higher education context. Therefore, there was a need for the Portuguese Naval Academy to adapt to the new legal framework of the national and military higher education.

The Naval Academy's mission is to prepare highly qualified navy officers to the level of Masters and Bachelor degrees in specialties required to perform the duties which are committed to them, conferring the right skills to fulfill the specific missions of the Navy and Armed Forces, and promote the self development in fulfilling their command, management and team leader tasks.

The present model of the military higher education provided at the Portuguese Naval Academy is characterized by the following three fundamental educational components: a technical and technological training to meet the professional qualifications required to the performance of technical duties; a behavioral training embodied in a solid military education, civic and moral in order to develop qualities of command, management and team leader; and a physical and military training essential for the missions' accomplishments.

Students, upon the completion of their courses embark in warships carrying out duties as junior officers, usually as team leaders. It is therefore essential that students develop their team leadership skills.

What team leadership skills, and how they can be developed by students at the Naval Academy are the fundamental questions. In order to answer them, the objectives of this paper are the following:

- First objective aims to identify the team leadership skills that are developed by the functional leadership model used by the Naval Academy.

- Second objective aims to evaluate the potential use of the Naval Academy navigation simulator in the development of such team leadership skills, by carrying out practical tasks of leadership.

2 TEAMS AND TEAM LEADERSHIP

In modern organizations, the use of teams formally established by the management to carry out relevant tasks to the fulfillment of its mission is a very present tendency. While the teams are related to how organizations are structured, the teamwork refers to the set of activities that are carried out by the team members. According to Gonçalves *et al.* (2014) setting up teams and encouraging leaders and team members to work together is a way to obtain effective teams.

A team can be defined as a set of two or more people who interact with each other in order to achieve common and interdependent goals, and have a collective perception of unity (Cunha *et al.*, 2006). Taking into consideration the importance of teams to the organizations, it is fundamental to evaluate which factors can influence the team effectiveness.

There are several models of team effectiveness suggested by many authors who continue devoting themselves to the study and research of this matter. The IPO model (input-processes-output) is the one that has been predominant in the investigation, although suffering criticism, focusing especially on its static character. In order to incorporate temporal aspects and the reflexivity of the team activity, important factors for adaptive and learning team processes, one of the models that has had more acceptance into the recent research is named IMOI (input-mediator-output-input) proposed by Ilgen *et al.* (2005). According to these models, the processes/mediators factors have to do with the teamwork that is designed so that the objectives are achieved or that task assigned to the team executed. Research has suggested that the team leadership is one of the most relevant behavioral processes that influence the team effectiveness.

Leadership has been one of the most studied and investigated phenomena of knowledge. Among the many definitions that can be found in the literature on this subject, the Global Leadership and Organizational Behavior Effectiveness project defines leadership as "an individual's capacity to influence, motivate, and enable others to contribute to the effectiveness and success of organizations" (House *et al.*, 1999, cited by Cunha *et. al.*, 2006). Although leadership has been discussed since ancient times up to nowadays, its systematic study only started from early 20th century. The study of leadership made it an object of a wide variety of approaches, many of them focused on the personality characteristics and behavior of the leader, and situational variables and values, which resulted in the development of several theories in which the leader, the situation and the relationship leader-team members are the main elements. However, the tendency for organizations to adopt teams as a working base unit led to the need to understand how team leadership contributed to its positive performance.

What does the leader have to do? What are his performance functions? These issues led to the designated functional approach of leadership, in which the team becomes the key element. In general, the functional approach sees leadership in terms of the functions that must be performed by the leader in order to meet the team's needs. In this sense Adair (1986) suggests that the leader's role is to support the team in achieving a common task, maintain a cohesive team and to ensure that each member does his best. Thus, and according to Burke *et al.* (2006), the team leadership can be generically made operational as a social and dynamic process of problem-solving, based on a set of behaviors promoted by the leader, according to what Fleishman *et al.* (1991) (cited by Zaccaro *et al.*, 2001) designated as leadership functions: search and preparation of relevant information to the team; use of this information in decision making; management of available human resource and management of material resources. On the other hand, Zaccaro *et al.* (2001) propose a functional leadership model in which the intervention of the team leader is addressed to the relevant interaction processes for carrying out the task, through the leadership functions proposed by Fleishman and collaborators.

3 FUNCTIONAL LEADERSHIP MODEL USED AT THE NAVAL ACADEMY

Faced with a variety of concepts about skills that can be found in the academic literature, according to Shippman *et al.* (2000) the skill defines the performance of an activity or task with success or the adequate knowledge of a certain area of knowledge (cited by Neves *et al.*, 2006).

As mentioned before, the young officers of the Naval Academy will take duties on board warships, usually as team leaders (for example as head of technical services or officer of the watch). A useful element to understand teamwork is thinking about the team members skills needs, in order to accomplish assigned tasks, to work within an interdependent team and to analyze vast, ambiguous and complex information. In addition to representing an important learning role throughout life, the development of these skills makes the future officers respond positively to the various situations and opportunities that they will have throughout their professional career. According to Stagl *et al.* (2007) the team leadership skills is one of the most important qualifications to achieve high performance and allows the leader to have the capacity to face internal and external challenges.

Taking into account the need and the importance of the development of leadership skills, a functional

leadership model was developed, which is presently used at the Naval Academy, based on the following assumptions:
- IPO/IMOI team effectiveness models suggest that team leadership is an important process that conditions the team effectiveness.
- The team leadership as a set of behaviors promoted by the leader, intervening at the relevant teams' interaction process levels in order to perform the tasks. These behaviors correspond to those functions established by Fleishman and colleagues functional leadership model, being that Santos *et al.* (2008) suggest to be designated as *clarification of the situation* (search and preparation of relevant information to the team), *clarification of the strategy* (use of information in decision making) and *coordination* (management of human resource and management of material resources) *skills*. Santos *et al.* (2008) also suggest that a new leadership skill be added to *facilitate team learning*, given the growing importance of the leaders in the teams learning process.
- Adopting a centralized team leadership perspective, which is characterized by the focusing of decisive power and practice of leadership functions in a single individual (formal leader), inserted into the team.
- The adoption of the episodic conception of teamwork. According to Marks *et al.* (2001) the teamwork is characterized by cycles of team performance, put into operation through phases. Santos *et al.* (2008) suggest that the tasks assigned to the teams have an episodic nature consisting of four phases referred to as *assessment of the situation, strategic structuring, action and reflection.*
- During the phase of the task performance cycles, certain interaction processes have more emphasis (Marks *et al.*, 2001).

3.1 *Functional leadership model*

The performance of more or less complex tasks and of strategic value for the organizations is the main reason for the teams' existence. When a task is assigned it means that the team is facing the problem, carrying it out in accordance with the established requirements and conditions (Essens *et al.*, 2005).

The functional leadership model currently used at the Naval Academy is made operational through the designated Practical Tasks of Leadership (PTL). The PTL requires high interaction and interdependence of team members, and consist in the execution of non-structured team tasks (not expected to have a specified and unique model for its execution, leaving the teams to resolve this issue), which imply the drawing up of a strategy leading to its implementation.

When the team is assigned a new task, the functional leadership model is developed through the following phases:
- Phase 1 (initial assessment of the situation) - the team leader receives relevant information concerning the mission/task to be performed, the goal to be achieved (how the task is considered complete), the means available (team constitution, material resources, maximum time available) and the constraints and limitations that must be respected upon its implementation. At this phase it is particularly relevant the collection and processing of information, which can allow the leader to clarify the situation.
- Phase 2 (assessment of the situation by the team) – this phase starts with the preparatory meeting of the team, which we call *briefing*, where the leader, through *clarification of the situation skill*, facilitates sharing, discussion and structuring of information within the team, considered relevant to the achievement of the task. This leadership skill, which triggers the development of a shared mental model of the situation within the team, is made operational through: (1) defined the task; (2) explained the purpose of the task; (3) set out the means available and the constraints or limitations; and (4) have checked whether the information was understood by all team members (i.e., if there is a collective understanding of the situation).
- Phase 3 (strategic structuring) – this phase, which takes place during the *briefing* and aims to develop on the team the way to the task resolution through the formulation of the strategy and planning of activities, is promoted by the leader through the *clarification of the strategy skill.* Team members are faced with the need to define a work plan and to allocate responsibilities. The planning of activities is thus the process of implementing a strategy. The participation of team members in the definition of the strategy and planning allows the development of shared mental models about the task and peers, which will facilitate the coordination and cooperation mechanisms of the team while performing the task. The stimulation of the strategy structuring, from the leader's point of view aims, in essence, that the team shares collectively the knowledge about the task, the team interactions and the material resources. The clarification of the strategy skill is made operational through: (1) presented a strategy; (2) asked and encouraged team members to submit suggestions for the development of the strategy or alternative strategies; (3) assign roles and mode of interaction and use of resources to team members; (4) integrates all the information in a final plan and explains it to the whole team; (5) checked whether the final plan was understood by

all the team members (that is, if there is a collective understanding of the planning to be followed, aiming the consolidation of shared mental models).

- Phase 4 (action) – at this phase, essentially of behavioral nature, the team is engaged in a set of processes that lead to the achievement of the task and planning done in the strategic structuring phase. We shall emphasize the monitoring compliance processes of the task, the team monitoring, coordination and cooperation of teamwork. In fact, during this phase the teams are faced with the need to coordinate their work effectively, reflect and constantly monitor the extent to which the results of the ongoing action are close to the objectives set out in the preparatory phases.

During the action phase the main role of the leader should be the facilitator of the team interaction processes, through the *coordination skill* which is made operational by: (1) coordinated the activities of the team members, in performing the task; (2) monitored and updated the evolution of performance, informing the team concerning the assessment of the situation and execution, and what is expected in the near future; (3) encouraged and promoted cooperation among members and provided assistance and helped the members in difficulty; (4) promoted a positive affective climate among the members, limiting the existence of conflicts and stimulating the motivation.

- Phase 5 (reflection) - after completing the task, or finished the time allocated to it, it follows the final evaluation phase, through an after-action meeting, which we called *debriefing*.

The reflection phase ends with the teams' collective analysis of the processes efficiency and effectiveness of the results reached. At this phase the leader's role is to encourage the team to analyses situations, reflect on these and extract learning and/or lessons learned that might be used later, through the *facilitate team learning skill*, which is made operational by: (1) promoted a reflection concerning the initial assessment of the situation and the strategy/plan followed; (2) promoted a reflection on the individual and team involvement in carrying out the task; (3) summarized the lessons learned.

3.2 *The PTL assessment instrument*

The effectiveness of leaders is measured by team productivity that is to what extent the teams reach the final goal set for each PTL.

The functional leadership skills (*clarification of the situation, clarification of the strategy, coordination and facilitate team learning*) are assessed through a behavioral observation grid, consisting of 16 items, using a Likert 5-point scale, ranging from ineffective to the extremely effective.

4 USE OF THE NAVIGATION SIMULATOR FOR THE DEVELOPMENT OF TEAM LEADERSHIP SKILLS

Once the profile of team leadership skills to be developed by the Naval Academy students is established, it is necessary to implement the education and training mechanisms of these skills. One way to develop team leadership skills is to provide opportunities for students to have leadership experiences while conducting practical exercises of seamanship and navigation, which can be performed in the context of simulation.

In this article we will focus on the use of Navigation and Maneuvering Simulator (NAVSIM) of the Naval Academy in the development of team leadership skills by carrying out PTL.

The simulation can be defined as a computer program or set of programs, specifically designed to simulate real situations and events, in order for knowledge and skills to be acquired. These situations and events represent real-life scenarios, on the assumption that students can learn from reconstruction of events and situations with which they will have to deal in their professional life. Simulations allow students to "practice" responses to different simulated situations and events, for example, they can have access to similar equipment and instruments that they will find on board Navy ships.

One of the great advantages of simulation is the possibility of students to make mistakes without having serious consequences. In this sense, the simulation can be considered as one of the best learning tools for students, since mistakes can serve as a learning mechanism. In fact, students to be involved in computer simulations can verify more quickly how to act and quickly understand the consequences of their actions.

In view of the great potential that the simulation can represent in the training of the Naval Academy students, the Portuguese Navy acquired in 2004 the NAVSIM from the Norwegian company Konsberg Maritime, with the possibility to simulate seven ship bridges, four of them being installed at the Naval Academy. Several consoles are available in each of the bridges allowing students to have access to navigation facilities such as: navigation Radar, ECDIS, AIS, NAVTEX, differential GPS, echo sounder, bottom log, anemometer, magnetic compass, gyroscopic compass and rudder angle indicator. Other information displays show the state of the propulsion system (diesel engines and turbines), if there is any fault in any of the on-board equipment or the engine itself, the state of anchors, the speed introduced and how the ship itself is maneuvering. Students also have at their disposal a desk to use paper nautical charts where they set up the navigation planning, and the respective positions of the ship. In addition a set of communications equipment is available, and complies with the requirements of the *Global Maritime Distress and Safety System (GMDSS)*.

With regards to the students training, the NAVSIM can be used, among other activities, to simulate navigation exercises in shallow waters; coastal navigation; maneuver of the ship (mooring, anchoring, etc.) and several naval exercises, as well as plan and execute search and rescue (SAR) operations. The objectives are as follows: practice and techniques procedures of navigation and maneuvering; operation of navigational and/or communications equipment; training officer of the watch; training a piloting team and conducting exercises and naval operations. All this complements the theoretical subjects that are taught in-class: Navigation, Seamanship, Communications, Tactical and Naval Operations.

In addition to the extensive use in training in the above areas, focused more on technical training, the NAVSIM can also simultaneously be used in the development of interpersonal skills in order to develop students` team leadership skills.

Based on the functional leadership skills model, the PTL was adapted to the simulation context provided by NAVSIM. Among many PTL that can be created, an example of a task carried out in the NAVSIM by the 4th year cadets is described, aiming the development of team leadership skills:

4.1 *Requirements and conditions of PTL*

Situation: a bridge team is onboard the frigate "Vasco da Gama". The ship is returning from a SAR mission and heading Lisbon Naval Base (LNB). It is 06:30 in the morning, local time. The ship is sailing over the leading line of the southern entrance of Lisbon harbor very close to *Bugio* lighthouse. It's raining and waves are about 2 meters high.

Task: Driving the ship safely from the position in which it is to moor in LNB at 07:30.

Team: Bridge team is formed by 5 members and one of them is appointed as team leader[1] (officer of the watch).

Material resources available: Navigation, command, driving and control electronic equipment, and other existing equipment in the bridges of simulated ships.

---

1 The functional leadership model and the NAVSIM can also be used successfully by teams formed by people with little or no knowledge and no naval and sailor experience, just by previously providing them with basic knowledge of navigation and seamanship, as well as handling and operation of main equipments in simulated bridges.

Constraints and limitations: Maximum speed 12 knots; time to complete the task 60 minutes; weather conditions can worsen/improve any time; the number of warships, merchant vessels and boats in the area may increase at any time; whenever appropriate use the International Regulations for Preventing Collisions.

PTL is performed according to the phases established by the functional leadership model already mentioned.

In the early phase of the situation assessment, the team leader receives information about the requirements and conditions of the task, in particular about the situation, task, team, material resources and the constraints and limitations to perform the task. Next, the leader meets with his team in one of the NAVSIM bridges and then starts the *briefing* for the situation assessment and strategic structuring. The plan to follow is defined and the roles of each member assigned. It is intended that during the *briefing* the leader develops the *clarification of the situation and clarification of the strategy skills*.

As mentioned above, the frigate "Vasco da Gama" is positioned over the leading line of the southern entrance of Lisbon harbor very close to *Bugio* lighthouse ready to be operated. It is intended that the ship sails for a while with some darkness, rain and 2 meters high waves. During the action phase the bridge team is faced with the following string of situations and events: navigation in restricted waters at night, several merchant ships and small boats in the area, a fog bank, an engine or rudder malfunction and a naval force leaving the Lisbon harbor. The team leader will receive, over the duration of the task, several written messages on different subjects, in order to increase the stress level. He or she will have to decide which information is relevant for the fulfillment of the task and to inform the team about it. The most relevant messages are concerned to the failure of the leading line lights, an anchored ship caused by an emergency situation, (which can be an obstruction to the navigation plan), and the change of the objective of the task no longer the ship will moor in LNB but will moor at *Alcântara* pier.

It is intended that during the action phase the team leader develops the *coordination skill* of the teamwork.

PTL ends with the Debriefing and it is intending that during this phase the team leader develops the *facilitate team learning skill.*

5 CONCLUSIONS

The complexity of the environments in which teams perform their tasks and deal with the unexpected makes relevant the existence of a leader able to facilitate effective teamwork.

The episodic perspective of the functional leadership model reflects a set of strategic skills that must be carried out by the team leader throughout the different phases of the task in order to ensure team effectiveness.

The development of the *team leadership skills* of the Naval Academy students, namely the *clarification of the situation, clarification of the strategy, coordination and facilitate team learning*, are considered fundamental in their preparation as future leaders and Navy officers.

The use of NAVSIM in the development of team leadership skills, besides considered one of the best learning tools of students learning, also has the advantage to simulate real life situations and events with which the future officers will face in their professional life.

REFERENCES

Adair, J. (1986). *Effective teambuilding.* London, UK: Pan Books.

Burke, C. S., Stagl, K. C., Klein, C., Goodwin, G. F., Salas, E., & Halpin, S. M. (2006). What type of leadership behaviours are functional in teams? A meta-analysis. *Leadership Quarterly, vol. 17*, pp. 288-307.

Cunha, M. P., Rego, A., Cunha, R. C., & Cabral-Cardoso, C. (2006). *Manual de comportamento organizacional e gestão, 5ª edição.* Lisboa, Portugal: RH, Lda.

Essens, P., Vogelaar, A., Mylle, J., Blendell, C., Paris, C., Halpin, S., et al. (2005). *Military command team effectiveness: Model and instrument for assessment and improvement (NATO no. ac/323 (HFM-087) TP/59).* NATO Research and Technology Institution.

Gonçalves, S. P., Braun, A. C., Antunes, A. C., Silva, A. D., Duarte, A. P., Oliveira, A. M., et al. (2014). *Psicossociologia do trabalho e das organizações.* Lisboa: Lidel - edições técnicas, Lda.

Ilgen, D. R., Hollenbeck, J. R., Johnson, M., & Jundt, D. (2005). Teams in organizations: From Input-Process-Output Models to IMOI Models. *Annual Review of Psychology, 56*, pp. 517-543.

Marks, M. A., Mathieu, J. E., & Zaccaro, S. J. (2001). A temporally based framework and taxonomy of team process. *Academy of Management Review, vol. 26(3),* pp. 356-376.

Neves, J. G., Garrido, M., & Simões, E. (2006). *Manual de competências.* Lisboa, Portugal: Edições Sílabo, Lda.

Santos, J., Caetano, A., & Jesuíno, J. C. (2008). As competências funcionais dos líderes e a eficácia das equipas. *Revista Portuguesa e Brasileira de Gestão, Edição especial 10 anos,* pp. 95-106.

Stagl, K. C., Salas, E., & Burke, C. S. (2007). Best practices in team leadership: What team leaders do to facilitate team effectiveness. In C. e. (Eds), *The practice of leadership: Developing the next generation of leaders.* (pp. 172-198). San Francisco, CA: Jossey-Bass.

Zaccaro, S. J., Rittman, A. L., & Marks, M. A. (2001). Team leadership. *The Leadership Quarterly, VOL. 12,* pp. 451-483.

Maritime Education and Training (MET)
Safety of Marine Transport – Marine Navigation and Safety of Sea Transportation – A. Weintrit & T. Neumann (eds.)

Paradigm Shift in Ship Handling and its Training

S.G. Seo
Southampton Solent University, Southampton, UK

K. Earl
Timsbury Shiphandling Centre, Warsash Maritime Academy,
Southampton Solent University, Southampton, UK

ABSTRACT: With the clearer exposition of the pivot point of a ship by some authors recently, together with the advancing technology of Global Positioning Systems, the accurate location of the pivot point is now available in real time. This information enables ship handling practitioners to perform more accurate and efficient manoeuvres within confined areas. The training of professionals should therefore reflect this change of scene. The manned model ship handling centres are the best places to accommodate this change quickly and implement new training schemes.
In this paper, the ship's pivot point is expounded in a fresh light, leading to an equation for the definition and others for the calculation of the pivot point location both in general and for specific examples. A number of exercises for both steady cases and unsteady cases have been suggested. Both cases of exercises are of practical value as shown in a real example of Southampton Container Port.

1 INTRODUCTION

Ship handling has been viewed by many as an '*art*', meaning that it cannot be performed by scientific calculations alone, but must also be relied upon one's own experience and intuition. One of the factors contributed to this view is the concept of 'pivot point' which has been the central and important tool in ship handling. It has, unfortunately however, been a rather ambiguous entity, resulting in some confusion and misuse amongst ship handlers. Yet practitioners has been trying to understand ships' motion in terms of it.

In recent years a number of authors gave clearer expositions of it - Artyszuk J. (2010), Seo (2011), Tseng C-Y. (1998) for example - and demonstrated what can be achieved with the correct understanding. These enabled the practitioners to have an unambiguous picture of the concept. Together with the advancing Global Positioning System technology, its usage can now be extended further. Its position can be calculated in real time showing the movement as it happens, as demonstrated in de Graauw (2012). This can be extrapolated into the near future, be it the next a few seconds or minutes. The ship handler can make a plan of action based on the calculated position and trend of the pivot point. This signals a change in the mental attitude of ship handling practitioners and thus in teaching and training the subject, too.

2 THE PIVOT POINT

The concept of pivot point has been an essential tool in ship handling. The knowledge about the position of the pivot point in a manoeuvring situation provides the ship handler with the information on the geometry of motion of the ship. Baudu (2014), Cauvier (2008), Clark I. (2005) and Rowe R.W. (2000). Hence, it is a requirement for the ship handler to understand how and why the ship behaves in a certain way. The pivot point now being available, a ship handler, using his knowledge of mechanics can control the pivot point to where he wants it to be, in order to take the next action in an effort to make the desired manoeuvre.

Ship's motion in a confined area can be modelled as a planar rigid body motion assuming no vertical movement of any point of the ship. This is justified for the relatively calm free surface in such an area.

Manoeuvring a ship ahead or astern (surge) does not pose much difficulty to the ship handler. One can easily make the required movement by making reference to some landmark alongside the ship. Figure 1.

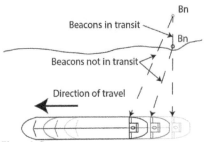

Figure 1. Surge

Sway motion alone can also be easily conducted by making reference to landmarks. Figure 2&3.

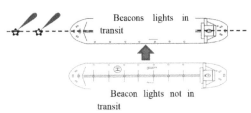

Figure 2. Sway

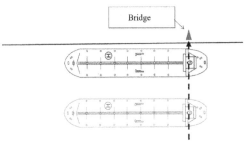

Figure 3. Sway

In real situations, making a rotation (yaw) is difficult to be precise, because a transverse motion (sway) is usually accompanied while making a rotation, the water being a yielding material. Figure 4.

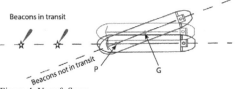

Figure 4. Yaw & Sway

When sway and yaw occur simultaneously, a ship handler can only perceive *the combined effect of drift and turn*, which gives him a false impression that only a rotation happened about a certain point on the ship's centreline. This seeming centre is called the *Pivot Point* of the ship. This is *a simplification of perception from two motions down*

to one motion, which is the very reason why the pivot point concept is so useful to ship handlers.

2.1 *The Pivot Point (How it is brought about)*

In Figure 5, the initial position of the ship is shown by the black outline (top). The ship now turns about the centre of mass (G1) to become the red hull (middle). While turning, the ship drifts at the same time into the blue hull (bottom). The two motions happening simultaneously, the centre of gravity moves from G1 to G2, and the pivot point moves from P1 to P2.

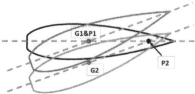

Figure 5. Turn and Drift

Had the drifting preceded the turning, as happens at every turn in zig-zag trial, the pivot point would have appeared at forward infinity and approached to P2, as shown in Figure 6.

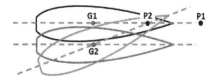

Figure 6. Drift and Turn

In continued zig-zag runs, at every turn of the rudder, the pivot point disappears into the forward infinity from P2 due to ship's momentum, and then reappear at aft infinity to approach back and settle at P2 until the next turn of the rudder.

2.2 *The Mathematical Definition of the Pivot Point*

Among all the points in the ship in planar motion, there is only one point on the centreline at which the sway and yaw completely cancel each other, thus making this point seem to be stationary. All other points appear to be turning about this point. This point is the Pivot Point. If the sway speed and yaw speed are known, the position of the pivot point can be obtained as the distance from the centre of mass (GP) using Equation (1). Tseng (1998).

$$v + (GP \times r) = 0 \qquad (1)$$

where, v(m/s) = sway speed of G; G = Centre of Gravity; P = Pivot Point; GP(m) = distance to P from G; r(rad/s) = yaw Speed.

2.3 The Calculation of the Pivot Point Position

Since the pivot point is defined on the centerline of the ship, only one dimensional coordinate system will suffice for our purpose. The vertical line through the centre of gravity is taken as the origin, one side of which is taken as positive direction, the other side negative direction. Figure 7.

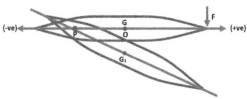

Figure 7. 1-D Coordinate System

When a force causes a ship to drift and turn, the centre of gravity will move due to the drift motion.

$$GG_1 = \frac{1}{2}\left(\frac{F}{\Delta}\right)t^2 \tag{2}$$

where, F is the force; Δ is the mass displacement of the ship; t is the time taken.

The arc drawn by G in an *imaginary* yaw motion with P as the pivotal point is:

$$(arc)GG_1 = GP \times \frac{1}{2}\left(\frac{F \times (-GF)}{I}\right)t^2 \tag{3}$$

where, GF is the distance from G to F, the negative sign indicating the other side of G (origin) from P; I is the second moment of mass of the ship about the origin.

For a small change of heading, GG_1 can be equated to $(arc)GG_1$ giving,

$$GP = \frac{I}{\Delta \times (-GF)} \tag{4}$$

Equation (4) gives the position of the pivotal point in terms of GP. This pivotal point (P) is naturally called the 'Pivot Point' of the ship even though it is *imaginary*. Under the assumption of a solid ship of uniform density with multiple number of controlling forces, Equation (4) becomes:

$$GP = \frac{-1}{V \times GF_c}\int r^2 dV \tag{5}$$

where, V is the volume of the ship; GF_C is the longitudinal distance along the centreline between G and Fc, the position of the resultant of all applied controlling forces; r in this equation is the radial distance of the infinitesimal volume from the origin.

Equation (5) reduces, for a box barge ($C_B=C_{lp}=1.0$), to an elegantly simple equation.

$$GP = -\frac{L^2 + B^2}{12GF_c} \tag{6}$$

Assuming GF$_c$ = -0.5L and B = L/7, Equation (6) gives GP = 0.170L. This means that the pivot point is at 0.330L from the bow. It is seen from Equation (6) that a bigger B will give a bigger GP, which implies that *a full-form ship (high C_B) with a bigger beam (as a fraction of L) will have the pivot point closer to the bow.*

A wall-sided hull could be defined by:

$$y = \frac{B}{2}\left\{1 - \left(\frac{2x}{L}\right)^2\right\} \tag{7}$$

where, x is the position along L and y is the half beam.

Again assuming GF$_c$ = -0.5L and B = L/7, Equations (5) and (7) give GP = 0.102L, which means that the pivot point is at 0.398L from the bow. This hull has a C_B=0.67 and a C_{lp}=0.67. By comparing the two hulls above, one can deduce that *a smaller block coefficient will cause the pivot point to be closer to G.*

Any applied force sets a ship into motion. The gradually increasing motion changes gradually the aerodynamic and hydrodynamic environment. The reactive forces increase until they balance with the active forces. By then the ship will have gained some momentum. This momentum adds further movement in the pivot point position, which settles down as the motion becomes steady. In reality, therefore, dealing with the unsteady process accurately is very difficult, if not impossible, particularly when various forces are involved. Fortunately, however, with a clearer understanding that the pivot point is the *result* of a ship's motion, and the advance of Global Positioning System technology, the pivot point location itself has become readily available by simply measuring the displacement of two fixed points, bow and stern for example, of a ship, as demonstrated by Arthur de Graauw (2012).

2.4 Interpretation of the formula for the Pivot Point Position

Two important aspects are noted from Equation (5).

Firstly, the minus (-) sign indicates that the pivot point appears *on the other side of G from Fc*. Secondly, a bigger GF$_c$ yields a smaller absolute GP, which means that *an external force farther away from G causes the pivot point to be closer to G.* These two findings are essential knowledge for the practitioners to proactively control the pivot point.

2.5 The Attribute of the Pivot Point

In deriving Equation (5), the ship was represented by her centre of gravity, G, fixed within the ship. And this point was taken as the actual centre of turning (yawing), rather than the pivot point which is just an imaginary point with ever-changing location and can even exist outside the ship's hull, even at the infinity forward or aft. The centre of gravity is a clearly better choice for the reasoning about the geometry of ship's motion. Taking the pivot point as the centre of turning would make manoeuvring problems needlessly difficult to solve. This will become obvious when one try to proactively control the pivot point.

The pivot point is defined by a purely *geometrical* consideration of a ship's motion. Two actual motions, sway and yaw, were replaced with one *imaginary* yaw motion in the process. This means that the pivot point is only an *imaginary geometrical* property. It is, therefore, wrong to associate the pivot point with any actual *physical* quantity, e.g. a turning moment. A mistake frequently seen is taking the moment arm from the pivot point in an effort to explain the direction of heading change when an external force is applied.

3 INTO THE FUTURE

The movement of the pivot point could be extrapolated into the immediate future showing the trend of movement. This extrapolation is justified by the fact that ships are operated in water which is a yielding matter. The large momentum of a ship in motion would prevent the ship from any jerky change of motion under normal manoeuvring circumstances. This would furnish a continuously differentiable pivot point movement except when it jumps from the forward infinity to the aft infinity as happens in zig-zag runs.

This brings about an important consequence that the pivot point location is now known not only for the past track but also for a foreseeable future. This in turn allows the ship handler to envisage the ships motion to be, and thus can make a plan or decision of manoeuvre. Or in a more proactive effort, he could try to control the pivot point location by using any available means such as tugs or bow thrusters, to achieve his manoeuvring objectives.

4 SOME BASIC EXERCISES TO ACTIVELY CONTROL THE PIVOT POINT

4.1 Ship Motion with Yaw but No Translation

This is the case when the ship is turning about the centre of yawing (S), normally taken at the centre of gravity. The ship has no translational motion (no

surge, no sway). In this case all three points coincide – the centre of yawing, the centre of circling (E) and the pivot point (P), Figure 8. This manoeuvre could be produced with the bow and stern thrusters, and tugs in combination.

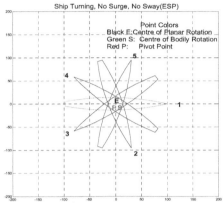

Figure 8. Yaw only

4.2 Ship Motion with Yaw and Sway only with the Pivot Point aft of Bow

In the absence of any longitudinal movement (no surge), if the ship drifts at the same time as turning, and if the pivot point is between S and the bow, the motion shown in Figure 9 will result. In this case the two points, E and P, will coincide.

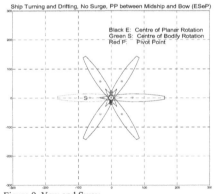

Figure 9. Yaw and Sway

4.3 Ship Motion with Yaw and Sway only with the Pivot Point forward of Bow

If the pivot point is ahead of the bow, the motion shown in Fig.10 will result.

The two points, E and P, are at the same location.

This manoeuvre could be produced using the stern thruster. In practice however, the same manoeuvre could be produced by a combination of all the three elemental motions – short surge followed by sway and yaw.

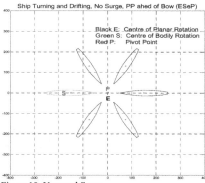

Figure 10. Yaw and Sway

4.4 *Ship Motion with Yaw and Surge only*

If the ship moves ahead while turning but without any sway motion, the resulting movement will look like the one shown in Figure 11. In this case, the pivot point will be on top of the point S. This manoeuvre could be produced with both bow and stern pods deflected. This manoeuvre without any drift causes no swing out of the stern, thus it may be a necessary manoeuvre in tightly restricted waters.

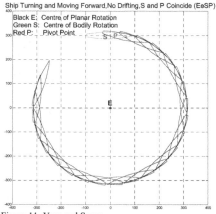

Figure 11. Yaw and Surge

4.5 *Ship Motion with Heading, Drifting and Turning*

When all the three motions (Surge, Sway, Yaw) are present, all the three distinctive points (E, S, P) will exist separately as shown in Figure 12. In this particular case, the stern swings out sweeping a bigger arc, as all skilled ship handlers are most conscious of. Ship motions in general fall in this category. The amount of swing out is directly related with the position of the pivot point.

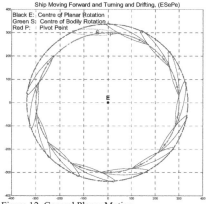

Figure 12. General Planar Motion

5 SOME PRACTICAL MANOEUVRES

5.1 *Turning Short Round*

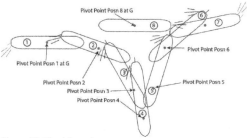

Figure 13. Short Round

1 Start manoeuvre at a slow speed, then hard to starboard with a kick ahead
2 Stop engines – rudder midships
3 Engine astern, transverse thrust continues to turn vessel.
4 Vessel at a stop over the ground, continue with engines astern. Transverse thrust still acting on vessel.
5 Engine still running astern
6 Engine still running astern, about to stop engine.
7 Rudder hard to starboard, engine ahead.
8 Vessel completed short round.
 Note: This is a simplified version rather than what would be required in reality.

5.2 Entering a Cut

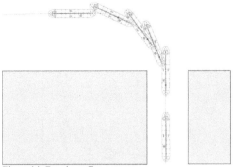

Figure 14. Entering a Cut

While preparing for the manoeuvre shown in Figure14, the clearance from the jetty and the longitudinal position are crucially important so as not to come into contact with any port structure during the manoeuvre.

The pivot point will initially appear near the centre of gravity, not nearer to the bow as normally quoted in ship handling literature, and then gradually move forward as the ship gains drifting momentum.

5.3 Southampton Container Port

The following sequence of screen shots have been taken from the PPU on the departure of the "CMA CGM Marco Polo" from Southampton. The pilots portable unit is an AD Navigation ADX-XR which includes RTK, giving a very precise position . The performance criteria are:-
- Position Accuracy: 1-2 cm (RTK mode)
  0.8 m with EGNOS/WAAS
  2 m uncorrected GPS/GLONASS
- Bow and Stern Speed: 1 cm/sec (0.02 knots)
- Vertical/Squat: 2-3 cm (RTK mode)
- Heading: 0.01 deg (20m POD separation)
- Rate of Turn: 0.1 deg/min

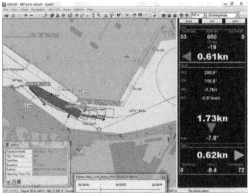

Figure 15.

The "CMA CGM Marco Polo" is clear of the berth, moving astern. The vectors for the bow and stern are indicated by the black arrow, whilst the predicted position of the vessel is outlined for 4 positions.

Figure 16.

The gray fill outline is the actual position of the vessel (400m LOA). The vessel has 3 tugs in attendance, and needs to swing within the swinging ground depicted by the purple circle.

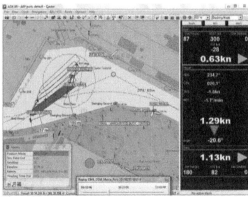

Figure 17.

The ship outlines with no fill, are the predicted positions of the vessel after a set time duration. This is set by the pilot / operator. The vector of the bow and stern can also be seen, indicated on the chart with the black arrow from the centre line, fore and aft respectively.

94

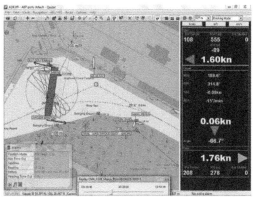

Figure 18.

This is the sort of manoeuvre as the exercise in Figure 8. The ship is using tugs, in combination with own engine, rudder and bow thruster so as to maintain a rotational movement about the vessels midship position. Bow velocity is depicted at the right hand side of the screen, in this instance 1.6knots to port, and depicts that the vessel is moving 0.06 knots astern. The stern velocity is presently 1.76 knots to starboard.

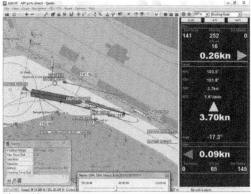

Figure 19.

This depicts the vessel having completed her swing 141m off 107 berth proceeding outwards at 3.7knots

Screen shots courtesy of ABP Southampton.

6 CONCLUSION

For some two hundred years or more the pivot point location in ship handling has been a rather ambiguous entity. Yet practitioners were taught and practiced to make out ships' motion in terms of it.

Now the location of pivot point can clearly be shown in real time. Even the future location can be shown if the ship's motion continued. This means the future situation can be envisaged and corrected if needed by proactively controlling the pivot point.

This dictates a change in the mentality of ship handlers from the passive use of the pivot point as a rather unclear clue to active use of controlling its position for their need to effectively, accurately and safely manoeuvre ships. The training and educating of ship handlers should thus be adjusted to reflect this change.

In other words, the clear understanding of the concept of the pivot point combined with the currently available GPS technology enable ship handlers to utilise more fully the convenient concept of the pivot point. This is signaling the forthcoming change of view point on the use of the pivot point concept - a *Paradigm Shift*.

REFERENCES & BIBLIOGRAPHY

Andy Chase G. 1999 Sailing Vessel Handling and Seamanship – The Moving Pivot Point *The Northern Mariner July 1999: 53-59*

Artyszuk J. 2010 Pivot Point in ship manoeuvring *Scientific Journals 2010:13-24*

Baudu H. 2014 Ship Handling *Dokmar Maritime Publishers*

Blackburn I. 1836 Naval Architecture *Longman, Rees, ORME, & Co. London*

Cauvier H. 2008 The Pivot Point *The Pilot October 2008*

Clark I. 2005 Ship Dynamics for Mariners *The Nautical Institute*

De Graauw, A. 2012. Where is my Pivot Point? *Seaways March 2012:* 23-24

Grassi C. R. 2000 A Task Analysis of Pier Side Ship-Handling for Virtual Environment Ship-Handling Simulator Scenario Development *Master's Thesis, Naval Postgraduate School*

Hwang W-Y. 1980 Application of System Identification to Ship Maneuvering *PhD Thesis, MIT*

Rowe R.W. 2000 The Shiphandler's Guide *The Nautical Institute*

Seo, Seong-Gi. 2011. The Use of Pivot Point in Ship Handling for Safer and More Accurate Ship Manoeuvring. *International Conference IMLA 19:* 7-10

Tseng C-Y. 1998 Analysis of the Pivot Point for a Turning Ship *Journal of Maritime Science and Technology*

Experimental Research with Neuroscience Tool in Maritime Education and Training (MET)

D. Papachristos & N. Nikitakos
Dept. Shipping, Trade and Transport, University of Aegean, Greece

ABSTRACT: The paper argues for the necessity to combine MMR methods (questionnaire, interview) and gaze tracking as neuroscience tool for personal satisfaction analysis at the maritime and training education (MET) and proposes a practical research approach for this purpose. The purpose of this paper is to compare the results from gaze tracker (Face analysis tool) of three experiments for satisfaction evaluation of the students-users' (subjective) satisfaction of the maritime education via user interface evaluation of several types of educational software (i.e. engine simulator, ECDIS, MATLAB). The experimental procedure presented here is a primary effort to research the emotion analysis (satisfaction) of the users-students in MET. The gaze tracking methodology appears to be one sufficient as evaluation tool. Finally, the ultimate goal of this research is to find and test the critical factors that influence the educational practice and user's satisfaction of MET modern educational tools (simulators, ECDIS etc.).

1 INTRODUCTION

In the shipping industry, the need for excellent education and on other hand, the usability evaluation of ship manipulation systems and engine management, leads to the use of new technologies in educational practice. Specifically, the Marine Education & Training (MET), the use of simulators (engine or ship's bridge) is fact. Various maritime educational standards (i.e. STCW, 95, Manila 2011) allow the simulators and other educational tools (i.e. educational software, MATLAB) use in educational practice.

The aim for the application of new technology (simulators, games etc.), in MET is the transport of capacity, i.e. to adapt the dexterities learned within the vessel operating training framework. We assume that the dexterities and the knowledge learned in the classroom can be applied effectively in real life similar situations (Tsoumas et al., 2004).

MET follows certain education standards (STCW'95/Manila 2011) for each specialty (Captain, Engineer) and for each level (A', B', C'). Its scope is the acquisition of basic scientific knowledge, dexterities on execution (navigation, route plotting, engineering etc.) as well as protecting the ship and crew (safety issues and environment protection issues)(IMO, 2003, Papachristos et al., 2012, Tsoukalas et al., 2008).

In MET, in particular, the user's satisfaction based on objective criteria poses an important research subject because via this we can determine the background explaining the satisfaction phenomena, recommending at the same time new considerations that will expand the up-to-date educational conclusions on the adult education in educational programs and software development (IMO, 2003, Papachristos, Nikitakos, 2010, 2011).

The paper argues for the necessity of a mixed approach to usability and educational evaluation at the engine room or Ship bridge simulation, and proposes a practical framework for this purpose. In particular, we use a multi-method approach for the usability and educational evaluation of maritime simulators and other educational tools that combines physiological data generated from gaze tracking data (neuroscience tool), questionnaires and interviews. The combination of these methods aims at the generation of measurable results of user experience complementary assessments (Papachristos et al., 2012).

Gaze tracking involves detecting and following the direction in which a person looks. The direction of the eye gaze can express the user's interests; it is a potential porthole into the current cognitive processes. Communication through the direction of the eyes is faster than any other mode of human communication. Gaze Tracking has been applied: in

Human Computer Interaction, Advertising, Communication for disabled, Virtual Reality, Improved image and video communication, Medical field and Human Behavior Study (Arpan, 2009).

Eye observation on handiness tests is a rather promising new field especially for system designers, as it may offer information on what may attract user attention and which are the problematic areas during system use. The research area on use of the optical recording tools is the quest for an exact interpretation of the optical measurements, their connection to the satisfaction and the learning effectiveness for users. Suggested research aims at this direction with the use of neuroscience methods in combination with the use of qualitative-quantitative researches aiming at the extraction of useful conclusion that will help simulator system designers to develop the systems (especially the interface, delivering & organizing education material), class designers to better organize material and modern tools use (better planed educational scenarios that thriftily develop the trainees abilities but also can offer a more objective evaluation of their abilities & function as future captains or mechanics) and finally the expansion of the adult education field by offering new conclusions regarding the use of e-learning (introduction modes, evaluation) and possible revision of maritime education models of the respective apposite organizations (IMO) (Dix et al., 2004, Papachristos, Nikitakos, 2010, 2011).

Generally, this approach is generic, in the sense that it can be the starting point for an integrated usability & educational evaluation of the interactive technologies during in-situ education, simulation and pragmatic ship operation management. Today, in total theapplication of neurosciences on education and especially gaze-tracking methods are an important research quest and expansion (Goswami, 2007, Papachristos, Nikitakos, 2010).

2 LITERATURE REVIEW AND SCOPE

As more information is integrated on board by implementing an e-navigation strategy plan in the future, graphic user interface (GUI) is likely to be more sophisticated. Such sophisticated equipment can enhance navigational safety if seafarers can operate equipment, access information and understand it properly. So, when seafarers misunderstand information, sophistication will not lead to navigational safety and rather may pose risks on the ship. Thus, it is important to establish a methodology for usability evaluation (with emphasis on user's satisfaction) navigational or engine management equipment (IMO, 2012).

Usability has been defined by ISO 9241 as "the extent to which a product can be used by specified users to achieve specified goals with effectiveness, efficiency, and satisfaction in a specified context of use". It is widely acknowledged that the efficiency and effectiveness can be measured in an objective manner, i.e. in specific contexts of use and with the participation of representative user groups. They are usually defined in terms of metrics like: task success, time-to-task, errors, learnability (in repetitive use tests), etc.; while personal satisfaction is subjective in nature and depends on the characteristics of the user groups addressed (Papachristos et al., 2012, Tullis and Albert, 2008, Kotzabasis, 2011). Usability testing procedures used in user-centered interaction are designed to evaluate a product by testing it on users. This can be seen as an irreplaceable usability practice, since it gives direct input on how real users use the system. Usability testing focuses on measuring a human-made product's capacity to meet its intended purpose (Dix et al., 2004, Nielsen, 1994). A number of usability methods have been developed and promoted by different researchers (Neilson and Mark, 1994, Norman, 2006, Ryu, 2005).

There is considerable work on the ergonomic & usability assessment of the human strain (Torner et al., 1994) and the design and arrangement of ship equipment. This work has few applications in shipping industry (Petersen et al., 2010) and has not yet resulted to well established evaluation methods and cases (Wang, 2001). More specifically, these studies tend to report on usage effects on health, safety and mental workload; however they offer little guidance on the evaluation methods and/or design of the respective technology and equipment (devices) with respect to usability (Papachristos et al., 2012).

Research in Human-Computer Interaction (HCI) has created many methods for improving usability during the design process as well as at the evaluation of interactive products. The study of usability itself is extended to include other aspects of the user experience like accessibility, aesthetics, emotion and affect and ergonomics (Papachristos et al., 2012).

The area of computer simulation has been successfully applied to the study and modeling of processes, applications and real-world objects (Rutten et al., 2012). The simulators constitute a category of educational software and follow a methodology of application in instructive practice (Crook, 1994, Solomonidou, 2001). According to de Jong and van Joolingen (1998) a computer simulation is "a program that contains a model of a system (natural or artificial; e.g., equipment) or a process". Their use in the science or technology education has the potential to generate higher learning outcomes in ways not previously possible (Akpan, 2001). In comparison with textbooks and lectures, a learning environment with a computer simulation has the advantages that students can systematically explore hypothetical situations, interact with a simplified version of a process or system, change the time-scale of events, and practice tasks and solve problems in a realistic environment without stress (van Berkum and de Jong, 1991). A student's discovery that predictions are confirmed by subsequent events in a simulation, when the student understands how these events are caused, can lead to refinement of the conceptual understanding of a phenomenon (Windschitl and Andre, 1998). Possible reasons instigating teachers to use computer simulations include: the saving of time, allowing them to devote more time to the students rather than setting up and supervising experimental equipment; the ease with which experimental variables can be manipulated, allowing for stating and testing hypotheses; and provision of ways to support understanding with varying representations, such as diagrams and graphs (Blake and Scanlon, 2007).

Specifically, the Maritime Engine Simulation (MES) allows the creation of real, dynamic situations that take place on a ship at sea in a controlled surrounding where naval machine officers are able to (Kluj, 2002; Tsoumas et al., 2004):

1 practice new techniques and dexterities
2 shape opinions from teachers and colleagues
3 transport the theory of a real situations in a safe operation
4 face several problems simultaneously rather than successively, can learn by giving priority to multiple objectives under high pressure situations and change situations accordingly.

Gaze interaction through eye tracking is an interface technology that has great potential. Eye tracking is a technology that provides analytical insights for studying human behavior and visual attention (Duchowski, 2007). Moreover, it is an intuitive human–computer interface that especially enables users with disabilities to interact with a computer (Nacke et al., 2011). Infrared monitor eye gaze tracking Human-Computer Interaction (HCI), which is limited by restrictions of user's head movement and frequent calibrations etc, is an important HCI method (Cheng et al., 2010, Hansen and Qiang, 2010). This method measuring the effect of personalization could be the relationship of users' actual behavior in a hypermedia environment with theories that raise the issue of individual preferences and differences (Tsianos et al., 2009). The notion that there are individual differences in eye movement behavior in information processing has already been supported at a cultural level (Rayner et al., 2007), at the level of gender differences (Mueller et al., 2008), and even in relation to cognitive style (verbal-analytic versus spatial-holistic) (Galin and Ornstein, 1974).

International bibliography provides many sources on the Eye-tracking research in education (Conati

and Merten 2007). In the field of learning and instruction, eye tracking used to be applied primarily in reading research with only a few exceptions in other areas such as text and picture comprehension and problem solving (Halsanova et al., 2009, Hannus and Hyona, 1999, Hagerty and Just, 1993, Hyona and Niemi, 1990, Just and Carpenter, 1980, Rayner, 1998, Van Cog and Scheiter, 2010, Verschaffel et al., 1992). However, this has changed over recent years, eye-tracking is starting to be applied more often, especially in studies on multimedia learning (Van Cog and Scheiter, 2010). Because eye tracking provides insights in the allocation of visual attention, it is highly suited tfor the study of differences in intentional processes evoked by different types of multimedia and multi-representational learning materials (Van Cog and Scheiter, 2010, Halsanova et al., 2009). For example, Qu and Johnson (2005), use eye-tracking for interaction adaptation within the Virtual Factory teaching systems (VFTS), an computer tutor for teaching engineering skills. Eye-tracking is used to discern the time the user spends reading something from the time the user spends thinking before taking action, with the goal of assessing and adapting to the motivational states of student effort and confusion.

The major idea of this paper is to compare the results from the gaze tracker (face analysis tool) of three experiments for satisfaction evaluation of the students-users' (subjective) satisfaction of the maritime education via user interface evaluation of several types of educational software (i.e. engine simulator, ECDIS, MATLAB). We use a combination of qualitative – quantitative methodology, on one hand, and the use of a neuroscience tool (use biometric tool –face analysis/gaze tracker), on the other hand. This aims at the combination of the positive aspects of the corresponding methodologies: aiming at countable results & variable check (quantitative, questionnaire use), interpretative, explanatory (qualitative, interview use) and more objective measurements by "observation" of the user's physiological data (gaze tracking use).

3 METHOD

The optical perception includes the stimulant's natural reception from the external world and the process/explication of that stimulant. The observation of eye movement is an established method in many years now. The eye movements are supposed to depict the level of cognitive process a screen demands and consequently the level of facility or difficulty of its process. Usually, optical measurement concentrates on the following: (a) the eyes' focus points, (b) the eyes' movement patterns and/or (c) the pupil's alterations (Dix et al., 2004; Duchowski, 2007).

The measurement methodology must fulfill all three requirements of the cognitive neuroscience (experiential verification, operational definition, repetition) and include data-tools: (a) Recording device: might include special glasses with the recording camera or a web camera, (b) Registration data process – analysis software and (c) data process software (Papachristos, Nikitakos, 2010). The following figure shows the optical data registration procedure:

Figure 1. The Gaze tracking process (as rich picture)

The elements of the proposed approach include (Fig.2)(Papachristos et al., 2013):

1 Registration and interpretation of user emotional states (questionnaires)
2 Optical recording (gaze tracker)
3 Usability/Satisfaction & Educational assessment questionnaires
4 Wrap-up interviews (emotional assessments).

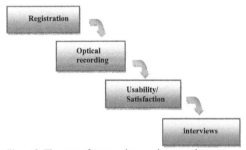

Figure 2. The steps of proposed research approach

In the experiment the optical data registration will be conducted by the "*Face Analysis*" software that was developed by the *IVML Lab of the National Technical University of Athens*, in connection with a Web camera set on the where the subject of the research (educational software i.e. MATLAB) (Asteriadis et al., 2009). That particular software records a large number of variables (42) that concern data on the form of the face as well but in the

present research we focus only 5 parameters that refer to the user's eyes and head movement. The next diagram shows the software's optical interface during the registration procedure (Fig.3) and a figure for tool's operation (Fig.4).

The formalistic presentation of the tool (Face Analysis) gives a total output with a parameters set of User's Visual Attention (VA):

$$VA=\{p_i\}, i\in[1..5] \tag{1}$$

who p_i: parameters of VA, as
- p_1 [time]: time recording
- p_2 [gv (h,v)]: gaze vector f(horizontal, vertical)
- p_3 [h_p]: head pose f (pitch, yaw)
- p_4 [d_m]: distance of monitor (metric)
- p_5 [h_r]: head roll (angle)

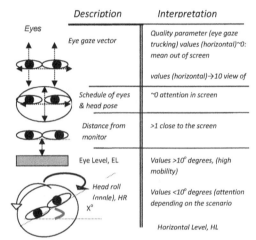

Figure 3. Biometric tool parameters interpretation ('Face Analysis')

Figure 4. 'Face Analysis' software in action

The personal satisfaction modeling contains 5 levels (Papachristos et al., 2013):

Very dissatisfied	dissatisfied	neutral	Somewhat satisfied	Very satisfied
LEVEL 1	LEVEL 2	LEVEL 3	LEVEL 4	LEVEL 5

Figure 5. The Satisfaction levels

From these levels, we design the Research Personal Satisfaction Framework (RPSF) (Fig.6) (Papachristos et al., 2013):
− *Negative Level* (Levels 1 & 2)
− *Neutral Level* (Level 3)
− Positive Level (Levels 4 & 5)

Negative Level	Neutral Level	Positive Level
Level 1 AND 2	Level 3	Level 4 AND 5

Figure 6. The Research Personal Satisfaction Framework (RPSF)

Finally, we use a combination of methods (questionnaires, interviews), because in the international bibliography, the use of multiple methods of educational evaluation in educational practice is more effective and the combinatorial use of quantitative and qualitative approaches confines their weaknesses (Brannen, 1995, Bryman, 1995, Patton, 1990, Retalis et al., 2005, Tsianos et al., 2009). Specifically, the Mixed Methods Research (MMR) employs a combination of qualitative and quantitative methods. It has been used as a distinct approach in the social and behavioral sciences for more than three decades. MMR is still generating discussions and debates about its definition, the method involved, and the standards for the quality. Although still evolving, MMR has become an establish approach. It is already considered the 3rd research approach, along with the quantitative and qualitative approaches, and has its own emerging world view, vocabulary, and techniques (Fidel, 2008).

The Personal (subjective) Satisfaction is a difficult measuring factor. For that, we use a mixed technique by using a gaze tracker (and language dimension/sentiment analysis) with MMR methods (questionnaire & interview), verifying measurements can be accomplish in order to extract safer conclusions. The size of samples is small because in experimental psychology by using equipment, are 20-30 participants usually. The size of sample depends from nature of research (Borg and Gall, 1979, Cohen et al. 2008, Papachristos et al., 2013a, 2013b, 2013c).

4 ANALYSIS

The data of this analysis come from three experiments (Fig.7)(Papachristos et al., 2013a, 2013b, 2013c):
− *Experiment-A(E-A)*: the execution a didactic scenario in a MATLAB environment that took place in Marine Academy of Aspropyrgos −

MAA (Merchant Faculty). The random sampling took place in January 2011 in the Computer Science Lab of MAA. The sample consists of 16 students (15 Male, 1 Female) that were subjected to the specific experimental procedure, completed the questionnaire and gave interviews (MMR approach). The scenario is based on the educational material (according to the STCW-95 corresponding standard) tutored in the 5th semester, aiming at the following educational goals:
− mathematic tool for control systems design,
− control systems modeling, and
− model analysis and simulation
The scenario involves the following activities:
− transfer functions (tf) to MATLAB:

$$G_1(s) = 1 / s+1 \qquad (2)$$
$$G_2(s) = 1 / s \qquad (3)$$

− and computation the total transfer function (Gtotal(s)). Furthermore, calculating the response (image) of the Gtotal(s) in which there is a unitary feedback (H=1) and a step function entrance.
The scenario combines educational goals with the use of simple implementation commands in the MATLAB environment. Video recording of 5 - 18 min per student. We use simple scale for usability assessment.
− *Experiment-B(E-B)*: the execution scenarios in e-navigation environments (ECDIS), aiming:
− to plan and display the ship's route for the intended voyage and to plot and monitor positions throughout voyage
− follow SOLAS V/19.2.1.4
The sampling was carried out on the January 2012 until May 2012, in the Information Technologies Lab of the National Marine Training Centre of Piraeus (NMTCP). Participated 3 Marine officers in experiment and they underwent a specific procedure (ships travels in different ports) in the ECDIS lab room with recording of 23 min per student. They completed the questionnaires and were interviewed follow the research methodology framework. We use SUS scale (SUS is a simple, ten-item scale giving a global view of subjective assessments of usability) for usability assessment (Brooke, 1996).
− *Experiment-C (E-C)*: sampling was carried out between May and June 2012, in the Marine Engine System Simulator (MESS) Laboratory of the National Marine Training Centre of Piraeus (NMTCP). The samples consisted of 13 professional (Merchant Marine officers) that were subjected to a specific experimental procedure (operation management) in engine room simulator ERS 5L90MCL11, (video recording ~23 minutes per student), completed the

questionnaires and gave interviews. We use SUS scale for usability assessment too.

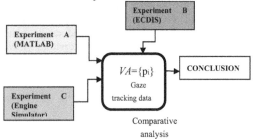

Figure 7. Diagram of Result Analysis

The result analysis:
- *E-A*: The VA parameters shows
  - the gaze vertical (p2) is 0 for a long time (mean, median and mode values for all satisfaction scales: Matlab & scenario), which means that users focus enough time out off screen,
  - in a distance from the monitor (dist_Monitor parameter) it is observed that approach the screen (>1) and keep a relatively close distance (values homogeneity),
  - Time$_{recording}$ parameter (video recording) is connected to the Satisfaction scale (grows in low scale to upper scale) (Fig.8).

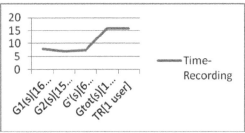

Figure 8. Time allocation relative success of didactic scenario (computing stages G1(s)/G2(s)/G'/Gtot/Time response)
- *E-B*: a relationship between Gaze parameter and Usability assessment of users. The gaze parameter depending from SUS score. It shows attention increases as assessment from ECDIS software, as shown the next table:

Table 1. Correlation between variables

Variable correlated	Spearman's rho	Sig. (2-tailed)	Remark
SUS Score – Gaze Vertical parameter (p2)	.393	.029	Positive

- *E-C*: we found the Visual Attention (VA) from the "Face Analysis tool" shows (Tab.2)
  - growing the attention as satisfaction scenario increase (mean grow high → very high) in dist

parameter (distance from monitor, >1 close to the screen),
- growing the attention as satisfaction scenario increase (mean grow high → very high) in Head Roll parameter (rolling of the head – eye angle from horizontal level, <10 attention depending on the scenario, >10 high mobility),
- growing the attention as satisfaction scenario increase (mean grow high → very high) in Gaze tracking parameter (Gaze vertical parameter >1 view the screen).

Table 2. Variation of VA parameters

Satisfaction Scenario (Average)	Very high (4 male)	High (8 male)	Medium (1 male)
Gaze Vertical	8.28	7.09	0.21
Dist_Monitor	1.12	1.07	0.99
Head Roll	1.9	0.49	-2.16

5 CONCLUSIONS

This experimental procedure is a primary effort to research the educational and usability evaluation with emotion analysis (satisfaction) of the users-students in maritime simulators.

The main purpose of this research, is the investigation of personal satisfaction of a user of MET equipment (Engine room simulator) via the assistance of Gaze tracker (Face Analysis tool) but also other methods like MMR (questionnaires-interviews).

The results from experiments (3) until now, are shows:
- the correlation between VA parameters and satisfaction:
  - time recorder (p1),
  - Gaze vertical (p2),
  - Distance Monitor (p4), and
  - Head Roll (p5),
- and the correlation between VA parameters and usability assessment (SUS scale)(ECDIS experiment).

The gaze tracking methodology appears to be one capable of satisfying evaluation tool. The research continues with the numeral increase of the sample and the total processing and evaluation of the research findings (qualitative and quantitative data). The proposed approach may require further adaptations to accommodate evaluation of particular interactive systems.

REFERENCES

Akpan, J. P. 2001. Issues associated with inserting computer simulations into biology instruction: a review of the literature. Electronic Journal of Science Education, 5(3), Retrieved from: http://ejse.southwestern.edu/article/viewArticle/7656/5423.

Arpan, S., 2009. "CSCI 8810 A report on Gaze Tracking", Retrieved from: http://www.docstoc.com/docs/80403505/CSCI-8810-C-Gaze-Tracking

Asteriadis, S. Tzouveli, P. Karpouzis, K. Kollias, S. 2009. Estimation of behavioral user state based on eye gaze and head pose—application in an e-learning environment, *Multimedia Tools and Applications, Springer* 2009;41:3:469-493.

Blake, C., and Scanlon, E. (2007). Reconsidering simulations in science education at a distance:features of effective use. *Journal of Computer Assisted Learning,* 23(6), 491–502.

Brannen, J. 1995. Combining qualitative and quantitative approaches: An overview, J. Brannen (ed.), *Mixing Methods: Qualitative and Quantitative Research.* UK:Avebury, 3-38.

Bryman, J. 1995. Quantitative and qualitative research:further reflections on their integration, *Mixing Methods: Qualitative and Quantitative Research.* UK:Avebury, 57-80.

Borg, W.R. and Gall, M.D. (1979) Educational Research: an Introduction (6th edition). NY: Longman.

Brooke, J. 1996. SUS: A "quick and dirty" usability scale. In: Jordan, P. W., Thomas, B., Weerdmeester, B. A., McClelland (eds.) Usability Evaluation in Industry, Taylor & Francis, London, UK pp. 189-194.

Cheng, D. Zhao, Z. Lu, J. Tu, D. 2010. A Kind of Modelling and Simulating Method for Eye Gaze Tracking HCI System, *Proceedings of 3rd International Congress on Image and Signal Processing* (CISP2010), IEEE, EMB, pp. 511-514.

Cohen, L. Manion, L. Morrison, K. 2008. *Research Methods in Education* (5th edtion). London: Routledge Falmer.

Conati, C. and Merten, C. 2007. Eye-tracking for user modelling in exploratory learning environments: An empirical evaluation. *Knowledge-Based Systems*;20:557-74.

Crook, C. 1994. *Computers and the collaborative experience of learning*. London, Routledge.

de Jong, T., and van Joolingen, W. R. (1998). Scientific discovery learning with computer simulations of conceptual domains. Review of Educational Research, 68(2), 179–201.

Dix, A. Finlay, J. Abowd, G. D. Beale, R. 2004. *Human-Computer Interaction*, UK:Pearson Education Limited.

Duchowski, A. T. 2007. *Eye tracking methodology: Theory and practice*, Springer, New York.

Fidel, R. 2008. Are we there yet?: Mixed methods research in library and information science, *Library & Information Science Research*, pp. 265-72.

Galin, D. and Ornstein, R. 1974. "Individual Differences in Cognitive Style—I. Reflective Eye Movements," *Neuropsychologia*, vol. 12, pp. 367-376.

Goswami. U. 2007. Neuroscience and education: from research to practice? *Nature Review Neuroscience*, 7:406-413.

Hagerty M, Just M A. 1993. Constructing mental models of machines from text and diagrams. *Journal of Memory and Language*; 32:71-42.

Hansen, D.W. Qiang, Ji 2010. In the Eye of the Beholder: A Survey of Models for Eyes and Gaze Pattern Analysis and Machine Intelligence, *IEEE Transactions on*, Vol.32, Is.3, pp.478-500.

Holsanova, J. Holmberg, N. Holmqvist, K. 2009. Reading information graphics: the role of spatial contiguity and dual attentional guidance. *Applied Cognitive Psychology* 2009; 23:1215-26.

Hyona J, Niemi P. 1990.Eye movements during repeated reading of a text. *Acta Psychologica*; 73: 259-80.

IMO-International, Maritime Organization, 2003. Issues for training seafarers resulting from the implementation on board technology, *STW 34/INF.6*.

IMO, 2012. "Development of an e-Navigation strategy implementation plan", NAV 58/INF.13, 27 April 2012.

ISO 9241, "Ergonomics of Human System Interaction", http://www.iso.org.

Just M A, Carpenter P A. A. 1980. Theory of reading: From eye fixations to comprehension. *Psychological Review* 1980;87: 329-55.

Kluj, S. 2002. Relationship between learning goals and proper simulator, *ICERRS5 Paper*.

Kotzabasis, P. 2011. *Human-Computer Interaction: Principles, methods and examples*, Athens, Kleidarithmos (in Greek).

Mueller, S.C. Jackson, C. P. T. and Skelton, R.W. 2008. "Sex Differences in a Virtual Water Maze: An Eye Tracking and Pupillometry Study". *Behavioural Brain Research, vol.* 193, pp. 209-215.

Nacke L.E. Stellmach, S. Sasse, D. Niesenhaus, J. Dachselt, R. 2011. LAIF: A logging and interaction framework for gaze-based interfaces in virtual entertainment environments, *Entertainment Computing* 2, pp. 265–273.

Nielsen, J. 1994. *Usability Engineering*, Academic Press Inc.

Nielsen, J, and Mack, R.L. (eds.) 1994. *Usability Inspection Methods*, New York, John Wiley.

Norman, K.l. 2006. Levels of Automation and User Participation in Usability Testing, Interacting with computers, Elsevier.

Papachristos D, Nikitakos N. 2010. *Application Methods and Tools of Neuroscience, in Marine Education*. Conference Proceedings "Marine Education & Marine Technology" (ELINT), 1 December 2010, Athens, Greece.

Papachristos, D. and Nikitakos, N., 2011. *Evaluation of Educational Software for Marine Training with the Aid of Neuroscience Methods and Tools*, Symposium Proceedings, TransNav2011 Symposium, Gdynia Maritime University, Poland, 15-17 June, 2011.

Papachristos, D. Koutsabasis, P. Nikitakos, N. 2012. Usability Evaluation at the Ship's Bridge: A Multi-Method Approach, In *Proceedings of 4th International Symposium on "Ships Operation, Management and Economics"-SOME12*, The Greek Section (SNAME), 8-9 November 2012, Eygenideio Foundation, Athens.

Papachristos, D. and Nikitakos, N., 2013a. Human Factor Evaluation for Marine Education by using Neuroscience Tools, In Proceedings 4th International Symposium of Maritime Safety Security and Environmental Protection, 30-31 May 2013, Athens, http://www.massep.gr/sponsorship-opportunities/.

Papachristos, D. Alafodimos, K. Lambrou, M. Kalogiannakis, M. Nikitakos, N. 2013b. Gaze tracking Method Use in the Satisfaction Evaluation (Matlab Environment) in Maritime Education. Information & IT Today, ISSN: 1339-147X, Vol.1, Issue 1, pp.11-18.

Papachristos, D. Alafodimos, K. Lambrou, M. 2013c. Marine e-Learning Evaluation: A Neuroscience Approach. International Journal of Marine Navigation and Safety of Sea Transportation, Volume 7, (3), Sept. 2013, DOI: 10.12716/10001.07.03.XX.

Patton, M. Q. 1990. *Qualitative Evaluation and Research Methods*. CA:Sage Publications.

Petersen, E.S. Dittman, K. Lützhöft, M. 2010. Making the Phantom Real: A Case of Applied Maritime Human Factors, Proceedings of SNAME SOME 2010,http://publications.lib.chalmers.se/cpl/record/I ndex.xsql?pubid=133364 (last access 18 November 2014).

Rayner, K. Xingshan, L. Williams, C.C. Kyle, R. C. and Arnold, W. D. 2007. "Eye Movements during Information Processing Tasks: Individual Differences and Cultural Effects," *Vision Research*, vol. 47, pp. 2714-2726.

Retalis, S. (eds.), 2005. Educational Technology. The advanced internet technologies in learning service, Athens:Kastaniotis Editions (in greek).

Rutten, N. Van Joolingen, W. R. Van de Veen, J. T. 2012. The learning effects of computer simulations in science education, *Computers & Education 58, pp. 136-153.*

Ryu, Y.S. 2005.Development of Usabilities Questionnaires for Electronic mobile Products and Decision Making Methods, Phd Thesis, Blacksburg, Virginia, USA, retrieving from http://scholar.lib.vt.edu/theses/available/etd- 08212005-234205/unrestricted/ETD_Ryu_Final.pdf (last access 19 November 2014).

Solomonidou, X. 2001. *Modern Educational Technology.* Saloniki, Kodikas (in Greek).

Tsianos, N., Lekkas, Z., Germanakos, P., Mourlas, C., Samaras, G 2009. An Experimental Assessment of the Use of Cognitive and Affective Factors in Adaptive Educational Hypermedia, *IEEE Transactions on Learning Technologies*, Vol. 2, No. 3, July-September 2009, pp. 249-258.

Tsoukalas, V. Papachristos, D. Mattheu, E. Tsoumas, N. 2008. Marine Engineers' Training: Educational Assessment of Engine Room Simulators, *WMU Journal of Maritime Affairs,* Vol.7, No.2, pp.429-448, ISSN 1651-436X, Current Awareness Bulletin, Vol. XX-No.10, Dec. 2008, IMO Maritime Knowledge Centre, pp.7.

Tsoumas, N. Papachristos, D. Matheou, E. Tsoukalas, V. 2004. Pedagogical Evaluation of the Ship's Engine Room Simulator, used in apprentice marine engineers' Instruction, 1st International Conference IT, Athens.

Tullis, T. and Albert, B. 2008. *Measuring the User Experience: Collecting Analysing and Presenting Usability Metrics*, Morgan Kaufmann.

Torner, M. Almstrom, C. Karlsson, R. Kadefors, R. 1994. Working on a moving surface—a biomechanical analysis of musculoskeletal load due to ship motions in combination with work, *Ergonomics*, Vol. 37.

van Berkum, J. J. A., and de Jong, T. 1991. Instructional environments for simulations. *Education & Computing,* 6, 305–358.

Van Gog T, Scheiter K. 2010. Eye tracking as a tool to study and enhance multimedia learning. *Learning and Instruction*; 20:95-99.

Verschaffel L, De Corte E, Pauwels A. 1992. Solving compare word problems: An eye movement test of Lewis and Mayer's consistency hypothesis. *Journal of Educational Psychology*; 84:85-94.

Wang, J. 2001. The current status and future aspects in formal ship safety assessment, *Safety Sciences 38*, pp. 19-30.

Windschitl, M., and Andre, T. 1998. Using computer simulations to enhance conceptual change: the roles of constructivist instruction and student epistemological beliefs. *Journal of Research in Science Teaching*, 35(2), 145–160.

Sea Ports and Harbours

Trends in Environmental Policy Instruments and Best Practices in Port Operations

O.-P. Brunila, V. Kunnaala-Hyrkki & E. Hämäläinen
University of Turku, Brahea Centre for Maritime Studies, Kotka, Finland

ABSTRACT: Environmental effects of port activities can be controlled and decreased with different kind of policy instruments. Policy instruments can be divided into regulatory instruments, economic instruments and information-based guidance. The term best environmental practice means the application of the most appropriate combination of environmental control measures and strategies. Nowadays also port operators are beginning to see the potential advantages of sharing ideas on best practice and trends. In this paper we conduct a literature review and a case study on environmental policy instruments and best practices in port operations. We briefly examine the strengths and weaknesses of the different policy instruments. In order to study the best practices in port operations we will examine for example previous projects, the ports' voluntary actions and the environmental initiatives in port development. In addition, we discuss in more depth about best practices in the Port of Helsinki in Finland.

1 INTRODUCTION

1.1 Background

Ports facilitate the movement of goods from one country of continent into another. However, ports are also sites of environmental pollution that originate from land-based activities, shipping and ports' own activities. With increasing regulations to control port pollution and intensified public debates, ports can no longer avoid environment concerns. There is also a pressure to increase services, modernize development and enhance economic efficiency in order to respond to the growing competition. In order to balance the competing needs, the port operations must be managed in a sustainable manner so that the economic growth in is balanced with environmental protection (Hiranandani 2014).

Environmental effects of port activities can be controlled and decreased in several ways. In Finland, both EU and national legislation regulate the port operations and set different kinds of economic incentives or disincentives to the operation. In addition, environmental effects can be controlled and decreased with voluntary actions. The ports can go even further in controlling and decreasing their environmental effects than required by law. Certifications, corporate social responsibility

and developing of best practices are means to effectively control and decrease the ports' environmental effects. The term best environmental practice means the application of the most appropriate combination of environmental control measures and strategies.

Nowadays, port operators are also beginning to see the potential advantages of sharing ideas on best practice and trends. Sharing knowledge and best practices allows the ports to enhance their operations and helps to choose the most cost-effective measures for decreasing their environmental impact.

1.2 Objectives

The first goal of this paper is to find out what kind of policy instruments affect the ports' environmental effects. In addition, the strengths and weaknesses of the different policy instruments based on their effectiveness are analysed. The second goal of this paper is to analyse the environmental best practices that are used in ports. Special attention is given to the Port of Helsinki in Finland.

1.3 Methodology

A literature review was conducted in order to study the applicable environmental policy instruments and best practices in port operations. In order to study the best practices in port operations we will examine for example previous projects, the ports' voluntary actions and the environmental initiatives in port development. In addition, we discuss in more depth about best practices in the Port of Helsinki in Finland. The case study discusses the port's recent environmental policies and best practices.

1.4 *Structure*

In the following chapter 2 we examine the applicable environmental policy instruments and their strengths and weaknesses. In chapter 3 we examine best practices in port operations and also the best practices in Port of Helsinki in more depth. Chapter 4 includes the discussion and conclusions of this paper.

2 ENVIRONMENTAL POLICY INTRUMENTS

2.1 *Different policy instruments*

Policy instruments can be divided for example based on the interests they aim to protect. The policy instruments can be directed to protect private goods, such as the competitiveness of companies, or public goods, which the market would otherwise neglect. These public goods include for example the protection of the environment from the harmful effects of port operations (Kuronen & Tapaninen 2010).

Policy instruments can be either preventive, such as a requirement for an environmental permit, or sanctions and consequences, such as financial liability. Both preventive measures and consequences can be either private or administrative measures (Kuronen & Tapaninen 2010).

Usually policy instruments are divided into regulatory instruments, economic instruments and information-based guidance. Environmental effects of port activities can be controlled and decreased with each kind of the policy instruments.

2.2 *Regulatory instruments*

Regulatory instruments include for example jurisdiction and law based decrees, restrictions and licenses (Kuronen & Tapaninen 2010). Regulatory instruments are effective and easy to enforce. Their weaknesses include their economic efficiency and public acceptance. Also their implementation and enactment can be expensive, difficult or practically impossible (Vieira et al. 2007). In addition, regulatory instruments may not promote changes or innovations, because there is no economic incentive (Klemmensen et al. 2007).

The EU has a lot of different regulations that influence the European ports and their management.

However, not all port related EU legislation affects the environment. The EU has directives for habitats, fauna and biodiversity and also different legislation on emissions, noise, soil, waste and air quality, and pollution from ships.

In Finland, all ports have strict environmental regulations. Ports have to follow national environmental policies and environmental management systems, they have to get environmental permits for the operation and go through an environmental impact assessment (EIA), and of course the Finnish ports have to follow the previously mentioned EU legislation.

2.3 *Economic instruments*

Economic instruments include for example taxes, subsidies and fees. Economic instruments can achieve environmental targets with good economic efficiency. However, they also often face acceptance difficulties, because their tendency to increase prices. If they have lateral effects or are used in combination with other policies, they can be more acceptable if the price increase is compensated by a price decrease of the other (Kuronen & Tapaninen 2013, Vieira et al. 2007).

In shipping, environmentally differentiated port fees are applied as a financial incentive to encourage shipping companies reduce environmental impact themselves. For example, The Swedish Maritime Administration in co-operation with the Swedish ports and Ship owners decided to implement a system of environmentally differentiated port charges in the year 1998. The aim was to create incentives to use low sulphur oil and to reduce NOx emissions. Similar systems are applied in other countries also (Breitzmann & Hytti 2013).

2.4 *Information guidance*

Information guidance is based on the idea that information can lead to a voluntarily change in behaviour and its effect is totally dependent on the interest of the operator. Information-based guidance includes for example information distribution, voluntary education, certifications and awards (Kuronen & Tapaninen 2010).

Information guidance is based on the idea that well-reasoned information can result in voluntary change in behaviour. While regulatory and economic instruments are usually based on legislation with consequences for non-conformity, information guidance is completely dependent on the actors' voluntary interests (Kuronen & Tapaninen 2010).

Because port activities have a significant effect on the environment, the environmental issues should be integrated into port management. National and international legislation is one of the most significant drivers that lead the ports to invest in environmental actions. Yet, the ports may also find motivation to reduce their environmental effects from their own driving force, from societal pressure, in order improve the port operations or in order to gain competitive advantage (Madjidian et al. 2013).

The policy of the European Sea Ports Organization (ESPO) is established to consist of the compliance with legislation and the achievement good environmental standards through voluntary, self-regulation, while reflecting the local circumstances of the individual port (Van Breemen et al. 2008).

2.5 *Criteria for effective policy instruments*

For example, Kuronen & Tapaninen (2010) have presented in their research criteria for effective policy instruments. Effectiveness refers to the potential improvement of the goal of the policy instrument and whether the instrument is appropriate and technically suitable for achieving the goal (Greiner et al. 2000, Vieira et al. 2007). Firstly, economic efficiency relates to effectiveness in terms of implementation costs and the economic efficiency of an instrument in a collective sense, in view of the total benefits against its total costs (Greiner et al. 2000, Vieira et al. 2007).

A second criterion is acceptability, which refers to the stakeholders' level of agreement on a new policy instrument and to the community acceptability of the instrument. Acceptability is required so that the policy can be durable (Greiner et al. 2000, Vieira et al. 2007).

The criterion of enforcement indicates how effectively the policy instrument can be implemented. Some policy instruments are difficult to implement, even though they would probably be quite effective (Kuronen & Tapaninen 2010). There can be several types of barriers for implementation, such as legal and institutional, financial, such as lack of resources, political and cultural, and technological (Greiner et al. 2000, Vieira et al. 2007).

While assessing the effectiveness of policy instruments, also the lateral effects of the instruments should be taken into consideration. They refer to possible spill over effects of an instrument into other sectors, for example reduction in air emissions can improve the health in the surrounding community and thus decrease health care expenses (Vieira et al. 2007).

The criterion for incentives and innovations requires that the instrument encourages innovation and change and provides incentives for improvements (Greiner et al. 2000).

As it was noted earlier, not all instruments are based on jurisdiction. Yet one aspect of the effectiveness of instruments based on jurisdiction is the consequences of non-compliance. The consequences of non-compliance should be severe

enough so that the temptation to non-compliance in minimized (Greiner et al. 2000).

3 BEST PRACTICES

3.1 *Best practices in port operations*

The society is expecting that the ports take responsibility of environmental protection and sustainable development. In addition, ports are often located in a close proximity of urban areas and may even be bounded by areas of special environmental significance, such as protected habitats. These issues have to be taken into consideration in the individual port's operations (Van Breemen et al. 2008).

One key element in the competition between the Baltic Sea ports now and in the future will be their environmental status and their capability to response to the challenges of sustainable development (Brunila & Anttila 2013).

The term best environmental practice means the application of the most appropriate environmental control measure or combination of measures that show results superior to those achieved with other means (e.g. GHD 2013).

Previous researches have produced a number of best practice planning lessons (e.g. Australian Government 2010, GHD 2013, PIANC 2013). During those researches it has been found that projects that develop from long term plans and also projects that have strong business cases are likely to be more successful. In addition, strong project governance means strong project delivery. In the studies, it was also highlighted that you should be open to learn the lessons from previous projects, since they can be very helpful (Australian Government 2010).

Thus nowadays, port operators are beginning to see the potential advantages of sharing ideas on best practice and trends. Sharing knowledge and best practices allows the ports to enhance their operations and helps to choose the most cost-effective measures for decreasing their environmental impact. In addition, there is a possibility to learn from others' mistakes.

Yet, it should be recognized that the ports are not the same. Each port and its surrounding area can be considered to be unique and the importance of different environmental aspects depends on the characteristics of each port, but yet they share several common environmental issues and face common environmental challenges (Hiranandani 2014). Thus not all best practices applied in one port are directly applicable in another.

Hiranandani (2014) has studied the drivers and constraints behind the ports sustainable development and also best practices. According to Hiranandani (2014), community pressure can be both a driving and a constraining factor. Yet, it should be noted that nowadays all marine operations and activities are under increasing scrutiny from nor only legislators but also other stakeholders with interests in the quality and condition of the environment itself (Madjidian et al. 2013).

According to Hiranandani (2014), one thing that impedes sustainable port activities is the lack and uncertainties of data. Data is necessary for the ports so that they can monitor their environmental performance. According to Hiranandani (2014), greater investments of time and resources are required, so that ports have adequate data and evidence to justify, plan, monitor and evaluate their environmental practices.

Because of the recent global economic recession, it can be considered that reducing environmental impacts in ports is too costly. Yet it should be noted that environmental initiatives by ports can also become a strong commercial argument and a competitive advantage. In addition, implementing state-of-the-art sustainable practices can reduce costs and enhance the port's operational efficiency (Hiranandani 2014).

According to Hiranandani (2014), when it comes to regulations and their implementation, they are both a driving and a constraining factor. In addition, the differences and constant changes in international, national and regional legislations can distort competiveness of local operators due to stricter local regulations (Hiranandani 2014).

3.2 Examples of best practices in ports

Ports utilize a range of sustainable practices. Hiranandani (2014) performed a multi-case study regarding sustainable development in seaports. The study also included examples of best practices in the case ports. Hiranandani (2014) established that air pollution was the biggest environmental concern in all case study ports. The ports had implemented three main strategies to reduce local emissions: alternative fuels, onshore electric power supplies and intermodal transport. Water quality, sediment and storm water management plans were proposed or implemented in all case study ports. Disposing dredge materials can also cause harm to the environment. The ports had used innovative practices such as recycling and re-use of dredged materials. Most case study ports have effective waste management programs, but two ports had implemented additional waste minimization and recycling and reuse programs (Hiranandani 2014).

The ports studied by Hiranandani (2014) had also carried out environmental impact assessments regarding their expansion and development projects. In three of the studied ports, sustainability was also expressed in resource conservation, efficient energy consumption (e.g. through co-siting and clustering of businesses), renewable energy, and environmentally friendly procurement policies and working practices that can reduce the overall negative environmental effects (Hiranandani 2014).

Best practices for ports have been developed in for example previous projects. For instance, in the Noise Management in European Ports (NoMEPorts) project in the Port of Amsterdam, a guide on port area noise management practices was developed (Van Breemen et al. 2008). One efficient way to reduce the environmental effects of mooring ships is electric power supply from shore to ship. The electrical connection eliminates the need to run the auxiliary engines and thus reduces the related air emissions. The project BSR InnoShip (Baltic Sea cooperation for reducing ship and port emissions through knowledge- & innovation-based competitiveness) produced technical solutions for shore-to-ship connections (Breitzmann & Hytti 2013). Also the project CLEANSHIP offered best practice guides for onshore supplies and port reception facilities for ship sewage.

Corson & Fisher's (2014) best management practices manual can be used by ports seeking to implement operational controls to reduce environmental impacts in port operations. The document is formatted according to the ISO 14001 Environmental Management System (EMS) Standard. It is designed for the use of port personnel to manage their port operations with consideration of the environment, in which those operations occur (Corson & Fisher 2014).

In addition, Port Equipment Manufacturers Association (PEMA) has published information on e.g. the energy saving and emissions reduction possibilities available in the design and operation of port equipment (Corbetta et al. 2011)

Best practices are closely linked to the concept of Green Port. The key elements in the Green Port concept include: long term vision, stakeholder participation, shift from sustainability as a legal obligation to sustainability as an economic driver, actively sharing knowledge with other ports and continuous strive towards innovation (PIANC 2013). The World Association for Waterborne Transport Infrastructure's (PIANC) Sustainable Ports –guide (PIANC 2013) provides best practices and challenge response options to for example the following issues: land and water area uses, air quality, water quality, dredging impacts, sound impacts, energy and climate change mitigation, habitat and species health, ship related waste management and sustainable resources management.

Best practices are also closely linked to environmental management systems or standards, such as Port Environmental Review System (PERS), ISO 14001 or the Eco-Management and Audit Scheme (EMAS). Environmental management systems and standards can include good practices that the ports can use in their operation. Environmental management systems also indicate the port's preparedness to actually comply with environmental legislation, and strive for environmental improvement and sustainable development (Madjidian et al. 2013).

3.3 *Best practices in Finnish ports - Case Port of Helsinki*

In Finland all ports have strict environmental regulations. Ports have to follow national environmental policies, environmental management systems, environmental permits, environmental impact assessment, and of course the previously mentioned EU legislation. Many Finnish ports have various independent environmental projects that are meant for enhancing their environmental status and for protecting the surrounding environment of the ports (Brunila & Anttila 2013).

Finnish Environmental Protection Act (EPA, 527/2014) requires that a permit is required for activities that pose a threat of environmental pollution (EPA 27§). In addition, it is a principle that appropriate and cost effective combinations of various methods shall be used in order to prevent pollution to the environment. This is called the principle of best environmental practice (EPA 20§).

Port of Helsinki is the most important general purpose port in Finland. It is one of the leading ports for unitized cargo and one of the busiest passenger ports in the Baltic Sea. The port comprises of three harbours – South Harbour, West Harbour and Vuosaari Harbour. Imports and exports at the Port of Helsinki are transported mainly in cargo units as containers, trucks, and trailers etc.

The Port of Helsinki in Finland frequently undertakes initiatives in various fields of social responsibility in order to practice sustainable port operations. The Port of Helsinki has an environment system that is based on the ISO 14001:2004 certification. Best practices are also a feature of the ISO 14001 standard. In addition, all three harbours of the Port of Helsinki have valid separate environmental permits (Port of Helsinki 2013).

The Port of Helsinki has been involved in several EU projects, such as Clean Baltic Sea Shipping, Baltic LNG and Penta (Port of Helsinki 2012a). The Clean Baltic Sea Shipping project aims to find means for harbours to affect environmental issues in the Baltic Sea area through for example environment-based harbour fees, development of waste reception procedures, increased use of onshore power supplies for ships, and shifting to the use of alternative ship fuels (Port of Helsinki 2011). The Baltic LNG project aims to find a shared solution for the refuelling of ships with liquefied natural gas or LNG in Baltic Sea ports (Port of Helsinki 2013).

The Port of Helsinki has achieved best practices in implementing environmental measures especially in port expansion. The Port of Helsinki places the utmost importance on complying with its responsibilities to the environment, and has proper procedures and measures to minimize environmental impacts caused from its operations. During the construction of Vuosaari, the Port of Helsinki had full consideration for the environment. Every decision was made with respect to future generations. For example road and railway traffic flows were constructed to operate through tulles so that no disturbance to the adjacent Natura 2000 conservation and residential areas would occur. In addition, several noise barriers were set in place during the construction. Every berth in Vuosaari Harbour was designed to have reception facilities for black and grey water. In addition, the most advanced cargo handling equipment is used at the harbour area in order to minimize emissions (Port of Helsinki 2012b).

There can be no doubt that best practices have been achieved in implementing environmental measures to accompany the port expansion of Vuosaari. The development of the Vuosaari Harbour area coincided with carefully considered ecological investments. The huge ecological investments are widely recognized as an international best practice in green port management (Van Hooydonk 2008).

There are several individual best practices or environmentally friendly actions used by the Port of Helsinki. In this chapter we will only discuss a few of them.

In the Port of Helsinki the majority of the waste waters received come from passenger ships. There are sewers in all of the quays in all of the three harbours. Waster waters are pumped from vessels with hoses and transferred into the Port of Helsinki sewers and from there into the city's general sewage system. The organic matter contained in the sludge produced in the treatment process is used by digesting the sludge, and the biogas generated in the digestion process is collected for further use. Because of the energy produced from biogas, the treatment plant is self-sufficient when it comes to heating and about 50% self-sufficient when it comes to electricity (Madjidian et al. 2013).

The environmentally friendly actions can consist also from little things. When it comes to noise from the harbour's operations that can affect the nearby residents, the Port of Helsinki has come up with a simple solution. The port updated its noise models and measured the noise levels of each new passenger vessel. The results influence the selection of berths and vessels that stayed overnight in Helsinki were placed at a berth further from housing (Port of Helsinki 2011).

Starting from the autumn of 2012, the vessels of the shipping company Viking Line operating on the

Stockholm connection have used onshore power supply when at the quay at Katajanokka. With the use of onshore power supplies, it is possible to minimize the noise and other emissions caused by the auxiliary engines of the ships. This is important especially when the harbour is situated in the city centre, close to residential areas (Port of Helsinki 2012a).

Even though the following best practice is not an environmental practice as such, it should be regarded, since the Port of Helsinki has been granted a reward based on it. The maintenance information system of Vuosaari Harbour won first prize in the Bentley Be Inspired competition in the Innovation in Government category in Amsterdam. The system is a comprehensive one, in which several systems have been integrated to work together. The system maintains the documents, maintenance schedules, and maintenance histories of buildings, other structures, and systems at the harbour area (Port of Helsinki 2011).

4 DISCUSSION AND CONCLUSION

Ports facilitate the movement of goods from one country of continent into another. However, ports are also sites of environmental pollution that originate from land-based activities, shipping and ports' own activities.

Environmental effects of port activities can be controlled and decreased in several ways. In Finland, both EU and national legislation regulate the port operations. In addition, environmental effects can be controlled and decreased with voluntary actions, such a developing best practices.

The first goal of this paper was to find out what kind of policy instruments affect the ports' environmental effects and what are the strengths and weaknesses of said instruments.

Based on the conducted literature review, policy instruments are traditionally divided into regulatory instruments, economic instruments and information-based guidance. Regulatory instruments are effective and easy to enforce. Their weaknesses include their economic efficiency, public acceptance and difficult implementation. The EU has a lot of different regulations that influence the European ports and their management and, in addition, all ports have strict environmental regulations in Finland. Economic instruments can achieve environmental targets with good economic efficiency. However, they also often face acceptance difficulties, because their tendency to increase prices. Information guidance is based on the idea that information can lead to a voluntarily change in behaviour. The instrument's effectiveness is completely dependent on the interest of the operator.

The second goal of this paper was to discuss the environmental best practices that are used in ports. The term best environmental practice means the application of the most appropriate environmental control measure or combination of measures that show results superior to those achieved with other means.

It can be deduced that ports have several reasons to develop and utilize best practices. Best practices can be used in fulfilling the ports' legal responsibilities. In addition, utilizing best practices can bring direct benefits in the form of more efficient operations and cost savings. It has been found that, in the future, one key element in the competition between ports will be their environmental status. Thus, developing and utilizing best practices can become a competitive advantage in a world, in which environmental issues are playing an increasingly important role.

During the literature review, several examples of best practices in ports and studies regarding best practices in port operations were found. In this paper we only discussed few of them. Best practices can be developed to different port operations. For example in order to reduce local emissions the ports can introduce alternative fuels, onshore electric power supplies and intermodal transportation. The number of best practices, port innovations and projects regarding best practices is probably going to increase rapidly in the next few years.

One important notion regarding best practices in ports was found. Even though the port operators are beginning to see the potential advantages of sharing best practice, it has to be recognized that all ports are not the same. Each port and its surrounding area can be considered to be unique, thus not all best practices applied in one port are directly applicable in another. Nevertheless, sharing ideas openly is still recommended, since ports still share several common environmental issues and face common environmental challenges and different practices are applicable in different parts of port operations and some best practices can be altered to suit each port's unique needs.

In this paper, a case study on the best practices of the Port of Helsinki was also conducted. It should be noted that in Finland, all ports have strict environmental regulations. In addition, the Finnish Environmental Protection Act requires that best environmental practices are used (EPA 20§).

The Port of Helsinki frequently undertakes initiatives in order to practice sustainable port operations. The port has an ISO 14001:2004 certification based environmental system and the port has also been involved in several projects, during which best practices and environmental practices have been developed.

In this paper, special attention was paid to the best practices achieved in port expansion in the Port of Helsinki, during which every decision was made with respect to the environment and future generations. This indicates the strong commitment of the Port of Helsinki to minimize environmental impacts caused from its operations and also that environmental issues are an important factor in the port's strategy and operational policy. Thus it is safe to assume that the Port of Helsinki has already discovered the potential benefits of developing and utilizing best practices and other sustainable practices. The huge ecological investments in the port expansion are also internationally recognized as a good best practice example.

REFERENCES

Australian Government – Department of Infrastructure and Transport. 2010. *Infrastructure Planning and Delivery: Best Practice Case Studies*. Australia. Available at: http://www.infrastructure.gov.au/infrastructure/publications /files/Best_Practice_Guide.pdf

Van Breemen, T., Popp, C., Witte, R., Wolkenfelt, F. & Wooldridge, C. (ed.) 2008. Good Practice Guide on Port Area Noise Mapping and Management. Developed by the partners of the NoMEPorts (Noise Management in European Ports) Project. Port of Amsterdam. Available at: http://ec.europa.eu/environment/life/project/Projects/index. cfm?fuseaction=home.showFile&rep=file&fil=NoMEports _GPG_PANMM1.pdf

Breitzmann, K.-H. & Hytti, M. (ed.) 2013. *Pan-Baltic Manual of Best Practices on Clean Shipping and Port Operations*. BSR InnoShip – Baltic Sea cooperation for reducing ship and port emissions through knowledge- & innovation-based competitiveness. Turku.

Brunila, O.-P. & Anttila, A. 2013. Green co-operation in the eastern Gulf of Finland. *Baltic Rim Economies, Quarterly Review* (3): 23.

Corbetta, L., Johannsson, S., Johanson, F., Linnartz, V. & Sanden, A. 2011. *Energy and Environmental Efficiency in Ports & Terminals*. A PEMA Information Paper. Port Equipment Manufacturers Association. Available at: http://www.pema.org/wp-content/uploads/downloads/2011/06/PEMA-IP2-Energy-and-Environmental-Efficiency-in-Ports-and-Terminals.pdf

Corson, L.A. & Fisher, S.A. 2014. *Manual of Best Management Practices for Port Operations and Model Environmental Management System*. Great Lakes Maritime Research Institute, a University of Wisconsin-Superior and University of Minnesota-Duluth Consortium. Available at: http://www.glmri.org/downloads/resources/manualBestMa nagementPorts.pdf

Finnish Environmental Protection Act (527/2014)

GHD 2013. *Environmental Best Practice Port Development: An Analysis of International Approaches*. Department of Sustainability, Environment, Water, Population and Communities. Canberra, Australia.

Greiner, R., Young, M.D., Macdonald, A.D. & Brooks, M. 2000. Incentive instruments for the sustainable use of marine resources. *Ocean & Coastal Management* 43: 29–50.

Hiranandani, V. 2014. Sustainable development in seaports: a multi-case study. *WMU Journal of Maritime Affairs* 13(1):127–172

Van Hooydonk, E. 2008. *Helsinki - North European Port Icon. Some considerations on Helsinki's identity as a port city*. Helsingin sataman julkaisuja sarja B 2008:8. Available at: http://www.portofhelsinki.fi/instancedata/prime_product_ju lkaisu/helsinginsatama/embeds/helsinginsatamawwwstructu re/13353_Helsinki_pages_2.pdf

Klemmensen, B., Pedersen, S., Dirkinck-Holmfeld, K., Marklund, A. & Rydén, L. 2007. *Environmental policy – Legal and economic instruments*. Uppsala: The Baltic University Press.

Kuronen, J., & Tapaninen, U. 2009. *Maritime safety in the Gulf of Finland – Review on policy instruments*. Publications from the Centre for Maritime Studies University of Turku A49.

Kuronen, J. & Tapaninen, U. 2010. Evaluation of Maritime Safety Instruments. *WMU Journal of Maritime Affairs* 9(1): 45–61.

Madjidian, J., Björk, S., Nilsson, A. & Halén, T. (ed.) 2013. *CLEANSHIP – Clean Baltic Sea Shipping*. Malmö. Sweden. Available at: http://www.clean-baltic-sea-shipping.com /uploads/files/CLEANSHIP_final_report_ for_download.pdf

PIANC. 2013. *Sustainable Ports. A Guidance for Port Authorities*. The World Association for Waterborne Transport Infrastructure. WG150 Issue 2013.04.28.

Port of Helsinki. 2011. *Annual Report 2011*. Available at: http://www.portofhelsinki.fi/instancedata/prime_product_ju lkaisu/helsinginsatama/embeds/helsinginsatamawwwstructu re/16066_HelSa_resume_2011_web.pdf

Port of Helsinki. 2012a. *Annual Report 2012*. Available at: http://www.portofhelsinki.fi/instancedata/prime_product_ju lkaisu/helsinginsatama/embeds/helsinginsatamawwwstructu re/15958_Helsa_Resume_2012_WEB.pdf

Port of Helsinki. 2012b. *Port of Helsinki: Vuosaari Harbour - The Capital Harbour of Finland*. Available at: http://www.portofhelsinki.fi/instancedata/prime_product_ju lkaisu/helsinginsatama/embeds/helsinginsatamawwwstructu re/13079_Vuosaari_the_Capital_Harbour_of_Finland_201 2.pdf)

Port of Helsinki. 2013. *Annual Report 2013*. Available at: http://www.portofhelsinki.fi/instancedata/prime_product_ju lkaisu/helsinginsatama/embeds/helsinginsatamawwwstructu re/16598_Helsa_Vsk_2013_resume_WEB.pdf

Vieira, J., Moura, F. & Viegas, J.M. 2007. Transport policy and environmental impacts: The importance of multi-instrumentality in policy integration. *Transport Policy* 14: 421–432.

Sea Ports and Harbours
Safety of Marine Transport – Marine Navigation and Safety of Sea Transportation – A. Weintrit & T. Neumann (eds.)

Decreasing Air Emissions in Ports – Case Studies in Ports

O.-P. Brunila, V. Kunnaala-Hyrkki & E. Hämäläinen
University of Turku, Brahea Centre for Maritime Studies, Kotka, Finland

ABSTRACT: Dense ship traffic, port operations and port related land transportation are the cause of emissions in harbour areas. Emissions to air are the most important factors that increase the greenhouse effect and climate change. The main source is exhaust gas from combustion engines that are used in marine and road traffic and partially in train traffic and working machines in the port area. In addition, emissions to air can cause health problems in the respiratory system and the eyes. In some cases, long-term exposure can cause cancer, e.g. leukaemia and cardiopulmonary mortality. It has been estimated that shipping causes approximately 60,000 deaths annually on a global scale. In Finland, emissions to air are controlled and measured trough environmental permits and environmental impact assessment procedures. Different regulations for SOx and NOx will decrease emissions; nevertheless, there is no revolutionary or suitable global technique to decrease emissions that would suit all ports.

1 INTRODUCTION

Dense ship traffic, port operations and port related land transportation cause a lot of different kinds of emissions to air and water. Especially ships produce waste, black and grey waters, ballast waters etc.

Emissions to air are the most important factors that increase the greenhouse effect and climate change. The main source is exhaust gas from combustion engines that are used in marine and road traffic and partially in train traffic as well as working machines in the port area. Diesel and petrol fuel are almost sulphur-free (0.01%), but the bunker that ships use had a sulphur content of 1.1% in 2009 (VTT 2009).

Sulphur dioxide reacts readily with water to form a chemical change that leads to acid rain. Acid rain corrodes metals and buildings. In addition, acid rain is found to be harmful to the health of humans as well as other organisms. In the event of acid rain, sulphur causes respiratory diseases and respiratory irritation to mucous membranes. (U.S. Environmental Protection Agency 2007)

In this study, the focus is on the air emissions of ships, working machines, port related transportation (turnover of trucks) and the recommendations that can be given to reduce air emissions in ports. Regarding ship originated emissions, the focus is on the different kinds of air emission calculation

methods. This study also contains methods to decrease ship originated air emissions in ports and an analysis of their strengths and weaknesses. The case study presented in this paper focuses on Finland`s biggest universal port, the port of HaminaKotka. The methods used in this study were literature review and interviews.

2 THE CURRENTS STATE ANALYSIS OF THE PORT OF HAMINAKOTKA

The Port of HaminaKotka is the largest universal port in Finland and it is located in the eastern part of the Gulf of Finland. There are 1,100 hectares of port land area and they are home to ten port operators and 170 other businesses. The port area is divided into 7 different port areas; Halla, Hamina, Hietanen, Hietanen South, City Terminal, Mussalo, Sunila and private quays. The Port of HaminaKotka is also the easternmost port in Finland and the distance to the Russian border is only 35 kilometres. The annual cargo volume of the port is approximately 15 million tons and there are around 3,000 ship calls annually. (Port of HaminaKotka 2013)

Every port in Finland must have an environmental permit and go through an environmental impact assessment (EIA) procedure. Dredging, disposal of sediments and port construction are not included in the port operation permit. These activities require a different permit based on water legislation. The operation of Finnish ports is regulated by official requirements such as the environmental permits for port operations but also by the voluntary commitments of the ports to their operations. Environmental information about operations must be collected regularly. There are mandatory and optional indicators that are recorded to determine the level of environmental protection. Optional environmental and quality management systems are also widely used. With the optionally provided information and the indicators and environmental management systems, the level of clarity, sincerity and comparability of the compiled information can be improved. (Port of HaminaKotka 2013)

When environmental information is collected, the basic principles for all the collected data are comparability, balance between problematic questions and opportunities, continuity, clarity and intelligibility. It is imperative that the gathered information is comparable so that changes in the level of environmental protection can be established. When compiling the environmental information that is common to all port operators, the level of environmental protection is improved by reporting. (Kujala 2010)

Port related emissions can be divided into two categories; direct and indirect effects. Direct emissions affect nature and humans directly. Climate change, emissions to air and water, noise and vibration can be categorized as direct impacts. Indirect impacts affect nature, humans, fauna, energy resources etc. in the long term. Indirect impacts include the infrastructure of ports, roads, railways, maintenance and the changes in biodiversity. Gases (such as SO_2, NO, NO_2, CO, O_2 and HC) and particulates ($PM10$, $PM 2.5$, $PM 1$, sulphate, nitrate and heavy metals) can also cause health problems in the respiratory system and the eyes. In some cases, long-term exposure can cause cancer, e.g. leukaemia. (Kalenoja & Kallberg 2005, Tenhunen 2008, U.S. Environmental Protection Agency 2000)

Possibly the most important measured and calculated parameters are emissions to air. Emissions to air are the most important factors that increase the greenhouse effect and climate change. The main source is exhaust gas from combustion engines that are used in marine and road traffic and partially in train traffic as well as working machines in the port area. In land-based transportation, the most common fuel is diesel. Diesel combustion exhaust gas composition is as follows: nitrogen (N_2) 66%, carbon dioxide (CO_2) 13%, water (H_2O) 11% and oxygen (O_2) 9%. In addition, the remaining one percent consists of other combinations of combustion, oxides of nitrogen (NOx), particulate matter, carbon monoxide (CO), hydrocarbons (HC), and sulphur dioxide (SO_2). (Kalenoja & Kallberg 2005)

In ship traffic, the most significant emissions are emissions of sulphur dioxide (SO_2), nitrogen oxides (NOx), carbon dioxide (CO_2) and particulate matter (PM). (Burel et al. 2013, Det Norske Veritas 2011) Up to 96% of the total transport emissions of sulphur, 50 % of nitrogen oxides and 22% of carbon dioxide are caused by water transport. In the fuel, sulphur is almost completely converted into sulphur oxides. In factories and power plants, the remaining sulphur can be removed nearly completely, but it is more difficult to clean from the exhaust gases of ships. Sulphur scrubbers are still in development and for the time being, the scrubbers are still quite expensive. (Lahtinen 2009)

Diesel and petrol fuel are almost sulphur-free (0.01%), but for the bunker that ships use, the sulphur content was 1.1% in 2009 (VTT 2009). Sulphur dioxide reacts readily with water to form a chemical change that leads to acid rain. Acid rain corrodes metals and buildings. In addition, acid rain is found to be harmful to the health of humans as well as other organisms. In the event of acid rain, sulphur causes respiratory diseases and respiratory irritation to mucous membranes. (U.S. Environmental Protection Agency 2007)

Since 2010, all ship types in EU harbour areas have been required to use low sulphur fuel (<0.1%) in port areas if the duration of their visit is longer than two hours. This requirement has a time limit because fuel switch from residual fuel oil - usually from heavy fuel oil (HFO) to distillate oil, marine gas oil (MGO) - has to be done gradually due to the large temperature difference. (Jalkanen et al. 2013)

The main sources of emissions to air are port operations and vessel traffic. To measure the emissions of vessel traffic, the readings are taken in the quay area and the measured unit is kg per year. The following emissions are measured from the ships: nitrogen oxides (NOx), carbon dioxide (CO_2), sulphur dioxide (SO_2), hydrocarbons (HC), methane (CH_4) and particulate matters such as soot (CO) and other small particulates ($PM 2.5$ and $PM 10$). The consumption of heavy fuel oil, marine diesel oil and gasoil is also calculated. Especially until 2015, a major part of the sulphur dioxide emissions comes from vessels' exhaust engines. The vessels' sulphur dioxide emissions also depend on the fuel type used. In port-related road transportation, the used diesel fuel does not contain any sulphur. (Kujala 2010, Port of Helsinki 2009)

Emission to air in port of HaminaKotka (tonnes/year)									
Parameter	2004	2005	2006	2007	2008	2009	2010	2011	2012
Dinitrogenoxide (N2O) (t)									0,89
Carbondioxide, FOSS (t)	38157	45065	55134,61	52705,58	56226,73	47184,99	45134	43888	44891
Carbonmonoxide (CO) (t)	88,9	109	130,24	178,91	181,64	129,18	111,4	118,1	112,7
Particles (t)	19,5	24,4	29,86	35,97	39,32	28,12	19,4	24,69	20,79
Methane (CH4) (kg)	1300	1300	1400	1600	1400	797	900	2620	2620
Other Volatile Organic Combounds (kg)	32,6	11028	48019,31	58420	58540	42075	39500	43700	42000
Suphur oxides (SOx/SO2) (t)	219,6	237	277,29	234,17	232,8	89,34	56,5	55,7	51,6
Nitrogen oxides (Nox/NO2) (t)	738	891	1082,17	1037,11	1078,86	837,46	850,8	772,8	766,8

Figure 1. Emission to air in the Port of HaminaKotka (2004-2012). (Värri 2013)

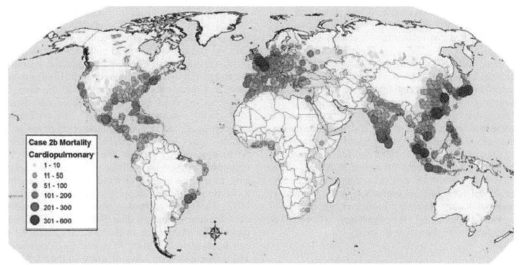

Figure 2. Cardiopulmonary mortality attributed to ship PM2.5 emissions worldwide. (Corbett et al. 2007).

Measured emissions to air in the Port of HaminaKotka are presented in Figure 1. Emission volumes were: dinitrogen oxide (N2O), carbon dioxide (CO2), carbon monoxide (CO), particles, methane, VOCs, sulphur oxides (SOx/SO2) and nitrogen oxides (NOx/NO2). Depending on the emission volume, measuring units were in tons or kg. As can be seen from Figure 1, all emission volumes have increased slightly between the years 2004 and 2008. The reason for this increase is the economic situation; handled volumes and ship calls in the port were rising along with the emissions. In the end of 2008, the economic recession began, which negatively affected the traffic and cargo volumes. Emission levels decreased immediately since transportation and emissions go hand in hand.

3 CALCULATION METHODS FOR SHIP ORIGINATED AIR EMISSIONS

In ship traffic, the most significant emissions are emissions of sulphur dioxide (SO2), nitrogen oxides (NOx), carbon dioxide (CO2) and particulate matter (PM). (Isakson et al. 2000) Up to 96% of the total transport emissions of sulphur, 50 % of nitrogen oxides and 22% of carbon dioxide are caused by

water transport (Lahtinen 2009). According to IMO (2014), it has been estimated that worldwide, 2.7% of CO2 emissions come from international shipping. In the fuel, sulphur is almost completely converted into sulphur oxides. In factories and power plants, the sulphur can be removed nearly completely, but it is more difficult to clean from the exhaust gases of ships. (Lahtinen 2009) Eyring et al. (2010) estimated that the contribution of ships' global emission of NOx is approximately 15%. In addition, 4-9% of SO2 emissions are from ships.

3.1 Effects of ships' air emissions

Many different kinds of studies have been made on the emissions of ships and how they affect fauna, flora and humans. According to Corbett et al. (2007), it has been found that shipping related particulate matters (PM 2.5) contribute 60,000 deaths annually on a global scale. Air pollution from ships does not have the direct cause and effect associated with, for example, an oil spill incident; rather, it causes a cumulative effect that contributes to the overall air quality problems encountered by populations in many areas, and it also affects the natural environment e.g. through acid rain. (MARPOL 2014)

The mortality of humans concentrates to areas where there is a large amount of ships and where ports are located nearby city centres. Figure 2 presents the cardiopulmonary mortality attributed to ships' particle matter emissions worldwide. As can be seen in Figure 2, there are a lot of cities that have ports in the city centre or nearby especially in coastal regions in Europe and Asia.

It should be noted that there are some uncertainties in the survey. For example, the absence of localized cell research (C-R) functions and incidence rates prevents the precise quantification of all anticipated PM-related health effects. (Corbett et al. 2007)

SOx and NOx emissions from ship engines increase the acidification of the environment and contribute to the eutrophication of terrestrial and aquatic environments. If ports are located near city centres like most of the big ports are in Europe, the combined emission of ships and urban emission increases the emission levels in quite small areas and so the risks for humans, flora and fauna increase as well. Well known health effects of SO_2, PM, ozone (O_3), NOX, and VOC (volatile organic compounds) at a local level include premature death from heart and pulmonary diseases. Moreover, ground level ozone is damaging to vegetation. (Corbett & Fischbeck 1997, Lonati et al. 2010)

3.2 Decreasing emissions of ships in the future

According to Agryros et al. (2014), the energy mix in shipping will be decreasingly conventional, which means that, depending on the scenario, heavy fuel oil will remain as the main energy source for the next twenty years in different ships. In the study, there were three different scenarios; status quo, global commons and competing nations. These scenarios represent alternative futures for the world and shipping in 2030, from business as usual to more globalisation or more localisation. Despite this, heavy fuel oil remains in the energy mix; however, LNG, bio-fuels and hydrogen will take a greater role in shipping. Additionally, a wide range of energy efficiency technologies and abatement solutions (including sulphur scrubbers and selective catalytic reduction for NOx emissions abatement) will be widely in use with ships.

Future trends in ship technology can be divided into two categories: low energy ships and green fuelled ships (Det Norske Veritas 2011). Low energy shipping technologies in shipbuilding reach for new materials, design and manufacturing processes. This means that ships must be designed and built in practice at a large scale and in a standardised way. Furthermore, there must be a low carbon development concept, and an emphasis on green technology research and building technology. (Wua et al. 2011) It is important to focus on drag

and water reduction, propulsion systems and whole scale energy efficiency. Holistic designs and the use of risk-based methods are necessitated by new technology and demanding targets in regard to emissions, efficiency, strength, and speed or cargo flexibility. The main triggers for development and innovations are the market forces, technological advances, technical development, safety issues, environmental regulations, high bunker costs, market realities and greener values. (Det Norske Veritas 2011)

Increasing fuel price, upcoming pollutant emission regulations, sulphur emission control areas (SECA) and coming nitrogen emission control areas (NECA) concern the whole shipping industry around the world. (Diaz-de-Baldasano et al. 2013) There are more and more interesting solutions in the form of green fuelled ships that refer to techniques which can meet increasingly strict environmental regulations and address rising bunker oil prices with natural gas and renewable energy sources. Alternative energy sources like LNG, bio fuels and renewable energy as well as more radical energy sources such as nuclear or wind energy can become a solution for future shipping. (Det Norske Veritas 2012) These new technologies and alternative fuels can decrease emissions like SOx, NOx, CO_2 and particulates. According to Burel et al. (2013), LNG can be used to cut SOx, NOx and CO_2 emissions and increase the efficiency of ships. In the study, the economic calculation showed that, depending on LNG system installation, the payback time would be around 3-8 years.

In 1997, International Maritime Organization (IMO) added a new annex to the International Convention for Prevention of Pollution from Ships (MARPOL). The new regulation for Prevention of Air Pollution from Ships (Annex VI) minimizes and decreases the airborne emissions from the ships. So far, 62 states have ratified this convention, which translates to 85% of the world's ship tonnage. In the convention, emission limits for SOx, NOx, ODS and VOCs were included. Moreover, new elements include regulations for equipment that contains chlorofluorocarbons, halogens and freons (e.g. CFC and HCFC). Annex VI entered to force in 2005 and significantly tightened emission limits were adopted in the end of 2008 and realized in 2010. (IMO 2014)

Annex VI was revised and the main changes included the progressive reduction of global emissions of SOx, NOx and particulate matter and emission control areas (ECAs). In these ECA areas, the limits for SOx and particulate matter were reduced to 1.00% starting from 1 July 2010 (from the original 1.50%). The limit will be further reduced to 0.10 %, effective from 1 January 2015.

3.3 Case studies in port areas

3.3.1 Port of Göteborg, Sweden

Port of Göteborg is one of the largest ports in the Nordic area. There are approximately 12,000 ship calls there every year and that is why emission levels are quite high. It has been estimated that 45% of total SO_2 emission and 15% of total NO_x emissions in the whole city area of Göteborg come from the port. In the study, emissions to air in the city of Göteborg were measured. A city as a measurement area is quite complex since many different sources produce emissions to the city area. These emission sources in the city area are for example traffic, heating/cooling, factories and other activities. That is, ships and the port area are only one emission source. When measuring ship emissions, time and wind direction were used as important parameters to identify emissions from ships. (Isakson et al. 2001)

The study and measurements showed that high concentrations of SO_2, NO and NO_2 are frequent along the ship lanes. The level of SO_2 and NO_2 emissions in the port area were much higher than in the city area. There was no difference whether it was summer or winter. The excess deposition of SO_2 compares well with model calculations for other large European ports. In the study, a positive correlation of NO, NO_2, SO_2 and metals V, Ni, Pb and Zn was also noted. The reason for correlation is the use of diesel fuel and exhaust gases from combustion engines. The levels of vanadium (V) were three times higher in the port of Göteborg than on the regional level. Sub-micrometer particles were measured. In ships' plumes, the number of ultra-fine micro particles was elevated compared to urban background concentration. (Isakson et al. 2001)

3.3.2 Case study in Danish ports

In this study, the operational meteorological air quality model was used to calculate the urban dispersion of air pollutants from ships in three Danish ports: Copenhagen, Elsinore and Køge. Measurements were taken from ships from two sources – manoeuvring and activity on the dock. Because of the different sizes and types of ships, an average size had to be determined and assumed for all ship types. NO_x levels may vary significantly depending on the engine type in a ship. Engine type does not affect the SO_2 emission, only the sulphur content of used fuel. (Saxe & Larsen 2004)

As a result from calculations and measurements, it can be observed that emissions of NO_x from ships in harbour areas could possibly induce health problems to people in Copenhagen and Elsinore sea ports. NO_x emitted by ships in the port of Copenhagen contributed substantially to the NO_x pollution level compared to central Copenhagen.

Ferry traffic in the port of Elsinore affects the NO_x pollution in the neighbourhood around the harbour area. The level of particles is increased by the combination of diesel combustion engines in road traffic and particles originating from ships. This may have an effect on people's health, especially for those in the port area in Copenhagen. The Port of Køge is small and the level of activity is low which means that ships do not affect the environment as significantly. (Saxe & Larsen 2004)

3.3.3 Studies in Finland

In Finland, there have been many different studies and projects which calculate, estimate and analyse the air emissions of ships. One project is the Shipping-induced NO_x and SO_x emissions - Operational monitoring network (SNOOP) –project. The main focus is to take the strategic evaluation of ships' emission effects to a new level by enlargement of the scope of NO_x, SO_x, PM, CO and CO_2 emissions to air. In addition, air quality management and the effect emissions have on human health in harbour areas are studied. (Finnish Meteorological Institute 2013)

In the SNOOP project, different measurement techniques and different measurement platforms for determining the ship sulphur content in the air were tested. Measurements were performed with the ship traffic emission assessment model (STEAM and STEAM2) which uses the automatic identification system (AIS). Previously, the proportion of sulphur was detected from ship fuel. The measurement process was also used to determine particulate emissions from ships as well as particulate size distribution and sulphur content from ship exhaust trails. Measurement points were in the ports of Helsinki and Turku. The Finnish Meteorological Institute measured emissions from ships at sea with a plane and a helicopter. Measurements were also carried out at a fixed station on the ship route. This was the first instance of using these kinds of direct outdoor measurements in Finland. These methods of measurement have been used previously in Netherlands and in Sweden. (Jalkanen et al. 2012)

In the SNOOP project, the used model enables examining the influences of relevant factors: accurate travel route of ships, ship speed, engine load, fuel sulphur content, multiengine setups, abatement methods and waves. The model includes the influence of engine load and sulphur content on the emissions. This methodology can be used to evaluate the total particle (PM) emissions and those of organic carbon, elemental carbon, ash, and hydrated sulphate. Particle (PM) and carbon monoxide (CO) emissions depend on engine load. This means that SO_x emissions vary depending on the sulphur content of fuel whereas CO emissions are sensitive to changes in engine operation or how large the "load" is for the ship engines. (Jalkanen et al. 2012)

Some uncertainties and probabilities arise in calculations and measurements in the used model. The largest challenge was that different fuel types are used in ships in different geographical areas. This uncertainty has been taken into account in the model with ship emission inventories. Another difficulty was the scarcity of detailed composition of measured data of PM emissions. This means that the chemical components of PM emissions should be analysed more comprehensively from the various engine loads and used fuels. In this study, it was not examined or modelled how emissions affect the environment, e.g. sea ice and marine currents. (Jalkanen et al. 2012)

3.4 *The current situation in the Port of HaminaKotka*

In the Port of HaminaKotka, the problem of emissions to air is well noted. As mentioned earlier in chapter 2, extensive emission measurements have been performed as obligated by environmental permits. In an oral interview, traffic director Markku Koskinen and development director Riitta Kajatkari mentioned that for the time being, there are no sufficient techniques to alleviate the emissions that are produced by ships in the port area in the Port of HaminaKotka. Worldwide, there are many different kinds of solutions to decrease ship related emissions from port areas. (Kajatkari & Koskinen 2014)

Ships' emissions of SOx, NOx, CO and PM cause real harm or danger, particularly in the areas where ports are located, usually in the neighbourhoods of populous urban agglomerations. Mobile-source, port-related emissions are generated by marine vessels and by land-based sources in ports. Depending on the type and size of a cargo ship, hotelling time can range from several hours to several days. One relatively clean technique is shore electricity. (Tarnapowicz & Tadeusz 2013)

In the Port of Helsinki in Finland, there is shore electricity for certain cruise ships but not for cargo ships. The largest problem according to interviews is that there are no common standards for plugs, adapters, voltage, power needs etc. that are needed to connect a ship to shore electricity. Also, the needed amount of power for ships might be equal to the consumption of a town or part of a city. For example in the port of Göteborg in Sweden, ships owned by Stora Enso use shore electricity in port. Annual CO2 emissions have been decreased by 2,500 tonnes and also noise levels and other emissions have dropped remarkably because auxiliary machines have been shut down during loading and unloading of cargo. (Martinsson 2010)

Finnish ports are comparatively small on the European scale, and for the moment, they cannot

provide the best available techniques (except choosing the fuel used) to decrease the emissions from ships in port areas. Decreasing land base emissions such as emissions from port operations or in port areas (including heating, water, waste, lighting, cooling etc.) is much easier than decreasing emissions in ships. Alternative bio fuels can be used to decrease the emissions to air in port areas. Onshore power is a very good solution, but it would take several years to make systems compatible worldwide, i.e. for all ship types to be connectable to all onshore power grids.

4 THE ECOLOGICALLY FRIENDLY PORT PROJECT

This study was conducted as a part of the Ecologically Friendly Port Ust-Luga (EFP) project, which was launched in December 2012. It is funded by the CBC ENPI program "South-East Finland-Russia" 2007-2013. The total budget for the project is 570,000 Euros. The duration of the project is 24 months. The project is led by the Russian State's Hydrometeorological University (RSHU). The project consortium includes Ust-Luga Company JSC, University of Turku Centre for Maritime Studies and Kymenlaakso University of Applied Science. The project associates are the Port of HaminaKotka, City of Kotka, Finnish Port Association and Administration of Leningrad Region Committee on Natural Resources.

The main objectives of the project are:
– Improving the environmental status of the eastern part of the Gulf of Finland
– Improving the ability of the ports to develop environmental protection and sustainable growth within the ports
– Improving the municipalities' ability to improve on environmental safety issues
– Increasing the awareness of citizens concerning green thinking
– Establishing a close bilateral cooperation between citizens and authorities on the basis of green values, green Economy and Ecological mentality, aimed at sustainable regional development
– Increasing the ecological knowledge and understanding of responsibility for the global challenge of climate change of Russian and Finnish ports and enterprises.

5 DISCUSSION AND CONCLUSIONS

Dense ship traffic, port operations and port related land transportation are the cause of various kinds of emissions in harbour areas. Annual statistics show that economic recession decreased emission levels in Finland. However, there are no new techniques or innovations in use that would decrease all ships' emissions to air. Emissions to air are the most important factors that increase the greenhouse effect and climate change. The main source is exhaust gas from combustion engines that are used in marine and road traffic and partially in train traffic as well as working machines in the port area.

Furthermore, emissions to air can cause health problems in the respiratory system and the eyes. In some cases, long-term exposure can cause cancer, e.g. leukaemia and cardiopulmonary mortality. It has been estimated that in shipping, Particulate matter (2.5) causes approximately 60,000 deaths annually on a global scale. In Finland, emissions to air are controlled and measured trough environmental permits and an environmental impact assessment (EIA) procedure.

Various regulations for SOx and NOx will decrease emissions to air in the future. IMO regulations for SOx come into force next year in the Baltic Sea, North Sea and Channel between U.K. and France. This will reduce the SOx emissions quite drastically. However, the regulation comes into force in the Mediterranean Sea only much later. Also, Russians have not implemented the IMO regulation, so they are not obligated to use low sulphur fuel in their ships. This skews the competition between ports as some countries have to pay much more for cleaner shipping than others. For ships, there are cleaner fuels like LNG, bio fuels, fuel cells etc. but there are not enough new ships that use these techniques, nor are there enough possibilities to refuel these ships. In the next 5 to 10 years the situation will hopefully have been improved.

Shore power can decrease emissions in port areas as portrayed by the case studies. Still, there are no universal plugging and power systems that would be compatible with all ships. The energy consumption of ships poses another problem; it has been estimated that one ship consumes the same amount of electricity as a small town. In the future, there should be standards for shipbuilding so that these technologies can be used on a global scale. As a result, there will be workable practices and suitable techniques to decrease air emissions in theory. In the future, considerable investments must be made so that emissions from ships can be cut on a larger scale.

REFERENCES

Agryros D., Raucci C., Sabio N. & Smith T. 2014. *Global Marine Fuel Trends 2030*. Loyds Register and UCL Energy Institute. Available at URL: < http://lr.org/Images/Global%20Marine%20Fuel%20Trends%202030%20single%20page%20v2_tcm155-249392.pdf>

Burel F., Taccani R., Zuliani N. 2013. Improving sustainability of maritime transport trough utilization of liquefied natural gas (LNG) for propulsion. *Energy journal* 57(2013) pp. 412-420.

Corbett J.J. & Fischbeck P. 1997. Emission from ships. *Science* 278(5339) pp. 823-824.

Corbett J.J., Winebrake J.J., Green E.H., Kasibhatla P.,Eyring V. & Axel Lauer A. 2007. Mortality from Ships Emissions: A Global Assessment. *Environ. Sci. Technol.* 41(24) pp. 8512-8518.

Det Norske Veritas 2012. *Technology Outlook 2020.* Research & Innovation. Available at: http://production.presstogo.com/fileroot/gallery/DNV/files/preview/9ec457bc750b4df9e040007f0100061c/9ec457bc75094df9e040007f0100061c.pdf.

Diaz-de-Baldasano M., Mateos F., Nunez-Rivas L., Leo T. 2013. Conceptual design of offshore platform supply vessel based on hybrid diesel generator-fuel cell power plant. *Applied Energy journal*. Available at URL: < http://ac.els-cdn.com/S0306261913009550/1-s2.0-S0306261913009550-main.pdf?_tid=c90caee6-aa91-11e3-8258-00000aab0f6c&acdnat=1394703067_4d2c47a4c3451b3c02db8d77c5a8477e>

Eyring V., Isaksen I.S.A., Bemtsen T., Collins W.J., Corbett J.J., Endersen O., Graigner R.C., Moldanova J., Schaler H. & Stevenson D.S. 2010. Transport impacts on atmosphere and climate: Shipping, *Atmosphehric Environment* 44 pp. 4735-4771.

Finnish Meteorological Institute 2013. *Shipping-induced NOx and SOx emissions–operational monitoring network (SNOOP).* Available at URL: < http://snoop.fmi.fi/?q=node/2>

Isakson J., Persson T. A. & Lindgren E. 2000. Identification and assessment of ships emissions and their effect in the harbor of Göteborg, Sweden. *Atmospheric Environment Journal* 35(2001) pp. 3659-3666.

IMO 2014. *Air Pollution and Greenhouse Gas Emissions.* Available at URL: < http://www.imo.org/OurWork/Environment/PollutionPrevention/AirPollution/Pages/Default.aspx>

Jalkanen, J.-P., Johansson, L., Kukkonen, J., Brink, A., Kalli, J. & Stipa, T. 2012. Extension of an assessment model of ship traffic exhaust emissions for particulate matter and carbon monoxide. *Atmospheric Chemistry and Physics* 12 pp. 2641-2659.

Jalkanen, J.-P., Johansson, L., Loven, K. & Rasila, T. 2013. Best practices and recommendations for ports and cities. 4.1 Current predicted exposure and health risk in Kotka/Hamina, Finland. *Pan-Baltic Manual of best practices on clean shipping and port operations.* ISBN 978-952-298-004-5.

Kajatkari, R. & Koskinen, M. 2014. Interview with the developing director Kajatkari and the traffic director Koskinen of the port of HaminaKotka. 13th of May 2014, Kotka.

Kalenoja, H. & Kallberg, H. 2005. *Liikenteen ympäristövaikutukset* (Environmental effects of transportation). Tampere University of Technology. Liikenne- ja kuljetustekniikka. Opetusmoniste 37. (Finnish)

Kujala, N. 2010. *Ports and environmental indicators room for cooperation?* Power point presentation.

Lahtinen, J. 2009. Rikkipesurit puhdistuvat laivojen pakokaasuja. (Sulphur scrubbers clean the ships' exhaust gases) University of Wasa. Available at URL: <http://www.uwasa.fi/midcom-serveattachmentguid-d2158d1e26a0d64f717e2d910aaf540b/su0911_artikkeli_lahtinen.pdf>

Lonati G., Cernuschi S. & Sidi S. (2010). Air quality impact assessment of at-berth ship emissions: Case-study for the project of a new freight port. *Science of the Total Environment* 409(2010) pp. 192-200.

Martinson, T. 2010. Göteborg kytki laivat puhtaaseen maasähköön. (Port of Gothenburg connected ships to clean shore base power). *Power & Automation magazine* 2/2010. Finnish translation available at URL: < http://www04.abb.com/global/fiabb/fiabb250.nsf!OpenDat abase&db=/global/fiabb/fiabb254.nsf&v=80EA&e=fi&url= /global/seitp/seitp202.nsf/0/F8C84D96E355726EC1257735 0035448D!OpenDocument>

Port of HaminaKotka 2013. *Basic information about port.* Available at URL: < http://www.haminakotka.fi/en/haminakotka-satama-oy>

Port of Helsinki 2009. *Environmental report.* (Finnish). Available at URL: < http://www.portofhelsinki.fi/download/13152_ymparistora portti.pdf>

Saxe H,. & Larsen T. 2004. Air pollution from ships in three Danish ports. *Atmospheric Environment* 38(2004) pp. 4057-4067.

Tarnapowicz, P. & Borkowski, T. 2013. Best practices and recommendations for ports and cities. 4.3 Electrical shore-to-ship connection, benefits and best practices. *Pan-Baltic Manual of best practices on clean shipping and port operations.* ISBN 978-952-298-004-5.

Tenhunen, J. 2008. *Tieliikenteen päästöt ilmaan.* (Air emissions from road transportation) SYKE. Available at URL:< www.ymparisto.fi/download.asp?contentid=86504>

U.S. Environmental Protection Agency (EPA) 2007. *Acid Rain.* Available at URL: < http://www.epa.gov/acidrain/what/index.html>

VTT 2009. *LIPASTO 2009 calculation system.* Available at URL: < http://lipasto.vtt.fi/lipasto_lask_tulokset.htm>

Värri, E. 2013. *Quality of Air emission reports from the port area of Kotka and Hamina.* E-mail inquiry.

Wua S., Cheng Y.-T. & Ma Q. 2011. Discussion on ship energy-saving in low carbon economy. Advanced in Control Engineering and Information Science. *Procedia Engineering journal.* Available at: http://ac.els-cdn.com/S1877705811024751/1-s2.0-S1877705811024751 -main.pdf?_tid=a6dd1564-aa13-11e3-abc0-00000aacb35e &acdnat=1394648893_86e648fc4fee70f0bb29a96740a86e 26

Port in a City – Effects of the Port

O.-P. Brunila, V. Kunnaala-Hyrkki & E. Hämäläinen
University of Turku, Brahea Centre for Maritime Studies, Kotka, Finland

ABSTRACT: Throughout the long history of ports, there have always been both negative and positive impacts and values. Values can usually be classified based on their socioeconomic importance – considerable, non-socioeconomic or soft values. Negative impacts have mostly been related to environmental impacts, land use, traffic and pollution. Positive impacts are also quite remarkable for the region. Benefits include economic benefits, employment and other direct and indirect catalysts, which help the surrounding business, enterprises and citizens. Ports may also have other values, like soft values. Soft values have gained a more and more important role in recent times.

1 INTRODUCTION

Many cities have been formed around ports, meaning that first a port was built in a good location with good connections to the sea or inland areas. Later, a town or city grew around the port area. Sea faring, ports and port operations have had a positive impact through the employment of citizens, and people have also moved to cities in search of better income. (Van Hooydonk 2007) In the year 1776, Adam Smith saw that ports are one of the stepping stones for economic growth and welfare (Haezendonck 2001).

Throughout history, ports have been nodal points connecting Europe, Asia, Australia and America (Girard 2013). The history of ports has usually had negative impacts on the environment and habitats, especially a few centuries ago. (Van Hooydonk 2007)

New ports are no longer located near city centres, but near good road and railway connections. Good examples of this are the Port of Ust-Luga in Russia, Maasvlakte 2 in Rotterdam, Netherlands, Port of HaminaKotka, Finland or Port of Helsinki, Vuosaari, Finland. All of these ports are quite new or still under construction. These ports are located close to good connections but near the city centres. Still older parts of these example ports are in city centres but traffic density has decreased a lot.

1.1 *Objectives of the research*

Throughout the long history of ports, there have always been both negative and positive impacts and values related to them. Those impacts can affect the cities and citizens in the proximity of the ports. In this study, the aim is to examine and discuss how the citizens and ports "live" together in the same area.

1.2 *Research methodology*

In this study, we conducted a literature review in order to examine the different positive and negative impacts of ports. In addition, we conducted a case study regarding the Port of HaminaKotka. In the case study, we examined what kind of best practice projects have been carried out or planned during the transformation of old industry areas to recreation parks. During the case study also interviews were carried out.

1.3 *Research structure*

In the following chapter 2 we discuss the positive and negative impacts of ports. In chapter 3 we examine the case study in the Port of HaminaKotka. Chapter 4 introduces the Ecologically Friendly Port project, to which this study relates to. Chapter 5 includes the discussion and conclusions of this study.

2 IMPACTS OF PORTS

Ports have affected cities and citizens in different, both positive and negative ways. On average, it can be said that positive impacts are related to economic matters and negative impacts to environmental issues, land use, traffic and pollution. (Merk 2013, Van Hooydonk 2007) Especially nowadays, when the logistics connections play a larger role for new ports than a location in the city centre, ports have other important values and new uses for the citizens and landlords of the port. Ports have different kinds of values, which usually have been classified according to their socioeconomic importance – considerable, non-socioeconomic or soft values. These soft values may be historical, archaeological, architectural, landscape, recreational, sociological and other cultural aspects. (Van Hooydonk 2007) Thus, ports have other meanings in addition to being an economic moneymaking machine or just spoiling the environment and landscape.

2.1 Benefits from ports

Despite the different values of ports, the most positive impacts are the economic benefits, employment and other direct and indirect catalysts, which help the surrounding business, enterprises and citizens. (Merk 2013) Especially, ports have been facilitators for the transport of goods and passengers. One good example of how important shipping and maritime transportation are for Finland's trade: 90% of exports and 70% of imports are transported using ships. That is why Finland can be called an island - we are so isolated from the rest of the Europe. (Hansen 2007) Northern Finland is connected to Sweden, Norway and Russia. In the South and West there is the Baltic Sea between Finland and other countries. The entire Eastern border of Finland is with Russia, which makes Russia an important trade partner. But the current situation between EU and Russia is not the best possible one due to economic sanctions.

Value can be added to ports through various positive values. One way to measure value is to compare port cluster value to regional GDP. Another way is to categorize values based on four different types of impact: direct, indirect, induced and catalytic impact. Direct impacts are concrete, such as jobs in the ports. Examples include people who are working in port construction and port operations. Indirect impacts are the employment suppliers of transit and other services, which are related to ports and port operations. Induced impact is the employment related to the direct and indirect jobs. This means that people who work in the port or port related jobs spend income on services that, in turn, create new jobs. Lastly, catalytic impacts are related to ports which work as boosters for new business

and companies, which create new jobs in the region. (Merk 2013 and Ferrari et al. 2012)

Employment is one of the main issues for the citizens. Ports have been and will be a major employer of citizens and nearby residents, but direct jobs are decreasing. Despite containerization, automation and port community systems have increased handled cargo volumes but the demand for labour has decreased. (Merk 2013, Haezendonck et al. 2011) According to Merk (2013), a meta-study of approximately 150 port impact studies was performed and as a result, the report indicates that on average one million tons of port throughput is associated with 800 jobs. In many big ports in Central Europe or in Asia, the cargo volumes exceed hundreds of millions of tons. For example, the volume of cargo handled in 2012 in the port of Shanghai was 744 million tons, and in the port of Rotterdam 441.5 million tons (Ship-technology 2013). These jobs include direct and indirect port jobs, but the assumption is an average estimation and based on port impact studies that use different definitions of ports and apply different methodologies. (Merk 2013, Haezendonck et al. 2001)

One interesting positive effect related to ports and cities is the development and innovation taking place in ports. Port-related maritime research is conducted especially in universities (Merk 2013). Through co-operation and different projects, port operation and logistics chains can be boosted, costs saved, alternative low emission fuels developed and the level of environmental status of ports increased.

Soft values can be non-socioeconomic values such as historical, sociological, psychological, artistic, cultural and moral or even religious functions (Van Hooydonk 2007). These values have played and will play a huge role in the residents' everyday life. Ports have been given spiritual values in mythologies or in legends and have been gateways to historic eras. Ports are also usually the first place where refuges or immigrants arrive in new areas to look for a better future. A good example is New York City, where the Statue of Liberty welcomes arrivals. From another point of view, ports have been a source of inspiration for art, comics, post stamps, books, stories and films. Especially in port areas there are a lot of different old cultural historic buildings, architecture and, in some cases, there are no longer practical uses for them, so in many cities these old buildings have found new purpose as clubs, museums, homes, office buildings, commercial buildings, cultural buildings etc. These kinds of urban planning and multipurpose buildings have become a very important factor to citizens and social welfare (Van Hooydonk 2007).

Almost 90 % of the EU's external freight trade is seaborne. Short distance sea shipping represents 40

% of intra-EU exchanges in terms of ton-kilometres (Mylly 2014). Maritime transport and ports are, thus, vital for the economy of EU and Europe. Various EU Directives like the Bird and Habitats directive have raised criticism in the port sector because these directives restrict the expansion of ports (Van Hooydonk 2007). On the other hand, for example in Finland and in Sweden, ports face stricter environment regulations than any other European ports. All Finnish and Swedish ports must have an Environmental Permit and go through an Environmental Impact Assessment (EIA) procedure. Finnish ports are driven by the official requirements, such as the environmental permits for port operations, but also by their voluntary commitments to improving their operations. The environmental permit also requires that environmental information about operations must be collected regularly and reported once a year (Brunila 2013). Despite the criticism that ports have received concerning environmental issues, ports have also put in effort to fix these problems by using best paractices. EU sector associations like ESPO, FEPORT and EcoPorts have helped ports to implement solutions to these issues by using best practices, open discussion with NGO's and developing and increasing the ports' environmental awareness (Van Hooydonk 2007).

2.2 Negative effects of ports

Ports also have a lot of negative impacts and these negative impacts often receive attention in the media. Non-governmental organizations have made protests concerning the arctic, oil drilling, antifouling paint, nuclear transportation etc. Environmental issues also get a lot of media coverage. (Van Hooydonk 2007, Wichmann 2012) Negative impacts in ports can be divided into different categories: environmental impacts, land use, traffic impacts and other impacts. (Merk 2013, Van Hooydonk 2007) Maybe the most significant negative port related impacts are the environmental impacts. There are emissions to air, water, soil, waste management, change of biodiversity, noise impacts, health impacts etc.

Emissions to air are the most important factors that increase the greenhouse effect and climate change and air emissions also represent a major port-related negative environmental impact - despite shipping on average being quite a clean mode of transportation. (Merk 2013) The main source of these emissions in ports is exhaust gas from combustion engines that are used both in marine and road traffic and, partially, in train traffic as well as working machines in the port area. Diesel and petrol fuel are almost sulphur-free (0.01 %), but the bunker that ships use had a sulphur content of 1.1 % in 2009 (VTT 2009).

Port-related emissions can be divided into two categories; direct and indirect effects. Direct emissions affect nature and humans directly. Climate change, emissions to air and water, noise and vibration can be categorized as direct impacts. Indirect impacts affect nature, humans, fauna, energy resources etc. in the long term. Indirect impacts include the infrastructure of ports, roads, railways, maintenance and the changes in biodiversity. Gases (such as SO_2, NO, NO_2, CO, O_2 and HC) and particulates ($PM10$, $PM\ 2.5$, $PM\ 1$, Sulphate, Nitrate and heavy metals) can also cause health problems in the respiratory system and the eyes as well as cardiac and pulmonary diseases. In some cases, long-term exposure can cause cancer, e.g. leukaemia. (Kalenoja & Kallberg 2005, Tenhunen 2008, U.S. Environmental Protection Agency 2000, Corbett & Fischbeck 1997, Lonati et al. 2010) According to Corbett et al. (2007), it has been found that shipping related particulate matters ($PM\ 2.5$) contribute to 60,000 deaths annually on a global scale. (Corbett et al. 2007)

In Finland, every port needs an environmental permit for port operations and environmental monitoring is carried out in connection to the environmental permit of the port and to the requirements set by legislation and other forms of regulatory measures (Port of HaminaKotka 2013, Finnish Port Association 2010). In Finland, all ports are under stricter environmental regulations than elsewhere in Europe. A similar environmental permit system for port operation is only used in Sweden. (Brunila 2013) Dredging, disposal of sediments and port construction are not included in this permit. They require a separate permit, usually based on water legislation. The permit is granted to the port authority controlling the port area. (Finnish Port Association 2010) The permit obligates every port to measure emissions of Dinitrogen oxide ($N2O$), Carbon dioxide ($CO2$), Carbon monoxide (CO), Particles, Methane, VOCs, Sulphur oxides ($SOx/SO2$) and Nitrogen oxides ($NOx/NO2$) to air, water monitoring, noise measurements, waste measurements and a waste management system. (Finnish Port Association 2010, Government resolution (480/1996), Government resolution (993/1992), Ministry of Environment 2009, Pulkkanen 2013, Värri 2013)

Changes in water quality have a more visible effect on the environment than air pollution. One major water pollution source is oil spills, but chemical spills are very dangerous for the environment and ecosystems as well. Especially if the chemical accident happens in the port area, it is plausible that the effects can be a cause of concern for the city residence as well (Häkkinen & Posti 2012). Illegal dumping, bilge water and ballast water discharge are quite common activities in the sea areas. As a result, trash, foreign species, oil and chemicals reach the shores, spoil them and have negative effects on flora and fauna. (Merk 2013)

3 CASE STUDIES IN THE PORT OF HAMINAKOTKA

A lot of different port transformation projects have been carried out all over the world using best practices. In Finland, the largest universal port, HaminaKotka Ltd, have given up for other purposes two of their old port sections. Both port sections are situated in the city centre of the town of Kotka. On a general scale, Finland's largest projects transforming old industry areas into recreation parks have been carried out in Kotka. Yet 25 years ago, the environmental issues and environment in general were last on the list of priorities.

3.1 From city harbour to culture harbour

The old city harbour is the oldest part of the port of HaminaKotka Ltd. Nowadays there is not much transportation activity, but it is still suitable for LoLo and RoRo transportation, especially for forest industry purposes. In the future, this part of the port will focus more and more on passenger traffic and cultural purposes. (Port of HaminaKotka 2014) In the year 1879, the first pier was opened for port operation and in the same place today is Finland's only maritime museum Vellamo, which was opened in the summer of 2008. The museum is very important to the citizens and of course to all of Finland, when measured based on soft values. In the museum there are the maritime museum sector, regional museum, maritime data/knowledge centre, meeting and teaching facilities, a restaurant and a store for souvenirs.

The future plans for this section of the port is to create a more multipurpose area that includes shopping centres, restaurants, hotels, concert areas and houses. An international development group has planned this project. Another important factor is a new cruise ship line from Kotka to St. Petersburg, Russia or possibly to a port in Estonia. (City of Kotka 2014) One prerequisite for cruise ship traffic is shore power and other ecologically friendly solutions and best practices. These techniques and best practices must be used for decreasing air emissions because the port is near the Kotka city centre.

In ship traffic, the most significant emissions are emissions of sulphur dioxide (SO_2), nitrogen oxides (NO_x), carbon dioxide (CO_2) and particulate matter (PM). (Burel et al. 2013, Det Norske Veritas 2011) Ships' emissions of SO_x, NO_x, CO and PM cause real harm or danger, particularly in the areas where the ports are located, usually in the neighbourhoods of populous urban agglomerations. Mobile-source,

port-related emissions are generated by marine vessels and by land-based sources in ports. Depending on the type and size of a ship, hotelling time can range from several hours to several days. (Tarnapowicz & Tadeusz 2013)

In the Port of HaminaKotka, the problem of emissions to air is well noted because extensive emission measurements have been performed as obligated by the environmental permit. In an interview it was found out that at the current moment, the Port of HaminaKotka does not have a suitable technique for decreasing the ships' air emissions in the port area. A lot of different techniques are available, but none of them fit various types of ships. (Kajatkari & Koskinen 2014) In the Port of Helsinki in Finland, there is shore electricity for certain cruise ships, but not for cargo ships. The largest problem, according to interviews, is that there are no common standards for plugs, adapters, voltage, power needs etc. that are required in order to connect a ship to shore electricity. In addition, the amount of power required for ships might be equal to the consumption of a town or a part of a city. (Martinsson 2010) These problems must be solved also in Kotka when the passenger cruise ships start to operate.

When the development project construction work starts, the economic investment in the first stage has been estimated to be 200 million euros. The project supports the Kotka-Hamina region industry, and the project will have a significant impact on employment. The construction phase will create almost a thousand person-years of direct employment and almost two thousand, when taking into account the indirect jobs. In the operational phase, there will be more than 700 direct permanent jobs and, in total, 900 jobs when indirect jobs are included. This also means that the municipality's economy will give a significant boost to the region, about 7 million euros in tax revenue. (City of Kotka 2014) Figure 1 presents an artists' depiction of what the new port would look like.

Figure 1. Picture of new port area in Kotka (City of Kotka 2014)

3.2 *From oil harbour to recreation park*

Originally, the current recreational park was zoned as a park but in the 1930's the construction of an oil port begun. Peak years were in the 1950s and the 1960s. In the area there were, in total, 56 storage units for oil and gasoline. Total capacity was 400.000 m³. The oil port lost its importance in the 1970s and the port was repurposed as a liquid cargo port. In the early 1990s, the safety regulations became stricter so all the liquid bulk cargo was moved to another part of the port. The old liquid bulk port came to an end in the year 2000. In the companies' contracts there was a deal, that they should remove and/or clean all contaminated and spoiled soil and remove the oil port facilities. (Southeast 135° 2013)

After the soil cleaning process, the land area was modified for different purposes. There are places for skateboarding, hammocks, and different exercise machines with sea views. The area is suitable for different kinds of events, concerts and barbeque parties. An example is the sand barriers - there are tables made from stone for 50 persons, or if someone wishes to sprinkle ashes into the sea. Every kind of activity has been thought of in sustainable development in park planning. (HS 2014)

Figure 2. Old oil port near the city centre of Kotka. (City of Kotka 2014).

Figure 3. Current Maritime Park Katariina. (City of Kotka 2014).

The following figures present the old and new version of the same port. Figure 2 is from the 1970s or 1980s, when the liquid bulk port was in active use. Figure 3 is from a few years ago, after the cleaning and land modification process. In Figure 3 it can be seen that the distance to the city centre is less than two kilometres. Nowadays, the new park is updated every year with new exhibitions and new activities for the citizens. These non-socioeconomic soft values in parks are very important to the citizens of Kotka and nearby towns.

4 ECOLOGICALLY FRIENDLY PORT-PROJECT

This study was conducted as a part of the Ecologically Friendly Port Ust-Luga (EFP) project, which was launched in December 2012. It is funded by the CBC ENPI program "South-East Finland-Russia" 2007-2013. The total budget for the project is 570,000 Euros. The duration of the project is 24 months. The project is led by the Russian State's Hydrometeorological University (RSHU). The project consortium includes Ust-Luga Company JSC, University of Turku Centre for Maritime Studies and Kymenlaakso University of Applied Science. The project associates are the Port of HaminaKotka, City of Kotka, Finnish Port Association and Administration of Leningrad Region Committee on Natural Resources.

The main objectives of the project are:
- Improving the environmental status of the eastern part of the Gulf of Finland
- Improving the ability of the ports to develop environmental protection and sustainable growth in the ports
- Improving the municipalities' ability to enhance environmental safety issues
- Increasing the awareness of citizens concerning green thinking
- Establishing a close bilateral cooperation between citizens and authorities on the basis of green values, green Economy and Ecological mentality, aimed at sustainable regional development
- Increasing the ecological knowledge and understanding of responsibility for the global challenge of climate change of Russian and Finnish ports and enterprises.

5 DISCUSSION AND CONCLUSIONS

Many ports are located near a town or a city. Usually, at first a port was built in a good location with good connections to the sea or inland areas. Later, a town or a city grew around the port area. Throughout history, ports have been vital connection points between continents; Europe, Asia, Australia, Africa and America. Ports have also raised a lot of criticism and have had both negative and positive impacts on the citizens and environment as well.

Throughout history, seafaring, ports and port operations have had a positive impact on the employment of the citizens and people have also moved to cities from elsewhere in search of a better income. Secondly, there have been negative impacts, which have been related to the environment, land use, traffic, air and water pollution and light and noise emissions. In addition, especially in the old days, seaman culture and seaman towns have had a negative impact on the residents of the city. Usually only negative impacts get media coverage, but the positive impacts are quite remarkable as well - especially as the ports work as an economic driver for the region. The benefits include: economic benefits, employment and other direct and indirect catalysts, which helps the surrounding business, enterprises and citizens.

Ports also have different kind of values. Values can usually be classified based on their socioeconomic importance – considerable, non-socioeconomic or soft values. These soft values have gained a more and more important role in recent times. Soft values include: historical, archaeological, architectural, landscape, recreational, sociological and other cultural aspects. New ports are no longer situated in the city centres like old ports were. For example, a lot of extra space and buildings are no longer in active use, therefore these can be repurposed.

A lot of different old port section transformation projects have been carried out all over the world using best practices. Generally, old buildings in ports or near ports have been transformed for new use as office buildings, residence use or other cultural purposes or parks. On a general scale, in Finland, the largest transformation work was carried out in Kotka when old industry areas were transformed into a recreational park and to cultural purposes. Finland's largest universal port, HaminaKotka Ltd, gave up two old port sections for other purposes. Both port parts are situated in the city centre of Kotka town. In another practical example, Shell Company`s old oil port was transformed to a recreational "sea park" and the old city port is now a "cultural port". The recreational park has been awarded nationally in Finland and the "cultural port" is home to Finland's only maritime museum. In these construction projects, the opinions of citizens have been heard and every year something new is created. These old port sections are now close to the citizens' everyday life and have an important role as cultural heritage.

REFERENCES

Brunila, O.-P. 2013. *The Environmental Status of port of HaminaKotka*. Publications of the Centre for Maritime Studies. University of Turku. A69.

Burel F., Taccani R. & Zuliani N. 2013 Improving sustainability of maritime transport trough utilization of Liquefied Natural Gas (LNG) for propulsion. *Energy journal* 57(2013) pp. 412-420.

City of Kotka 2014. City of Kotka www-pages. In Finnish. Updated 21 October 2014. Available at URL: <http://www.kotka.fi/asukkaalle/ajankohtaista_kotkassa/10 1/0/kantasatama-hanke_etenee_kotkassa>

Corbett J.J. & Fischbeck P. 1997. Emission from ships. *Science* 278(5339) pp. 823-824.

Corbett J.J., Winebrake J.J., Green E.H., Kasibhatla P.,Eyring V. & Axel Lauer A. 2007. Morality from Ships Emissions: A Global Assessment. *Environ. Sci. Technol*. 41(24) pp. 8512-8518.

Det Norske Veritas 2012. *Technology Outlook 2020*. Research & Innovation. Available at URL: <http://production.presstogo.com/fileroot/gallery/DNV/file s/preview/9ec457bc750b4df9e040007f0100061c/9ec457bc 75094df9e040007f0100061c.pdf.>

Finnish Port Association 2010. *Port operations and environment*. (Only in Finnish). Available at URL: < http://www.finnports.com/fin/tietopankki/ymparisto/>

Ferrari, C., Merk, O., Bottasso, A., Conti, M. & Tei, A. 2012. *Ports and Regional Development: a European Perspective*. OECD Regional Development Working Papers. 2012/07. OECD Publishing. Available at URL: http://dx.doi.org/10.1787/5k92z71jsrs6-en

Girard, L. F. 2013. Towards a Smart Sustainable Development of Port Cities/Areas: The Role of the "Historic Urban Landscape" Approach. *Sustainability* 5, pp. 4329-4348.

Government resolution (480/1996). *Valtioneuvoston päätös ilmanlaadun ohjearvoista ja rikkilaskeuman tavoitearvosta*. (Resolution on reference values of air quality and sulphur deposition target values). FINLEX – Valtion säädöstöpankki. Available at URL: < http://www.finlex.fi/fi/laki/alkup/1996/19960480>

Government rsolution (993/1992). *Valtioneuvoston päätös melutason ohjearvoista*. (Resolution on the reference values of noise level). FINLEX – Valtion säädöstöpankki. Available at URL: < http://www.finlex.fi/fi/laki/alkup/1992/19920993>

Haezendonck, E. 2001. *Essays on strategy analysis for seaport*. Elivira Haezendonck & Garant Publishers. ISBN 90-441-1153-1.

Hansen, S.-O. 2007. Finnish merchant maritime transportation is on crisis. *Turun Sanomat* (newspaper). Available at URL: <http://www.ts.fi/uutiset/talous/1074189278/Suomen +kauppamerenkulku+kriisissa>

Van Hooydonk, E. 2007. Soft values of seaports. *A strategy for the restoration of public support for seaports*. pp. 192. ISBN 978-90-441-2148-3.

HS 2014. Helsingin Sanomat - Park theme issue. *Mistä lähtien puistoissa on saanut tallata nurmikoita?* (Since when it has been allowed to walk on the lawn in parks?) (Only in Finnish). Available at URL: < http://www.hs.fi/sunnuntai/a1408687167075 >

Häkkinen, J. & Posti, A. 2012. *Survey of transportation of liquid bulk chemicals in the Baltic Sea*. Publications from the University of Turku Centre for Maritime Studies. ISBN 978-951-29-5000-3.

Kajatkari, R. & Koskinen, M. 2014. Interview with the developing director Kajatkari and the traffic director Koskinen of the port of HaminaKotka. 13th of May 2014, Kotka.

Kalenoja, H. & Kallberg, H. 2005. *Liikenteen ympäristövaikutukset* (Environmental effects of transportation). Tampere University of Technology. Liikenne- ja kuljetustekniikka. Opetusmoniste 37. (Only in finnish)

Lonati G., Cernuschi S. & Sidi S. 2010. Air quality impact assessment of at-berth ship emissions: Case-study for the project of a new freight port. *Science of the Total Environment* 409(2010) pp. 192-200.

Martinson, T. 2010. Göteborg kytki laivat puhtaaseen maasähköön. (Port of Gothenburg connects ships to clean shore base power). *Power & Automation magazine* 2/2010. Finnish translation available at URL: < http://www04.abb.com/global/fiabb/fiabb250.nsf!OpenDat abase&db=/global/fiabb/fiabb254.nsf&v=80EA&e=fi&url

Merk, O. 2013. *The Competitiviness of Global Port-Cities Synthesis Report*. OECD Port in the cities case studies series. pp. 183. Available at URL: < http://www.oecd.org/gov/regional-policy/Competitiveness-of-Global-Port-Cities-Synthesis-Report.pdf >

Ministry of Environment 2009. *Liikenteen melu ja tärinä*. (Noise and vibration from transport). Available at URL: http://www.ymparisto.fi/default.asp?contentid=139310&la n=fi

Mylly, M. 2014. *EMSA on maritime safety throughout Europe*. Available at URL: < http://www.adjacentgovernment.co.uk /farming-environment-marine-sustainable-news/importance -maritime-safety-throughout-europe-2/>

Port of HaminaKotka 2013. *Basic information about port*. Available at URL: < http://www.haminakotka.fi/en/ haminakotka-satama-oy>

Port of HaminaKotka 2014. *Harbours- Kantasatama*. Available at URL: < http://www.haminakotka.fi/en/kantasatama >

Pulkkanen, P. 2013. *Waste management reports from HaminaKotka and Emission to air reporting from HaminaKotka*. E-mail inquiry.

Ship-technology 2013. *The world's 10 biggest ports*. Available at URL:< http://www.ship-technology.com/features /feature-the-worlds-10-biggest-ports/>

Southeast 135° 2013. *South-East Finland tourist information*. Available at URL: < http://www.southeast135.fi/en/ >

Tarnapowicz, P. & Borkowski, T. 2013. Best practices and recommendations for ports and cities. 4.3 Electrical shore-to-ship connection, benefits and best practices. *Pan-Baltic Manual of best practices on clean shipping and port operations*. ISBN 978-952-298-004-5.

Tenhunen, J. 2008. *Tieliikenteen päästöt ilmaan*. (Air emissions from road transportation) SYKE. Available at URL: < www.ymparisto.fi/download.asp?contentid=86504>

U.S. Environmental Protection Agency (EPA) 2007. *Acid Rain*. Available at URL: < http://www.epa.gov/ acidrain/what/index.html>

VTT 2009. *LIPASTO 2009 calculation system*. Available at URL: < http://lipasto.vtt.fi/lipasto_lask_tulokset.htm>

Värri, E. 2013. *Quality of Air emission reports from the port area of Kotka and Hamina*. E-mail inquiry.

Wichmann, W. A. 2012. *An analysis of protest carried out by ships (PCS) Should PCS be regulated by a new IMO instruments?* World Maritime University, Sweden. Available at URL : <http://dlib.wmu.se/jspui/ bitstream/123456789/829/1/76408.pdf>

The Influence of Internalizing the External Cost on the Competiveness of Sea Ports in the Same Container Loop

E. van Hassel, H. Meersman, E. Van de Voorde & T. Vanelslander
University of Antwerp, Antwerp, Belgium

ABSTRACT: A lot of research has been done on the effects of internalizing the external costs on the maritime part of the transport chain or on the hinterland part of that chain. The unanswered question is what the effect is of such internalization on the competitive position of sea ports which are part of the same container loop. A derived question is how the internalization of the external costs over the total logistics chain affect the cost structure of the total chain.

1 INTRODUCTION

This paper researches the effects of the internalisation of the external costs in the total logistics chain.

The internalisation of external costs may have an impact on the hinterland distribution between the various ports. Ports that are located more inland will have shorter hinterland distances (and therefore cost) and could therefore have a competitive advantage over their competitors which are located directly at sea. This effect could be increased if the external costs in the hinterland are internalised. This could then lead to a situation where port competition may be affected due to the internalizing the external costs. Within this paper, it is examined whether sea ports that are located more inland (such as Antwerp or Hamburg) will have an advantage over their competitors that are located directly at sea when external costs are internalized.

In this paper, a model is applied that allows calculating the total generalized chain costs for different container loops (including the external cost). This model was first developed in van Hassel et al. (2014). In this model, the total supply chain, including maritime transport, the port process and hinterland transport is taken into account. The main reason to lay the emphasis on the supply chain is that container liners and ports will compete along these supply chains. This is illustrated in figure 1, where the chain with the lowest overall generalised cost will be the most successful chain.

In order to answer the research questions, first, a literature review is made. Secondly, the developed chain model is explained, which allows calculating the generalised cost of a several chains. This model will be applied to a container loop (U.S.- EU) in the third section. This logistics chain relates to the transport of a container from a port of origin at the East Coast of the US, via a port in the Hamburg – Le Havre range to any of 250 European hinterland areas. Finally, conclusions will be drawn.

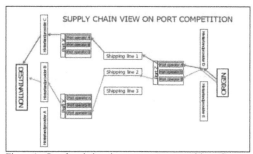

Figure 1. Supply chain view on port competition, Source: Meersman and Van de Voorde (2012)

2 LITERATURE REVIEW

Several sources of literature can be found regarding research into external costs. The most important ones that are used in this research are briefly described in this section.

According to Blauwens et al (2009), the main cost components of the external costs are: congestion, infrastructure, environmental (air quality, climate and noise) and accidents.

Miola et al (2009) give an overview of the environmental impacts of maritime transport (both at sea and in ports). The main cost components are those related to air pollution and climate change.

With respect to the external cost for land modes (hinterland), use is made of the research performed by CE Delft (2008). In this research, an overview is given of the monetary cost for the relevant external cost items for road, rail and inland waterway transport.

These external cost insights were incorporated in the maritime chain model that was developed in van Hassel et al (2014). With this model, the different port hinterlands can be calculated. In Kronbak and Cullinane (2011), also a visual representation of different port hinterlands is given. In that research, also a part of the maritime chain section and the port process is taken into account. For the hinterland transportation, only road haulage is taken into account, while in this paper, also rail and inland navigation transport are considered.

3 MODELLING APPROACH

In this section, an overview is given of the model that is being applied, and more in particular of its different components: maritime, port and hinterland.

3.1 Overview of the methodology

In this section, a short overview of the main components of the model is given. In van Hassel et al (2014), the total modelling approach is more extensively explained. Basically, a model was developed that allows calculating the generalised chain cost from a selected point of origin, via a predefined container loop to destination point. The model was coded in C# and uses Microsoft Excel (data) and JMP11 (maps) as output formats. In order to calculate the chain cost, first a container loop has to be defined. A loop is defined as a circle route of a ship from one port to the next (it has no beginning nor an end). This loop will determine the maritime part of the chain. In figure 2, the general overview of the developed model is given.

The model is built up as a route builder for ships. This route builder connects different aggregated hinterlands via a route of ports (bold lines). The aggregated hinterlands are defined as a summation of different smaller geographical areas, which in Europe correspond to NUTS-2 areas. In the aggregated hinterland, at least one, but mostly more ports are located that can serve the same set of hinterland areas. Examples of aggregated hinterlands are mainland Europe or the United States.

Once a ship has been selected, the main dimensions and related costs of that ship are known. Based on the physical characteristics of the ship, a set of ports which can accommodate the selected ship are available. At current, the model encompasses 42 ports in total, on which the loop can be set up (section 3.3).

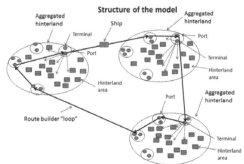

Figure 2. Structure of the model, Source: Van Hassel et al (2014)

Each port has a set of terminals which in their turn have an own set of characteristics, such as: allowable draught, navigation channel to enter the port, locks (if available), number of container cranes, etc. Per terminal, the total port entering cost can be determined. The reason to determine these costs at a terminal level is that port dues, tug boat cost, pilotage cost, etc. can differ between the different terminals in the same port (see also section 3.4).

From each terminal in a port, the hinterland distances via road, rail and inland waterways (if available) are incorporated in the model. The hinterland areas are defined as NUTS-2 areas in Europe and in Great Brittan. Using the hinterland distances, it is possible to calculate the hinterland cost per mode from a terminal to a hinterland destination (see also section 3.5).

A chain is defined as a route from a hinterland area in a specific aggregated hinterland to another hinterland area in another aggregated hinterland. A chain therefore has a beginning and an end. In order to calculate the chain cost from a point a origin to a destination point, the model must not only calculate the total cost of the ship, but it must also incorporate the cost of transporting a container from a hinterland area to a port on both ends of the chain, the cost of a container in the port phase (port dues, pilotage, container handling, etc.) on both chain sides, and the cost of transporting via sea the container from the port of loading to a port of unloading. In Figure 3, the overview of the model is given when a chain calculation is made.

Not yet all the aggregated hinterlands are fully developed. At this stage, the hinterlands of mainland Europe and the UK are fully developed. In the example of Figure 3, the left side aggregated hinterland has no hinterland areas in the model.

Therefore, it is not possible to select a point of origin there for now. In order to solve this problem in the current model structure, a port in that specific aggregated hinterland has to be chosen as a point of origin (for example Hong Kong in the aggregated hinterland of Asia). The aggregated hinterland in which the origin of the chain lies is called the aggregated *from* hinterland, whereas the hinterland in which the end of the chain is located, is called the aggregated *to* hinterland. If a aggregated hinterland is part of the selected loop but is not selected to be an aggregated *from* or a *to* hinterland, then only the maritime cost of sailing to the different ports in that aggregated hinterland are taken into account. In the aggregated *to* hinterland, all the possible chains are calculated. This means that from all the ports, that are part of the loop and that are located in the aggregated *to* hinterland, the port cost and the hinterland cost from the ports to all hinterland areas are calculated (250 NUTS-2 area in Europe). Due to the fact that all possible combinations are calculated, it is also possible to determine the lowest chain cost from a port of origin to all the different hinterland areas, including which ships, sailing routes, ports of call and hinterland modes are to be chosen to achieve that lowest cost.

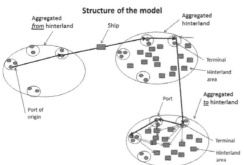

Figure 3. Definition of a chain, Source: Van Hassel et al (2014)

3.2 *Input parameters*

The input for the chain model consists of four main elements. The first desired input is a selection of an existing container loop. Secondly, a specific vessel needs to be selected that will sail within the selected loop. Thirdly, the size of the European hinterland must be chosen. There is a choice for a core hinterland (direct hinterland of the ports in the Hamburg – Le Havre range) and the total hinterland. The last element which requires input is the value of the goods transported in the containers. All these elements are further elaborated in the following sections. Based on this input, it is possible to simulate an existing container loop or to build one by oneself.

3.3 *Ship model*

The maritime model consists of three main parts: a routing module, a design(technical) module and a cost module. These three components are merged into an integrated maritime ship model. In the routing module, a maritime distances database has been built to connect all the 42 ports to each other, based on the AXSMarine (2013) distances calculator. The design module is used to determine the technical parameters of the ship. The technical model makes use of Holtrop and Mennen (1978) for resistance calculations, propeller calculations based on the Wageningen B-series propellers (Oosterveld en van Oossanen, 1975) and the engine characteristics (Wärtsila data). Based on the results from the routing and the design module, the transport cost can be calculated in the cost module. In the latter module, all the relevant cost components are determined, split up into three main sub-groups (Drewry, 2005): operational, voyage and capital. All the cost data used in the model has been updated to 2012 values. In van Hassel et al (2014), more detail is given about the technical and maritime cost calculations and data sources.

The external costs arise from air pollution, congestion, infrastructure and accidents. The air pollution costs will be calculated on the basis of the fuel consumption of the ship. For every ton of fuel burned, it is known how many tonnes of emissions is produced. These emissions are divided into CO_2, NO_x, SO_x and PM_{10}. The produced emissions per burned ton of fuel oil is taken from Dijkstra (2001), while the monetary values are taken from Minola et al. (2009) and CE Delft (2008).

The other external costs are set to zero because no external congestion occurs at sea (this is only possible at the Suez or Panama Canal and those channels are not passed in the proposed case study). Moreover, the external congestion costs, which occur in the port, will be included in the port model. There are no external infrastructure and accident costs. There will be accidents at sea, but these costs are covered by insurance. An exception might be when a ship sinks in territorial waters and a large oil spoil has occurred and the owner of the ship, through a complicated ownership structure, tries to escape its liability. In that case, the cost is borne by the coastal state and can only then be considered an external accident cost.

3.4 *Port model*

All the relevant cost components that a ship owner will bear when a ship will enter a port are taken into account in the port model. The total cost is built up of three different cost components: port shipping, port authority and third party.

The port shipping cost is related to operating the vessel in the port. The port shipping costs are composed of the crew costs and the operating and maintenance costs of the ship while the ship is at port, i.e. fuel consumption during the port stage and crew cost while the ship in the port. All these costs are made a function of the size and type of the ships that will enter the port.

The port authority costs are the dues that have to be paid by the ship owner to the port authority. Each port has its own system of setting charges, and hence, their value and the way of calculating them will vary from port to port (Wilmsmeier, 2007). For each of the ports, these costs were modelled.

The last cost element are the third party costs. Four different types of third party cost are used in this research. First, there is the price to be paid for the use of a tug boat (depending on vessel type and size). Secondly, there is the price to be paid for the use of pilotage. Third, there is the cargo handling cost. This is the price that has to be paid for the handling of cargo in the port. Lastly, also the costs of mooring and unmooring are taken into account.

A generic port model was developed which is built up from different container terminals. These terminals are connected by shared infrastructure (locks and waterways). The different processes that take place in the port, and that are important for the treatment of a ship, are included in the model. To model the port process, queuing theory was applied. In the model, there is queuing to pass the tidal window, to pass a lock and at the terminals. In reality, there will be no real queue in front of the locks. The ship's speed will be adjusted so that the ship will arrive at the lock when it can be handled. If there are a lot of ships that have to pass a lock, the speed has to be reduced further which will imply congestion. This additional time (congestion) will be modeled via a queue model. At the terminals, the process of loading and unloading the ships is modeled. First, the ship has to sail from the entrance of the terminal to the quay wall. The ship is moored and containers will be handled. The handling time is related to the number of cranes deployed per ship length and the nominal handling rate of the cranes. When the ship has been handled, it sails back to the buoy. So, with this model it is possible to calculate the total time and cost per ship for every terminal in the port. Also for the reverse process (export), this time and cost can be calculated. The cost data is in 2012 values.

The external cost during the port phase is determined by the external emission cost during a port stay. In the model, the ships are set to use MDO as a fuel. Therefore, the emission factors of MDO are used to determine the external cost. Besides the external costs that are related to emissions, there are the external costs that are related to infrastructure, accidents and noise. The last two cost items are, in principle, borne by the originator, but they are typically subject to insurance and are therefore not external. Noise costs are hard to estimate, but all studies available indicate that noise effects of port activities are very limited, so we can disregard their marginal cost effect (Meersman et al, 2004).

3.5 Hinterland model

A hinterland model is developed that will be used to calculate the hinterland transport cost from the selected container terminals in the selected ports to in total 250 European hinterland areas. The generalised costs of three different transportation options (road, rail and inland waterways) are calculated.

In this model, the distances are determined from the various container terminals in the selected ports to the hinterland destinations. These hinterland destinations are defined at NUTS 2 level. NUTS 2 areas have usually the size of a provinces in one of the European member states. For all these areas, the distances for the direct road connections (to the centre of the area) and the inland shipping and rail connections to the various inland terminals are determined. Also distances from the inland terminals to the centre of the various NUTS 2 areas are determined. All land distances were composed based on Port of Antwerp (2014). For all three modes of hinterland transport (road, rail and inland navigation), the generalised transportation costs are calculated. Because three different values of generalised cost are calculated (one for each mode of transportation) these three values must be combined in to one single value. That is done firstly by calculating the modal split per hinterland area based on the calculated generalised cost per mode of transport by making use of logit type modelling approach. The logit model is fit based on mode split data from Port of Antwerp (2014) for the different hinterland areas. Secondly, the total hinterland costs are determined by applying the following formula:

$$GHC_{q,j} = \sum_{k}^{n} GC_{k,q,j} . MS_k \qquad (1)$$

where $GHC_{q,j}$ = the generalised hinterland cost from port q to hinterland area j; $GC_{k,q,j}$ = the hinterland cost from port q to area j using hinterland mode k and MS_k = the modal share of mode k.

All costs are again in 2012 values. The basic cost structures of the different hinterland modes is taken from Grosso (2012). For a more extensive description on the way how the calculations are made, reference is made to van Hassel et al (2014).

For each of the hinterland transport modes, the external costs are calculated. In Table 1, the monetary values of the different external cost elements are given.

Air quality, climate and noise cost can be grouped together as the environmental cost. The cost data presented in Table 1 is from 2006, so that the data must be updated to 2012 values. In Arcadis et al. (2009), a set of index figures is given. With that data, it is possible to calculate the external cost of the hinterland modes. It is also possible to calculate new mode shares if the external costs are internalised.

Table 1. External cost of hinterland modes

	Road €cent/veh.km	Rail €cent/ton.km	IWT €cent/ton.km
Air Quality	5,12	1,26 (0,03)	0,60
Climate	0,91	0,13 (0,04)	0,06
Accidents	7,68	0,11	0,01
Noise	3,04	0,09	0,00
Congestion	42,23	0,02	0,00
Infrastructure	0,15	1,63	0,00

Source: MOW, 2009
() are values corresponding to electrical trains

4 AN APPLICATION TO A LOOP FROM THE US TO EUROPE

The loop that is being analysed in this paper, is a loop from the East Coast of the U.S. to Europe. For this case study, the CMA-CGM loop (Liberty bridge loop) is used (CMA-CGM, 2015). On this loop, ships with an average slot capacity of 4,600 TEU are deployed (CMA-CGM, 2013). For the calculations, the ship speed is set at 22 knots and the ship is loaded up to 80% of its slot capacity. In the analyses, the port of origin is set at Miami and all the calculations are made for TEU as a unit. The ports of call in Europe in this loop are Antwerp, Hamburg, Rotterdam and Le Havre (in that order) and in those ports, 20% of the total cargo volume aboard the ship is unloaded and loaded.

For this loop, two main calculations will be made: one for the import flows (from the US to Europe) and one for the export flows (from Europe to the US). For both transport flows, the hinterland distribution between the four European ports will be calculated.

The chain cost from a point of origin in the aggregated *from* hinterland (i) via port p in the same *from* hinterland and further on via port q in the aggregated *to* hinterland to a hinterland area in the same *to* hinterland (j) can be calculated with the following formula (fig 3):

$$GC_{i,p,q,j} = GHC_{i,p} + GPC_p + GMC_{p,q} + GPC_q + GHC_{q,j} \quad (2)$$

where $GC_{i,p,q,j}$ = the generalised chain cost from hinterland area i via port q to hinterland area j via port p; GHC = the generalised hinterland cost from a terminal in a port to a hinterland area; GPC = the

generalised port costs in ports p and q and GMC = the generalised cost from port p to q.

A hinterland area j is assigned to a specific port q if the generalised cost via that port is the lowest:

$$\underset{q=1}{\overset{n}{Min}}\left(GC_{i,p,q,j}\right) \quad (3)$$

where n is the number of ports in the same aggregated hinterland which can access hinterland area j. In our case study this is four, while the point of origin is set as the port of Miami so p is one.

When the difference in total chain cost in a hinterland area between two ports is less than 5%, then that area is assigned to both ports (common hinterland).

4.1 *Base case scenario*

The result of the hinterland split calculation for the import flows between the four European ports can be seen in Figure 4. In this graph, the external costs are not included in the calculations and are therefore used as the reference scenario. Also in this graph, only the core hinterland is shown and not to total European hinterland. The green colour represents the hinterland of the port of Antwerp, blue belongs to Hamburg, and purple to Le Havre. The orange colour represents the hinterland where the difference in generalised chain cost between Rotterdam and Antwerp is less than 5%.

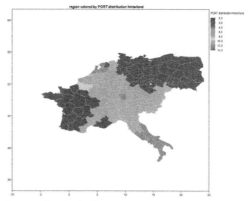

Figure 4. Hinterland split import flows from the US

From Figure 4, it can be observed that the port of Antwerp will take up a large part of the hinterland, while the influence of Rotterdam is very limited. This is due to the fact that Antwerp is the first port of call in Europe in this specific loop. Because the generalised cost is used as the basis for the hinterland split, also the effect of time is cooperated. Therefore, the opportunity cost of a longer sailing time will influence the total chain cost and therefore also the hinterland split between the different ports.

Also for the export flows, the hinterland split is calculated. In Figure 5, the results can be observed. The same colour codes as for the import flows apply.

From Figure 5, it can be observed that Rotterdam takes up a very large part of the hinterland (light blue) and the influence of Antwerp is lower. This is also due to the fact that from an export point of view (from Europe to the US) the port of Rotterdam has an advantage over Antwerp because it is called later in the loop. There are also area with a light green colour (east of Poland) and in between the hinterland of Rotterdam (light blue) and Hamburg (dark blue). These areas are the areas where the difference between Rotterdam and Hamburg is less than 5%.

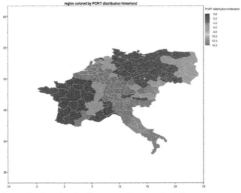

Figure 5. Hinterland split export flows to the US

The areas that are coloured red are the areas where the difference between the four ports is less than 5%. So generally speaking, it can be observed that in this loop, the port of Antwerp functions as the import port and the port of Rotterdam as export port, while the hinterland of Hamburg and Le Havre is more or less the same for the import and export flows.

4.2 Internalizing the external cost

Now that the reference scenario has been developed, the model is again applied to calculate the hinterland split between the four considered European container ports when the external costs are internalized. In the calculations, the external cost are included over the total chain (during the maritime transport, in the port and during the hinterland transport). In Figure 6, this new hinterland split can be found for the import flows.

If Figure 6 is compared to Figure 4, it can be observed that the influence of the port of Rotterdam is increasing. There is a larger area where the difference in total chain cost between Antwerp and Rotterdam is less than 5% (orange area). This can be explained by the fact that due to the internalization

of external costs, more use will be made of the alternative modes (especially inland waterway transport for the ports of Antwerp and Rotterdam). So the areas that are located around the river Rhine in Germany will be more served by inland waterways.

The reason why the difference between Antwerp and Rotterdam is decreasing is that the generalized hinterland cost via inland navigation is almost the same. This is due to the fact that the sailing distance from Antwerp to, for instance Manheim, is almost the same as for Rotterdam. For areas that are more located towards the ports (shorter hinterland distances), road transport is still the most used mode of transport, and for these areas, the port of Antwerp is still the most dominant port.

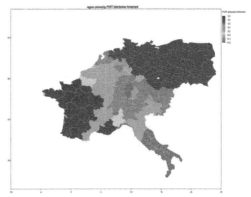

Figure 6. Hinterland split import flows to Europe including external cost

What also can be observed is that the hinterland of Hamburg is growing (more blue in Figure 6 than in Figure 4). This is the port gaining extra hinterland areas when the external costs are internalized.

The hinterland split for the export flows (from Europe to the US) is given in Figure 7.

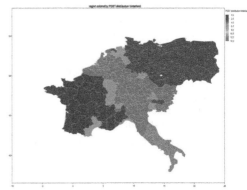

Figure 7. Hinterland split export flows to Europe including external cost

In Figure 7, it can be observed that the total hinterland is now split between Hamburg in the north (dark blue), Rotterdam in the middle (light blue) and Le Havre in the south (purple). Only in the very south of Italy, the port of Antwerp still has some influence (orange). This means that the good hinterland connections of the port of Rotterdam via the inland waterway network and the fact that Rotterdam is called at a later stage in the loop make that for the export flows, Rotterdam takes up a large part of the hinterland.

4.3 Effect of internalizing the external cost on the chain cost structure

When the cost structure of the chain for this loop has to be analyzed, it is necessary to determine the "average" chain cost. This is needed because from one point of origin (in this case the port of Miami), there are 250 different destinations (and thus chains). The average chain costs are determined with the following formula from van Hassel et al (2014):

$$AGC_i = \sum_{j=1}^{n} f_j GC_{i,j} \qquad (4)$$

where GGC_i = the average generalised chain costs fom origin i to a total aggregated hinterland; $GC_{i,j}$ = the generalised chain cost from origin i to hinterland area j and f_j = a weighing factor related to hinterland area j. This weighing factor f_j is calculated based on the total container hinterland distribution of the six main container ports in the Hamburg-Le Havre range. So, areas with a lot of containers are the important hinterland areas and therefore they will have a larger contribution to the average chain cost. In Figure 8, the total hinterland container distribution is given, in which the more dark the hinterland areas are, the more containers have an origin or destination in that area.

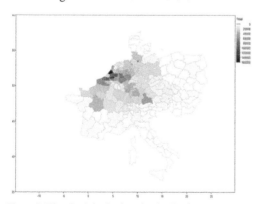

Figure 8. Hinterland destinations for the Hamburg – Le Havre range ports, source: Port of Antwerp, 2014

In Figure 8, it can be observed that most of the containers which are handled in the six main container ports have a destination relativity close to the ports.

By applying Formula 4, the average chain cost for the total hinterland can be calculated. Three different calculations are made. One where none of the external costs are internalized, one in which only the external cost at the hinterland side are internalized, and one in which the external costs over the complete chain are internalized. In Figure 9, the results of the calculations can be found.

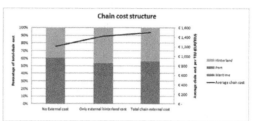

Figure 9. Cost structure of the average chain cost

In Figure 9, it can be observed that if no external costs are internalized, roughly 40% of the average chain costs are determined by the maritime component, another 40% by the hinterland and 20% of the chain cost are incurred during the port phase of the chain.

If only the external costs at the hinterland side are taken into account, 46% of the average chain costs are determined by the hinterland while the contribution of the maritime part is 36%, making the hinterland part of the chain the dominant factor. Equally, the average chain cost increases by 17% from €1.200 /TEU to €1.400/TEU. When all external costs are taken into account, the hinterland part of the chain will have the largest contribution to the total chain cost (44%). Also the average chain cost increases again, only now by 5%, up to €1.500/TEU. Thus the impact of internalizing the external cost at the hinterland has a larger impact on the average chain cost than the internalization of the external cost in the maritime part of the chain.

5 CONCLUSIONS AND FURTHER RESEARCH

In this paper, the effect of internalizing the external cost on the competiveness of the ports has been analyzed. The analyses have shown that by internalizing the external cost, the hinterland split between ports that are in the same loop will be influenced. Hereby, not only the location of the port has an influence (inland position of a port such as Hamburg and Antwerp), but also the position of the port in a loop and the hinterland connection of a port via the inland waterways or rail network . By internalizing the external cost, more use will be made of alternative modes of hinterland transport (inland waterway transport and rail), making that their modal shares will increase. So, although the port of Rotterdam, with the Maasvlakte I and II directly located at sea, can also benefit from internalizing the external cost due to its good connections to the inland waterway network.

It has to be mentioned that in the calculations of the inland waterway cost, the water levels of the Rhine are kept constant (i.e. all inland ships can be loaded up to their maximum capacity). However, it is noted that the water levels can fluctuate much and in case of low water levels, less cargo can be transported with inland ships due to draught restrictions. These fluctuations could increase in the future due to climate change (more wet winters and dryer summers) (Jonkeren et al., 2013). Normally, this effect is less for the transportation of containers due its lower mass per volume ratio than for bulk cargo. However, the draught restriction can play a role if the weight of the containers is higher than 10 tons per TEU (van Hassel & Vanelslander, 2014). When the payload of inland ships is decreased due to draught restrictions, other modes of transport will gain importance (rail and road). Therefore also the mass of the container can play an important role in the choice in hinterland transport mode. Therefore, in future research, this aspect has to be taken into account. Equally, the model will be extended with more ports and fully developed aggregated hinterlands.

REFERENCES

Arcadis et al. 2009; Maatschappelijke Kosten-batenanalyse voor de vervanging van drie 600 ton-sluizen door een of meer sluizen van min. 1350 ton. (in Dutch)

AXSMarine, 2013; http://www.axsmarine.com/distance/

Blauwens, De Baere & Van de Voorde, 2012; Transport economics, Uitgeverij De Boeck NV, Antwerpen

CE Delft, 2008; M. Maibach, C. Schreyer, D. Sutter (INFRAS),H.P. van Essen, B.H. Boon, R. Smokers, A. Schroten (CE Delft), C. Doll (Fraunhofer Gesellschaft – ISI), B. Pawlowska, M. Bak (University of Gdansk), Handbook on estimation of external costs in the transport

sector Internalisation Measures and Policies for All external Cost of Transport (IMPACT), Version 1.1

CMA-CGM, 2015; http://www.cma-cgm.com/products-services/line-services/flyer/LIBERTY

Dijkstra, ir. W.J.,2001; Emissiefactoren fijn stof van de scheepvaart Rapport Delft (in Dutch)

Grosso, 2011; improving the competitiveness of intermodal transport: applications on European corridors PhD – thesis, Antwerp University

Holtrop en Mennen, 1978; A statical power prediction method, international shipbuilding progres, Vol 25, The Hague

Jonkeren O., Rietveld P. van Ommeren J., te Linde A., 2013; Climate change and economic consequences for inland waterway transport in Europe, Springer-Verlag Berlin Heidelberg 2013

Kronbak and Cullinane, 2011; Captive and contestable Port Hinterland: modelling and Visualozation using GIS, in International Handbook of Maritime Economics (pp 348-362), Cheltenham: Edward Elgar

Meersman, Van de Voorde (2012) a new approach to structuring port competition, presentation at MIT, Boston (VS).

Meersman, H., Van de Voorde, E. & Vanelslander, T. (2010). Port competition revisited, Review of business and economics - ISSN 2031-1761 - 55:2(2010), p. 210-232

Meersman H., Van de Voorde E. and Vanelslander T. (2004); Port Pricing. Considerations on Economic Principles and Marginal Costs, EJTIR, 3, no. 4 (2003), pp. 371-386

MOW (2006); De opmaak van een standaardmethodiek MKBA voor socio-economische verantwoording van grote infrastructuurprojecten in de Vlaamse zeehavens, studie uitgevoerd in opdracht van MOW door Resource Analysis en Universiteit Antwerpen, ITMMA (Dutch)

Miola, A, Paccagnan V., Mannino I., Massarutto A., Perujo A.& Turvani M., 2009; External costs of Transportation, Case study: maritime transport; European Commission Joint Research Centre Institute for Environment and Sustainability

Oosterveld & van Oossanen, 1975; Further computer-analyzed data of the Wageningen B-Screw series; International shipbuilding progress Rotterdam –Holland Vol.22-No.252 1975

Port of Antwerp, 2014; Container hinterland distribution in the European hinterland provided by the port of Antwerp

Van Hassel E. , Meersman H., Van de Voorde E., Vanelslander T., 2014; .Impact of scale increase of container ships on the generalized chain cost, IAME conference 2014

Van Hassel E. & Vanelslander T. 2014; The impact of climate change on the water levels on the Rhine: What are the consequences for the hinterland connections via the inland waterways from Antwerp?, Presentation at the Scheldt-conference 8 May 2014, Antwerp

Wilmsmeier, 2007; Changing practices in European Ports – user reaction on differentiation of charges, ETC conference 2007

Development of Dry Ports: Significance of Maritime Logistics on Improving the Iranian Dry Ports and Transit

A.H. Pour & H. Yousefi
Khoramshahr University of Marine Science and Technology, Khorramshahr, Iran

ABSTRACT: This paper analyzes the strategic position of Iran as land-bridge for container transit at the Persian Gulf; therefore it is essential to develop the Iranian intermodal dry ports. Rodrigue & Notteboom,(2012) stated that Inland port development everywhere has grown substantially as a byproduct of the growth of intermodal freight transportation, driven by changing requirements of supply chains in global markets and distribution systems. This growth is described under various names including hinterland networks, dry ports, inland ports, intermodal centers, inland logistics centers, inland freight centers, and inland freight terminals. In this paper, the author is going to investigate the concept of Maritime Logistic which is a mode of management for supplying chain and science of managing and controlling the flow of cargos, information and other resources like energy and persons between the point of origin and the point of consumption in order to meet customers' requirements. It involves the integration of information, transportation, inventory, warehousing, material handling, and packaging of goods. The main part of this paper is dedicated to evaluate the role of container transit from the Iranian west transport corridor with a focus on Iran-Iraq-turkey-Syria corridor via Khoramshahr and Basreh port. The research methodology of this paper will be designated to consider SERVEQUAL method in order to grade the involved factors of improving the Iranian Dry Ports and as a result of that increasing volume of container transit for Iranian Southern ports. It should be noted that as railways play a key role in transporting intermodal containers or other commodities, so some wider information on rail transport has been included through this paper too.

1 MARITIME LOGISTICS

A dry port is an inland intermodal terminal directly connected by road or rail to a seaport and operating as a centre for the transshipment of sea cargo to inland destinations. In addition to their role in cargo transshipment, dry ports may also include facilities for storage and consolidation of goods, maintenance for road or rail cargo carriers and customs clearance services. The location of these facilities at a dry port relieves competition for storage and customs space at the seaport itself. Roso (2009b, p.308) has defined the dry port concept as: The dry port concept is based on a seaport directly connected by rail to inland intermodal terminals, where shippers can leave and/or collect their goods in intermodal loading units as if directly at the seaport. In addition to the transshipment that a conventional inland intermodal terminal provides, services such as storage, consolidation, depot, maintenance of containers and customs clearance are also available at dry ports.

Regardless of the terminology used, three fundamental characteristics are related to an inland node:

- An intermodal terminal, either rail or barge that has been built or expanded.
- A connection with a port terminal through rail, barge or truck services, often through a high capacity corridor.
- An array of logistical activities that support and organize the freight transited, often co-located with the intermodal terminal.

It can thus be seen that the functional specialization of dry ports has been linked with the clustering of logistical activities in the vicinity and have become excellent locations for consolidating a range of ancillary activities and logistics companies. In recent years, the dynamics in logistics networks have created the right conditions for a large-scale development of such logistics zones. (Dr. Jean-Paul Rodrigue, Hofstra University, New York, USA, & Dr. Theo Notteboom, President of ITMMA, University of Antwerp, Antwerp, Belgium)

Panayides (2006) stated that for a better understanding and ultimate definition of the term, the starting point should be to consider the underlying scope and characteristics of the two areas making up the term 'maritime transport' and 'logistics and supply chain management'.

A supply chain is composed of a series of activities and organizations that move materials such as raw materials and information on their journey from initial suppliers to final customers. Supply chain management involves the integration of all key business operations across the supply chain. In general, logistics and supply chain management relate to the coordinated management of the various functions in charge of the flow of materials from suppliers to an organization through a number of operations across and within the organizations, and then reaching out to its consumers (Harrison and van Hoek, 2011).

2 DEFINITION OF A DRY PORT

There is no officially agreed definition of a dry port. However, the working definition is that "a dry port provides services for the handling and temporary storage of containers, general and/or bulk cargoes that enters or leaves the dry port by any mode of transport such as road, railways, inland waterways or airports" (Iannone.F, 2011). As originally conceived, a 'dry port' was defined as an inland terminal to and from which shipping lines could issue their bills of lading, with the concept being initially envisaged as applicable to all types of cargo (UNCTAD, 1982). In both theory and practice, however, the concept has evolved not only to be closely associated with the rapid expansion of containerization and related changes in cargo handling (UNCTAD, 1991), but also to be applied in a variety of different contexts having the common characteristic of relating simply to 'a place inland that fulfils original port functions' (Cullinane and Wilmsmeier, 2011). In other words, the main objectives of a dry port are: to provide an additional hinterland terminal to which a seaport can outsource its workload; to improve the efficiency of the logistics chain; and to facilitate the trans-shipment of cargo to another mode of transportation.(Imai.A, et al, 2001(

However, since dry ports are expected to perform several different functions, the working processes at a dry port tend to be complex. Dry ports are bidirectional logistics systems - goods coming from seaports are received and transferred to modes of land transportation, while freight arriving by rail or by road is received and subsequently delivered to seaports. As a rule, dry ports should have a direct road or rail connection to a seaport, have a high capacity for processing traffic and offer the same type of facilities as those found at a seaport (Imai.A, et al 2006).

3 ANALYSIS OF DRY PORTS

Within the Western and Central Asian intermodal transport network, dry ports play a crucial role in shifting trade flows from one mode of transportation to another. For landlocked countries in the region such as Afghanistan, which have no direct access to the sea, dry ports are crucial nodal points of commercial trade. In accordance with this alignment of port development to the Product Life Cycle (Cullinane and Wilmsmeier, 2011), the 'dry port' concept can be implemented to extend the product life cycle of a port; specifically, by elongating the maturity phase and deferring a port's entry into a state of decline. In consequence, any required expansion of a port is redirected from the seaward to an inland location (UNCTAD, 1991). Of course, for this to work, the 'dry port' option must be practically feasible, with available and suitable physical site locations and the appropriate means of connectivity to the port itself either already present or potentially implementable. They also assist in the development of a hinterland zone, which can be particularly beneficial in creating jobs and promoting economic development in less commercially active areas within Western and Central Asian countries. There are also benefits in terms of accessing the existing hinterland, expanding a port's hinterland and the capturing of cargo closer to source and/or further up the supply chain. It is also reasonable to recognize, however, that on some occasions, the economic case may require the receipt of some form of subsidy (for example, large infrastructure grants are sometimes available from public sector authorities and agencies) and that, of course, there also certain sets of circumstances where there exists no realistic level of subsidy that will prompt the adoption of the concept (Bergqvist et al, 2010). The possible impacts of this emergent strategic freight network within Europe are described as being: a reduction in road freight transport into and out of ports; the loss of value-added logistics activities from seaport locations and the reversion of seaports to focusing solely on port-related activities; the relocation of many ancillary activities back into the hinterland – most critically, customs clearance and; greater externalities (particularly, congestion and pollution) occurring in and around the inland location of hinterland hubs. Currently, there are 48 dry ports in the entire Western and Central Asian region. The majority are Located in Afghanistan and are situated along its borders. Dry ports located near the border with Pakistan, at Torkham in Afghanistan (along the Kabul–Peshawar highway) and Chaman in Pakistan.

The author to conclude that rail and dry port capacity utilization needs to be improved through the imposition of more streamlined regulation and by better organization and administration, all of which should be supported by logistics marketing initiatives that spread the improvements achieved.

4 DRY PORTS AS LOGISTIC POINTS FOR CONTAINER TRANSIT

At first, it is better to understand the concept of a dry port. Mrs.Violeta Roso senior lecturer of Chalmers University in Sweden stated in this regard that "A "dry port" is defined as "an inland intermodal terminal directly connected to a seaport, with high capacity traffic modes, where customers can leave/collect their goods in intermodal loading units, as if directly at the seaport". And also, H.Yousefi (2011) expressed that A dry port is generally a rail terminal situated in an inland area with rail connections to one or more container seaports. The development of dry ports has become possible owing to the increase in multi-modal transit of goods utilizing road, rail and sea. This in turn has become increasingly common due to the spread of containerization which has facilitated the quick transfer of freight from sea to rail or from rail to road. So, Dry ports can therefore play an important part in ensuring the efficient transit of goods from a factory in their country of origin to a retail distribution point in the country of destination (RosoV 2009).

The Persian Gulf has an area of approximately 240,000 km2 and is very shallow, averaging just 50m-80m (1994; 1997), with only one opening– the Strait of Hormuz linking the Persian Gulf with the Arabian Sea. There are eight littoral Gulf States – Iran, Iraq, Kuwait, Saudi Arabia, Emirates, Bahrain, Qatar and Oman. The establishment of a shared place as dry port for all the above Gulf States will improve maritime transportation at the Persian Gulf. Based on IMO and WTO and the other relevant International regulations, it is necessary to consider the experiences of the container terminals operation at the Persian Gulf ports. It is useful for specifying the hub of container terminals at the Persian Gulf for further consideration. The dry port concept is an intermodal transportation system. The dry port itself is an inland intermodal terminal with additional services located inland. It is directly connected by rail to seaport or in some cases two or more seaports. In a dry port concept the maximum possible amount of freight transportation is accomplished by rail between the dry port and the seaport. Only the final leg of the door-to-door transportation is carried out by road transport. In an optimal dry port implementation the whole freight transportation between seaport and dry port is carried out by rail. However, that is usually not possible due to capacity of rail connection. (Roso.V, 2009)

The dry port offers value-creating services (e.g. consolidation, storage, depot, maintenance of containers and customs clearance) to actors which operate within the transportation system i.e. there is a whole range of administrative activities that could be moved inland with implementation of a dry port. Outsourcing activities from seaport to dry port relieves seaport, and hence seaport can concentrate in its core tasks and competencies.

According to most recent literature, dry ports are categorized into three different categories. They are close dry port, midrange dry port and distant dry port. The distances of different dry port are their location from seaport. Close dry ports are located approximately 50 kilometers from seaport. Distant dry ports are located 500 km or over from seaport.

4.1 Transit trade trends

There are insufficient official statistics available to establish the real volume of Afghan or Iranian transit trade via either country. However, according to local traders, there has been a steady increase in transit trade between the Islamic Republic of Iran and Afghanistan in recent years. Although Pakistani transit routes are shorter and somewhat cheaper for trade, Afghan traders have frequently favored the Iranian transit trade route, since it presents fewer potential dangers (Roso.V, et al, 2002).

There are no accurate Afghan-Iranian transit trade figures available. However, overall levels of Afghan transit trade via the Islamic Republic of Iran can be extrapolated from trade flows between Afghanistan and those markets that it gains access to as a result of the Afghan-Iranian transit agreement. For instance, Afghanistan conducts a significant amount of trade with countries in the Persian Gulf, China and Europe by transiting Turkey and using Iranian seaports at Bandar Abbas and Chabahar. Between 2004 and 2010, Afghan exports to China surged from US$ 860,000 to US$ 3.3 million.91 Exports to countries in the Middle East and North Africa rose from US$ 7.6 million in 2004 to US$ 18.5 million in 2010 and exports to the euro-zone increased from US$ 29.8 million in 2004 to US$ 38.7 million in 2010.92 Imports from each of those regions and countries to Afghanistan have also increased since the implementation of the new Afghan-Iranian transit trade agreement. For example, imports from the euro-zone more than doubled between 2004 and 2010, while Chinese imports more than tripled during the same period. It can therefore be assumed that Afghan transit trade levels via the Islamic Republic of Iran have also increased since the implementation of the agreement in 2005 (Ng, A.K.Y.and Tongzon, J.L, 2010).

4.2 *Transportation and customs regulations*

In accordance with the transit trade agreement, trucks from Afghanistan and the Islamic Republic of Iran are fully permitted to transit both countries' respective territories and are not required to pay any fees at road passes. while transit trucks are not required to pay customs duties at BCP in Afghanistan, they are instead required to pay tariffs at Herat. However, in order to avoid customs duties, transit trucks from the Islamic Republic of Iran often bypass Herat or simply offload their goods once they have crossed the Afghan border. As a result, transit trucks also avoid any additional cargo inspections within Afghanistan and thereby enhance the risk of chemical precursors being successfully smuggled into the country via the Islamic Republic of Iran. The only information that Afghan and Iranian transit trucks are required to provide is country of origin and country of export information on customs declaration forms. In Afghanistan, Herat is the main entry point for Iranian transit goods. Afghan and Iranian transit goods that are sealed in containers at Herat must be checked at all Afghan border crossings. At Bandar Abbas, only documents for outgoing containers are monitored (Padilha.P, et al, 2011).

4.3 *Dry ports along transit routes*

Overall, there are six major dry ports located along the three transit trade routes from the Islamic Republic of Iran to Afghanistan. Those are at Bandar Abbas, Chabahar, Iranshahr and Zahedan in the Islamic Republic of Iran and Dogharoun and Nimroz in Afghanistan.

Chabahar dry port receives 300-350 trucks and containers daily. They mainly carry foodstuffs, fresh fruit, cloths, cars, spare parts, fuel and construction materials such as cement and chalk. Most of those goods come from countries in the Persian Gulf. Foodstuffs, cloths, rugs, medicine, construction materials and spare parts for motorbikes are imported along the route to Afghanistan from China. At Dogharoun dry port, containers are checked extensively for drugs, with the help of sniffer dogs, x-ray machines, acetic anhydride test kits and night vision equipment (Roso.V, et al 2010).

5 AN EVALUATION OF SERVQUAL METHOD

This model developed by Cronin & Taylor, (1992) which is good to measure service quality but does not provide information on how customers will prefer service to be in order for service providers to make improvements. Teas, (1993), developed the Evaluated Performance model which measures the gap between perceived performance and the ideal amount of a dimension of service quality, rather than the customer's expectation. This was to solve some of the criticism of some previous models Gronroos, (1984); Parasuraman et al., (1985, 1988). Parasuraman et al., (1985), developed a model of service quality after carrying out a study on four service settings: retail banking, credit card services, repair and maintenance of electrical appliances, and long-distance telephone services. The SERVQUAL model represents service quality as the discrepancy between a customer's expectations of service offering and the customer's perceptions of the service received Parasuraman et al., (1985).

SERVQUAL was developed in the mid-1980s by Zeithaml et al. SERVQUAL means to measure the scale of Quality in the service sectors. The SERVQUAL authors originally identified 10 elements of service quality. (1) reliability; (2) responsiveness; (3) competence; (4) access; (5) courtesy; (6) communication; (7) credibility; (8) security; (9) understanding/knowing the customer; (10) tangibles. Later on the ten elements were minimized into the following five factors: (1) Reliability. (2) Assurance. (3) Tangibles. (4) Empathy. (5) Responsiveness. The willingness to help customers and to provide prompt service Businesses using SERVQUAL to measure and manage service quality deploy a questionnaire that measures customer expectations of service quality in terms of these 5 dimensions, and their perceptions of the service they receive. When customer expectations are greater than their perceptions of received delivery, service quality is deemed low. In this research, a regular SERVQUAL questioner consisting of two segments of individual characteristics and the five service quality dimensions which used as data collection. The SERVQUAL dimensions/items are main variables used in this study, and then the author coded these dimensions/items in order to ease the analysis of data collected. Demographic information was collected from respondents and these variables have to be coded as well for analysis.

6 THE STRATEGY OF CONTROL OVER TRANSIT CONTAINERS IS REINFORCED BY IRAN

Bandar Abbas Shaheed Rajaee port officially joined the UNODC-WCO brokered Container Control Program in February 2012, following the establishment of a Container Control Unit and training of personnel on profiling high risk containers. Shaheed Rajaee is the country's biggest port, handling more than fifty percent of the country's trade, including incoming, and outgoing and transit containers. As a result of its geographic location, it represents a strategic hub for transit containers going to and coming from Afghanistan.

The Container Control Program (CCP) assists governments to establish specialized units in seaports and dry ports with the aim of improving control over containers carrying illicit cargos such as drugs, precursors, explosives, and counterfeits. The consistent increase in use of containers worldwide makes it extremely important to focus on high risk maritime cargos while facilitating the legitimate flow of trade. For doing so, the best way to stem the illegal traffic is to maximize the efficiency of enforcement in picking out high risk containers for inspection. This goal can be achieved through the establishment of inter-agency container profiling units, which CCP seeks to promote, in seaports and dry ports.

In the framework of its program of Technical Cooperation on Drugs and Crime in the Islamic Republic of Iran (2011-2014), UNODC and WCO jointly organized a two-week training course for the personnel of the Bandar Abbas Container Control Unit on profiling of containers. During the training, participants were familiarized with various risk indicators related to the trade in containers, and learned how to backtrack and profile suspicious containers using open source data. In addition, the participants were trained on how to identify the high risk container, evaluate the frequency of the risk, look for options to treat the risk, and finally treat the risk with the best possible option using the available resources.

"Bandar Abbas Container Control Unit will hopefully contribute to increase considerably the drug seizures of Iranian law enforcement" said Mr.Antonino De Leo, UNODC Representative in Iran during the closing ceremony. In 2009, Iranian law enforcement seized around 89% of total opium seized in the world. The Heroin seizure of Iran also stands at a record of 41% of global heroin seizure in 2009. The regional counter-narcotics cooperation is expected to get enhanced by establishing new profiling units in the region and empowering the existing ones. Capable staffs of experienced units are often used to train the personnel of newborn units. This is an outstanding example of sustainable capacity building in law enforcement cooperation (United Nation Office on Drugs and Crime Report, 2011).

The Container Control Program has operational profiling units in countries such as Cape Verde, Costa Rica, Ecuador, Ghana, Pakistan, Panama, Senegal, Turkmenistan, and is being expanded to 21 additional countries in four continents. The Iran section of the program, which falls within the EC-ECO project "Fight against illicit drug trafficking to/from Afghanistan", is financed by the European Commission.

7 CONCLUSIONS

International intermodal terminals and freight villages usually have warehouse and distribution centers in Iran based along with the railway stations. Therefore, the cost of space, handling and associated warehousing costs involved with the actual storage of the product. It should be noted that the 'dry port' concept will continue to evolve as it is increasingly applied across the globe as a response to the challenges facing modern logistics in general, and ports and their hinterlands in particular. This study illustrates lack of support for the discriminate validity of SERVQUAL which is related to the factor analysis. In addition, the SERVQUAL model offered a satisfactory level of overall reliability. From the gap score analysis implemented, it was found that, the overall service quality is medium as alleged by consumers of the Iranian dry ports. Data from the study show that, the Iranian dry ports have to improve performance on all the dimensions of service quality in order to increase customer satisfaction in order to let them maintain high level of competitiveness.

The following four development stages need to be considered for improving the Iranian dry ports and Transit activities in international logistics. First, they would be arranging to receive containers or to work with foreign freight forwarders; second, export shipments, or negotiating with overseas suppliers of transportation services; third, overseas distribution, setting up just-in-time (JIT) suppliers; or production scheduling, and finally, would be integrating operations planning and control worldwide or participating in strategic decisions worldwide.

REFERENCES

1. Bergqvist.R, et al (2010) Establishing intermodal terminals. International Journal of World Review of Intermodal Transportation Research (WRITR) 3(3): 285–302.
2. Bergqvist, R. et al. (2011) The development of hinterland transport by rail – The story of Scandinavia and the Port of Gothenburg. Journal of Interdisciplinary Economics 23(2): 161–177.
3. Cordeau.J, et al (2001) A unified tabu search heuristic for vehicle routing problems with time windows. Journal of the Operational Research Society 52: 928–936.
4. Cullinane.K.P.B. (2010) Revisiting the productivity and efficiency of ports and terminals: Methods and applications. In: C. Grammenos (ed.) Handbook of Maritime Economics and Business. London: Informal Publications, pp. 907–946.
5. Do, N.H, et al (2011) A consideration for developing a dry port system in Indochina area. Maritime Policy and Management 38(1): 1–9.
6. Haralambides.H, et al (2011) On balancing supply chain efficiency and environmental impacts: An eco-DEA model applied to the dry port sector of India. In: K.P.B. Cullinane, R. Bergqvist and G. Wilmsmeier (eds.) Maritime

Economics and Logistics, Special Issue on Dry ports 16: 122–137.

7. Iannone.F, (2011) A model optimizing the port-hinterland logistics of containers: The case of the Campania region in Southern Italy. In: K.P.B. Cullinane, R. Bergqvist and G. Wilmsmeier (eds.) Maritime Economics and Logistics, Special Issue on Dry ports 40: 33–72.

8. Ng, A.K.Y. and Tongzon, J.L. (2010) Transportation improvements as a catalyst for regional development in India: The role of dry ports. Eurasian Geography and Economics 51(5): 669–682.

9. Padilha.P, et al (2011) the spatial evolution of dry ports in developing economies: R. Bergqvist and G. Wilmsmeier (eds.) Maritime Economics and Logistics, Special Issue on Dry ports 23: 99–121.

10. Roso.V, et al (2010) A reviews of dry ports. Maritime Economics and Logistics 12(2): 196–213.

11. Roso.V, et al (2009) the dry port concept: Connecting container seaports with the hinterland. Journal of Transport Geography 17(5): 338–345.

12. Sofia Isberg (2010), Using the SERVQUAL Model to assess Service Quality and Customer Satisfaction, Umea School of Business.

13. UNCTAD. (1991) Handbook on the Management and Operation of Dry Ports. Geneva: United Nations Conference on Trade and Development.

14. UN ECE. (1998) UN/LOCODE – Code for Ports and Other Locations. Recommendation 16, Geneva.

15. UNESCAP. (2009) Development of Dry Ports. Bangkok: UNESCAP Transport and Communications Bulletin for Asia and the Pacific.

16. Yousefi.H, et al (2011), Balanced Scorecard: A Tool for Measuring Competitive Advantage of Ports with Focus on Container Terminal, published in Vol.2, No.6, December 2011, IJTEF.

A Study on Rapid Left-turn of Ship's Head of Laden Cape-size Ore Carriers while Using Astern Engine in Harbor

T.G. Jeong
Korea Maritime & Ocean University, Busan, South Korea

K.H. Son
Association of Pohang Harbor, Pohang, South Korea

S.W. Hong
Graduate School, Korea Maritime & Ocean University, Busan, South Korea

ABSTRACT: Ships have been getting progressively larger due to the rapid development of ship-building technology and the increase in the quantity of cargoes transported. Because large-sized ore carriers entering the port of Pohang are getting even larger, shipping companies and owners of cargo request that the minimum allowable under-keel clearance be smaller than before. At the present time, the maximum size of ships entering Pohang harbor is 250,000 deadweight tonnes. This is expected to increase in the near future. The authors and other pilots in the Pohang harbor area have experienced that when cape-size ore carriers approach the No. 10 berth at Pohang Port, they tend to turn left rapidly whenever the astern engine is used. This phenomenon is quite different from that of the text book maneuvering, which would suggest that the ship's head should turn starboard.

This paper seeks to discern the causes of the left-steering phenomenon by considering the factors that have an effect on the ship's mobility. The main factor is little under-keel clearance. The left-circle discharging current caused by the reverse turn of the propeller does not flow under the hull bottom and accumulates at the sea bottom. The repulsive or repelling force by the accumulation of discharging current is much larger than the later force of the right-circle discharging current. The variables of increasing the repulsive force are the block coefficient of ship's stern, the speed used at that time, the engine order used, and the thrust produced from the engine. The bank effect can also be considered because of the short distance between the ship's side and the pier.

A future study requires using a water tank or computer simulation to test various under keel clearances. i.e. h/d.

1 INTRODUCTION

Ships have been getting progressively larger due to the rapid development of ship-building technology and the increase in the quantity of cargoes transported. Because large-sized ore carriers entering the port of Pohang are getting even larger, shipping companies and owners of cargo request that the minimum allowable under-keel clearance be smaller than before. At the present time, the maximum size of ships entering Pohang harbor is 250,000 deadweight tonnes. This is expected to increase in the near future.

The authors and other pilots in the Pohang harbor area have experienced that when cape-size ore carriers approach the No. 10 berth at Pohang Port, they tend to turn left rapidly whenever the astern engine is used. This phenomenon is quite different from that of the text book maneuvering, which would suggest that the ship's head should turn starboard.(Honda, 2008; Inoue, 2011; Rowe,1996; Yun, 2012).

This paper seeks to discern the causes of the left-steering phenomenon by considering the factors that have an effect on the ship's mobility. That is, this paper is to just suggest the possible causes of rapid left-turn preceded by astern engine based on the circumstances.

This paper is to deal with the overview of the environmental conditions including the cape-size ore carrier and berth concerned and the investigation of each causal factor that can contribute to the rapid left-turn.

2 SHIP'S LEFT-TURN AFTER USING ASTERN ENGINE

The authors selected a case in which ore carriers approaching the No. 10 berth at Pohang Port the rapid left-turn of ore carrier while they approached the No.10 berth for iron ore.

2.1 Ship's particulars of ore carriers

The ship's particulars of ore carriers entering the port of Pohang are given in Table 1. The ore carriers were cape-size, whose deadweight tonnage ranged from 186,330~245,609 metric tons. The maximum draft was 17.40 m. The stated depth of Pohang port is 19.50m. Thus, the under-keel clearance was 2.10m. However, we expect that the actual UKC is less than 2.10m because dredging has not been done for a long time.

Table 1. Ship's particulars of ore carriers

Name	DWT	Draft	LOA	Beam
HJ Melbourne	188,125	18.02	291.50	48.00
HJ Gladstone	207,390	18.02	309.00	50.00
O. Universe	245,609	19.22	316.30	53.00
K. Camellia	207,874	18.01	311.97	50.00
K. Cosmos	242,300	19.02	315.50	58.00
HD. Olympia	186,330	18.01	291.50	48.00

2.2 Docking at No. 10 berth & left-turn

Figure 1 shows that when an ore carrier approaches the No. 10 berth, the ship is turns to the left rapidly when using astern engine.

At position No.1 the ship approaches at a speed of 5 or 6 knots, where the rudder can have effect of controlling the ship. At the moment four tugs are taken to the ship; two fore, and two aft. To reduce ship's forward thrust all tugs are pulling the ship in 6 o'clock direction. At position No.2 the ship takes the starboard rudder and then stops the engine to proceed to position No.3.

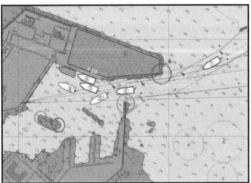

Figure 1. An ore carrier docking at No. 10 berth in the Port of Pohang and her left-turn

Because the distance from the ship to the berth is very short, it is essential to use the astern engine. Therefore, when proceeding to position No.3, the ship uses the engine order of 'Slow Astern' or 'Half Astern'. Astern engine being used, the ship's head

then turns to the right (starboard), as it should, due to the effect of the starboard rudder. Moments later, the ship stops turning starboard and then turns to the left rapidly.

Turning to the left, the ship moves to position No. 3' (in red) and is put in such danger that she cannot control herself, relying entirely on the tugs. Therefore the ship stops the engine at once. And then, using the engine order of 'Slow Ahead' or 'Half Ahead', the ship uses the starboard rudder. Finally the ship stops turning to the left and is placed at position No.4' (in red).

2.3 Conditions of left-turn occurrence

According to the experiences which authors have had so far, the left-turn has occurred in the following conditions.

1 The allowance of depth
   The smaller the under-keel clearance is, the more frequently the unwanted left-turn occurs. When depth/draft =1.1, the situation happens more often. Since 2012, at the time of completing dredge, the left-turn phenomenon seldom happens.

2 The speed of ship
   When the ship's speed is more than 4 knots at position No. 2 of Figure 1, the left-turn phenomenon occurs. While the ship's speed is less than 3 knots, the phenomenon seldom occurs.

3 The engine used
   When the engine used is more than 'Half Astern', the left-turn phenomenon happens frequently. Under low speed, or less than 'Slow Astern', the unwater left-turn seldom occurs.

3 INVESTIGATION OF CAUSE OF RAPID LEFT-TURN BY USE OF ASTERN ENGINE

We can consider the cause of rapid left-turn dependent on the use of astern engine, including the conditions described in the preceding subchapter. The key factors of the cause are bank effect and small allowance of depth. The following factors are only subsidiary factors of describing the two key factors.

1 The ship's speed
2 The engine used
3 The thrust produced
4 The shape of stern bottom
5 The starboard rudder

Therefore we can investigate the possible causes of the rapid left-turn which occurs when the cape-sized ore carrier uses the astern engine.

3.1 Bank effect and shallow effect

Waterways that are shallow compared to the ship's draft and beam, and waterways such as narrow rivers or narrow canals are called 'restricted waterways.' Whenever a ship makes way through a restricted waterway, the motion of the ship is principally affected by shallow water. The effect of shallow water becomes larger by the narrower width. And also the bank effect occurs by the interaction between the ship's side and the side wall of the canal or bank of the river.

While a large ship is passing near the side wall at a slow speed, the hydrodynamic force of the bank effect, the magnitude of which is dependent upon ship's speed, the shape of the wall, the ratio of depth and draft (h/d), and the longitudinal and transverse distance of the wall. This bank effect, we expect, is one of main factors causing the unwanted left-turn phenomenon.

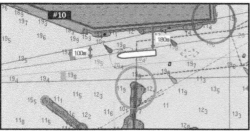

Figure 2. The transverse distance between the ship's side and the wall

As shown in Figure 2, the ship approaches No.10 berth at a transverse distance of 180m. The depth is very shallow compared to the draft, i.e. less than h/d=1.2. When the ship passes by the wall at a speed of more than 4 knots, the ship can turn to the left as a result of bank effect.

3.2 Small allowance of depth

The allowance of depth is given by the depth (h) and the draft (d). In this paper the draft is 17.40m and the depth is 19.50m. The ratio of depth and draft (h/d) is given as

h/d = 1.12

At the places where the marginal depth is too small, we can consider the repelling force by the discharging current of reverse propeller as follows.

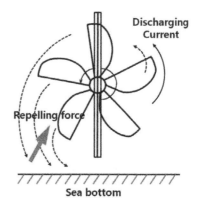

Figure 3. Repelling force by accumulation of discharging current

As shown in Figure 3, the discharging current of the right side of the screw cannot go forward because of the forward speed of 4 knots, and almost stops above the right of propeller. Whereas the discharging current of the left side of the screw strikes the sea bottom nearly at a right angle and can be accumulated at the sea bottom. The accumulated repelling force of discharging current acts on the port (left) bottom hull and propeller and the stern turns to the right. That is, the ship's head turns to the left. In fact we suggest that this accumulated discharging current is the main factor that the following sub-sections describes.

3.3 Ship's speed

At the ship's speed of more than 4 knots the flow of discharging current produced by reverse propeller cannot reach the starboard quarter and the lateral force of discharging current cannot be given to the starboard quarter.

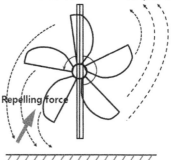

Figure 4. Discharging current on the right above the propeller

However, at the speed of less than 3 knots the flow of discharging current of reverse propeller can go forward and hit the starboard quarter. As shown

147

in the text book of maneuvering, the discharging current acts upon the starboard quarter and the stern turns to the left. That is, the ship's head turns to the right.

3.4 *Engine used*

If the engine used on the ore carrier is more than 'Half Astern', the thrust produced by reverse propeller is strong enough to make the flow of the discharging current hit the sea bottom on the left at a right angle. If it is less than 'Half Astern', the discharging current by reverse propeller flows afterward and does not accumulate.

3.5 *Thrust of engine produced*

The thrust of the engine can be a factor of deciding if a flow of the discharging current of reverse propeller is fast. If the thrust at the given engine order is large enough (as that of container ship) the majority of the discharging current flow can reach the quarter faster than the sea bottom. It is natural that the ship's head turns to the right.

However, in the case of small thrust at the given order of engine (as in the case of a bulk or ore carrier), the discharging current of reverse propeller accumulates at the sea bottom and causes the repelling force to turn the stern to the right, and thus the ship's head to the left.

3.6 *Shape of stern bottom*

In a ship with large block coefficient like an ore carrier, the discharging current does not smoothly flow along the hull bottom forward, but rather hits the sea bottom on the left. The discharging current causes a repelling force which in turn causes the stern to turn to the right.

Figure 6. Shape of hull bottom of ore carrier

Figure 7. Shape of hull bottom of container ship

3.7 *Starboard rudder used*

The rudder used is a factor of turning the stern to the right. Using starboard rudder while going forward, the ship turns to the right. And then, even if the ship uses the astern engine, the ship goes forward. However the normal rudder pressure forms around the rudder by reverse propeller. The composition of force vectors of forward thrust and normal pressure in the ship's stern is given by Figure 7. Finally the ship turns to the left.

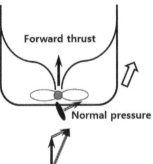

Figure 5. Ship's turning of starboard rudder used

4 DISCUSSION

In Section 3 we investigated the seven possible factors of unwanted or unintended rapid left-turns preceded by astern engine, based on the factors present when this phenomenon occurs. Of them we can expect that the main factors are bank effect and accumulated repelling force caused by the screw's discharging current. The bank effect is shown in the text book of maneuvering. The repelling or repulsive force is suggested only here. It is not yet verified by experiment or simulation. Other factors are to describe the two main factors, especially the repelling or repulsive force by accumulation of discharging current.

5 CONCLUSIONS

This paper was to investigate the phenomenon of rapid left-turn of cape-sized ore carriers in the port of Pohang. To do it, we, the authors, made a thorough investigation into the conditions the phenomenon when these unwanted rapid left-turns occurred. Based on the conditions, the causes of the rapid left-turns were analyzed and this hypothesis was developed. As a result of this analysis, the main factors have been determined to be bank effect (and shallow effect) and accumulated repelling force of the screw's discharging current. We suggest that the latter is the more important of the two factors. The following are variables that constitute the repelling force.

1 The ship's speed
2 The engine used
3 The thrust of engine produced
4 The shape of stern bottom
5 The starboard rudder

Therefore the phenomenon of rapid left-turn is composed of two reasonable simple factors that are themselves composed of several variables. The phenomenon of rapid left is quite different from that as shown in the traditional text books. It therefore requires further study, including testing models in a water tank or computer simulation, in order to explain why the ships violate normal laws of motion by rapidly turning the ship's head to the left when the ship's rudder, and normal expectation, indicate that the ship's head should turn right.

REFERENCES

[1] Inoue, K. Z.(2011), Theory and Practice of Ship Handling, Seisando, pp. 79~80.
[2] Honda, K, N.(2008), Introduction to Ship's Maneuvering, Seisando, pp. 46~47.
[3] Rowe R.W.(1996), THE SHIPHANDLER'S GUIDE for Masters and Navigating Officers, Pilots and Tug Masters, THE NAUTICAL INSTITUTE, p.23.
[4] Yoon, J.D.(2012), Theory and Practice of Ship's Maneuvering, Sejong Publisher, pp. 128~132.

Port Facilities

Port Facilities
Safety of Marine Transport – Marine Navigation and Safety of Sea Transportation – A. Weintrit & T. Neumann (eds.)

A Geographical Perspective on LNG Facility Development in the Eastern Baltic Sea

D. Gritsenko
Center for Maritime Studies, Brahea Center, University of Turku, Finland

A. Serry
Centre d'études pour le développement des territoires et l'environnement, University of Orleans, France

ABSTRACT: This paper presents an overview of the liquefied natural gas (LNG) facilities development in the eastern Baltic Sea. The mapping of gas supply networks, LNG facilities and traffic patterns establish originality of this research. The paper shows that factors motivating the development of LNG terminals in eastern Baltic ports come from areas of energy and maritime policy. For the first area, willingness to diversify national gas supply stemming from the energy policy prompts the development of LNG import facilities. For the other, the need to provide SECA-compliant marine fuel favors the uptake of LNG in ports. The investigation shows that to date the development of LNG terminals is a priority at the national (state) rather than local (port) scale. Moreover, in future the emerging LNG infrastructure may have an effect upon port competition in the eastern Baltic range.

1 INTRODUCTION

This paper aims at giving an overview of the current situation and near-future developments of the LNG facilities in the eastern Baltic port range. The investigation draws upon the body of institutional theory devoted to examining how outcomes generated in one policy area can influence or change the rules under which policy outcomes are generated within the other policy situation (McGinnis, 2011). Since LNG infrastructure in ports can potentially serve two sectors (energy and maritime industries), the incentives to develop LNG terminals derived from one sector may affect the technology uptake in the other. In particular, the paper seeks to understand whether the emerging terminals reflect the maritime trade patterns in the region and have a potential to support a network of LNG bunkering in the Baltic Sea.

The Baltic Sea is very transport-intense. Different types of cargoes and maritime transport technologies can be found in Baltic trade (Figure 1), although some restrictions stem from navigational limitations (shallow depths, archipelago, sea ice, high number of vessels at sea). The eastern Baltic port range has its own characteristics and a specific organization, in which the ports have linked similarities with the combination of competition, complementarities and cooperation (Serry, 2014). Port classification is emerging in this range dominated by three Russian

ports: multipurpose ports of St. Petersburg and Ust-Luga, and a specialized oil port in Primorsk. Except the Russian ports, the ports of Riga and Klaipeda are the most dynamic in the region, using the Russian proximity, but also different hinterland possibilities toward Belarus, Ukraine and Central Asia. Whereas ports of the Baltic States (Estonia, Latvia, Lithuania) are positioned as a gateway between Russia and Europe (Masane-Ose, 2013), the Finnish segment consists of two major ports (Helsinki and Kotka-Hamina) and a number of specialized harbors (Merk et al., 2012).

One of the latest developments within the eastern Baltic port range is an increased interest towards expansion of liquefied natural gas (LNG) facilities (Rozmarynowska, 2010; Liuhto, 2012). Whereas elsewhere in Europe and particularly in the North Sea LNG terminals are widely spread, the eastern Baltic ports have started planning activities only in 2010s. The first terminal in Klaipeda has become operational in late 2014. LNG terminal development seems favorable for the eastern Baltic Sea countries, as it will allow decreasing energy dependency by increasing gas supply diversification, as well as providing regional shipping with an alternative type of low-emission marine fuel.

A geographical approach is used to investigate where future LNG terminals will be constructed and how their location may affect the dynamics between the ports in the range. The mapping of gas supply networks, LNG facilities and traffic patterns establis originality of this research. Mapped maritime and port reality is a useful tool to put under scrutiny commonly accepted assumptions regarding the LNG facility development and provide a visual image of connecting relation between energy and shipping in the Baltic Sea.

The paper proceeds as follows. Section 2 presents background information on LNG technology, LNG terminal development and LNG for maritime use. Section 3 introduces method and data. Section 4 focuses on the eastern Baltic port range and presents paper's findings. Section 5 discusses the current situation and near-future outlook, and concludes.

2 BACKGROUND

Natural gas is an important source of energy in Europe in general and in the Baltic Sea region in particular (Eurogas, 2014). In the eastern part of the Baltic Sea, Russia is a net gas exporter, whereas the four other countries (Estonia, Finland, Latvia, and Lithuania) are net gas importers, until recently fully dependent on Russian supplies (Table 1).

The recurrent EU-Russia energy crises make the issue of energy dependency particularly relevant and the desire for diversification particularly strong (Liuhto, 2013). By building capability to import LNG, the gas importing countries will have an opportunity to create complex multi-supplier portfolios (Rozmarynowska, 2012), including gas import from Qatar, Malaysia, Algeria and other major LNG producers (IGU, 2013).

Table 1. Gas supply in the Baltic States, Finland, and EU 28*

Country	Norway	Russia	other	total net
Estonia	0	7	0	7
Finland	0	36	0	36
Latvia	0	15	0	15
Lithuania	0	28	0	28
EU 28	1699	1332	917	4996

* Source: Eurogas statistical report 2014.

Today the logistics of natural gas in Europe is largely dependent on grid connections (gas pipelines) (Figure 2). Gas pipelines remain cost efficient way of gas transportation, but they require extensive initial investment. Moreover, they are spatially bound and do not allow flexibility in supply choices. The liquefied natural gas, an increasingly popular form of storage, transportation and use of natural gas, has a potential to revolutionize the future gas logistics (Hayes, 2007). LNG allows a greater volume to be stored at smaller facilities as it takes up about 1 : 600 the volume of natural gas and requires only 1/3 of the volume occupied by compressed natural gas (CNG), the traditional form of gas storage.

LNG technology also offers an alternative way to transport natural gas. LNG can be transported in tanks by truck, train, or ship, subsequently stored and/or regasified at the final destination. Finally, LNG can be used as transportation fuel that offers environmental advantages in comparison to oil-based fuels (i.a., lower air emissions, Figure 3).

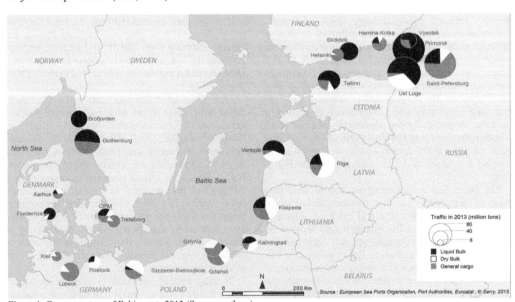

Figure 1. Cargo structure of Baltic ports 2013 (Source: authors)

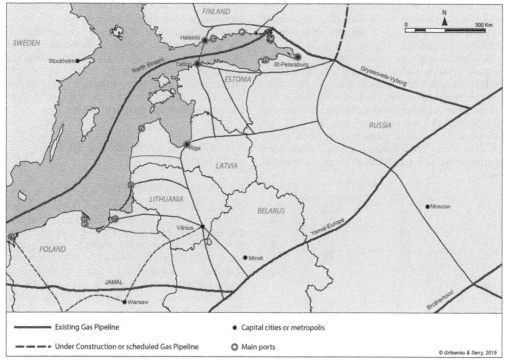

Figure 2. Gas pipelines in Europe (Source: authors)

Though LNG technology has been available for several decades and LNG has been in regular use since 1960s, its role in natural gas logistics remained marginal until the so-called 'shale gas revolution' (Stevens, 2010). In Europe, the escalation of political tensions between major gas-producing and gas-consuming countries increased attention to investment into the LNG technology. Though LNG outperforms pipeline grids when it comes to easy flexible transportation of variable volumes and convenience of storage, the development of LNG infrastructure is costly (Bengtsson et al. 2011).

As most part of LNG is supplied by sea, construction of LNG terminals in ports has become central to development of LNG infrastructure all over the world (Andresson et al., 2010). LNG terminals are complex infrastructural installations with multiple functions (reception of the LNG tankers, discharging of the LNG cargos, tanking and storage, regasification and injection to pipeline system, supplying for further use as fuel, including bunkering). Some LNG terminals are built as extension of existing port facilities, whereas others are built purposefully as new ports with focus upon LNG import/export functions. Types of LNG terminals range from onshore installations (tanks and (re)gasification facilities) to offshore solutions (floating storage barges and (re)gasification plants on an artificial island).

Overall, the outlook for LNG trade is positive as global consumption prompted by several potential LNG uses is set to increase. The UNCTAD Maritime Transport review 2014 indicated that global LNG shipments are expected to rise by 5.0% in 2015.

The analysis of literature has drawn attention to two major drivers for the development of LNG facilities in the eastern Baltic Sea ports. First, LNG is seen as a way out of energy dependence by diversification of gas supply through allowing more flexibility in natural gas logistics (Liuhto, 2013). The second rationale for increased interest to LNG in the Baltic Sea is its potential to serve as a marine fuel compliant with the latest air emission regulations (Burel et al., 2013). From 1.1.2015 Baltic and North Seas were designated as Sulphur Emission Control Areas (SECA) under MARPOL Annex VI, resulting in more stringent regulation on air emissions applied in these areas. Also the EU sulphur directive (2012/33/EU) aims at ensuring a substantial reduction of SO_x in ship exhaust to the benefit of coastal communities and the marine environment. In order to meet the SECA regulation, several compliance options have been proposed: a) use of low-sulphur fuel (MDO/MGO); b) use of exhaust gas cleaning systems (scrubber); (c) use of LNG (Kalli et al., 2009, EMSA, 2010, Bengtson et al., 2011). LNG seems to be a viable option as it offers a number of advantages with respect to conventional maritime fuels, including significant reduction of SO_x, NO_x, CO_2 and PM emissions in ship exhaust (Figure 3).

Notwithstanding the environmental benefits, academic studies indicate several problems with LNG as an alternative marine fuel. In particular, needs for vessel retrofitting, lack of infrastructure, as well as other regulatory and technical uncertainties are identified (Acciaro and Gritsenko, 2014). It also shall be noted that some ports may be better suited than others for LNG bunkering (e.g., emergency response might prove difficult in densely populated areas).

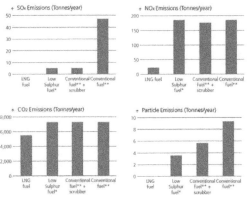

Figure 3. Comparison of exhaust emissions from fuels (Source: DNV, 2010).

The regulatory uncertainties are expected to improve with the recent adoption of the IGF Code that provides mandatory provisions for the arrangement, installation, control and monitoring of machinery, equipment and systems using low-flashpoint fuels, among others LNG (World Maritime News, 27.11.2014). The IFG Code will help to minimize the risk to the ship, its crew and the environment, having regard to the nature of the fuels involved. Also the EU has prepared LNG rules, and companies worked out their codes. Also some of the technical issues (including safety and efficiency issues in liquefaction/regasification, bunkering process and vessel modifications) are being rapidly resolved (Järvi, 2010; Einang, 2011).

3 DATA AND METHOD

The research was carried out as a combination of desktop research (review of existing literature, collection of publicly available data on transport/handling volumes, energy interdependence, and LNG facilities in eastern Baltic Sea port range) and of geographical visualization approach (mapping the links between infrastructural objects and the traffic flows in establishing LNG-related networks).

Methods used also include cartographic and spatial contextual analysis. Academic journals, official published statistical data, periodicals, newspapers and Internet sites of different stakeholders (e.g., ports, operators and terminals) have been used as sources of information. Software used is Mapinfo and Adobe Illustrator.

4 LNG IN THE EASTERN BALTIC PORT RANGE

4.1 *The eastern Baltic port range*

The maritime activity in the Baltic Sea grew considerably since the mid-1990s. The overall traffic has almost doubled since 1997 (Figure 1), whereas the growth of the world seaborne trade over the same period was approximately 65%. Increase in maritime transportation has prompted an unprecedented growth of the Baltic ports. The compound annual growth rate in cargo handling from 1995 to 2010 was 3.2%.

Four features define the dynamics of maritime traffic in the eastern Baltic port range: (1) increasing traffic of oil based on Russian exports, (2) the upward movement of containers, (3) the growth of the intra-Baltic Ro-ro flows, and (4) the concentration of traffic in some ports (in particular, in Russia). Liquid bulk (primarily, crude oil, oil products, and chemicals) represent almost 40% of all Baltic traffic, thereof nearly 60% is export of Russian-origin oil through the eastern Baltic ports (Primorsk, St. Petersburg, Ust-Luga and Vysotsk, but also in transit through Tallinn/Muuga, Ventspils and Klaipeda).

The second largest group of cargoes is dry bulk (coal, iron ore, grain as well as fertilizers) which account for almost than 30% of all transports and has a particularly important role in the structure of the port traffic on the Eastern shore. In 2012, the number of containers handled among Baltic Sea ports amounted to 9.4 mln TEUs, which equals to an annual growth of 7%.The amount of containers shipped in the Baltic Sea is determined by the proximity of consumer markets with Russia being the key destination point of containerized cargo. To encourage the container traffic, ports and railway companies are organizing block trains between the Baltic States and the CIS, but also with Afghanistan and China. Moreover, due to geography of demand, most Baltic ports cannot reach the volumes of large Western Europe ports, which act as transshipment hubs and the leading position in container segment is thus taken by the port of St. Petersburg (Figure 4).

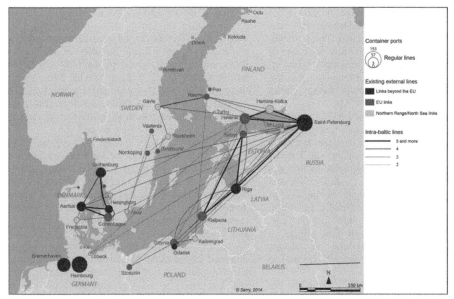

Figure 4. Baltic Sea containerization network in January 2014 (Source: authors).

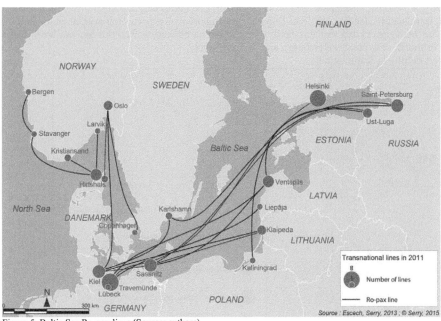

Figure 5. Baltic Sea Ro-pax lines (Source: authors).

In eastern Baltic ports short sea shipping lines and feeder ships, which transport cargo from/to the hub ports of Europe (Rotterdam, Antwerp, Bremerhaven, and Hamburg), are common. Intra-Baltic shipping, mainly relying on Ro-ro vessels and ferries which cover up to 80% of total internal Baltic traffic, constitutes an important part of regional maritime activity (Figure 5). Intra-Baltic services clearly mark a paradox: whereas the Baltic economy becomes increasingly globalized, its transportation is regionalizing.

4.2 Mapping the LNG infrastructure in the eastern Baltic Sea

Figure 6 presents a map of LNG facilities in eastern Baltic ports as of 1 January 2015. It shows that LNG terminals are a new development in this region. In

157

autumn 2014 the first terminal, the floating LNG barge with a symbolic name "Independence", was commissioned in Klaipeda, Lithuania. The terminal, leased from Norway's Hoegh LNG at a daily rate of USD 189,000, has the capacity to supply 4 bln m³ gas annually, covering not only Lithuanian needs but also about 80% of the total Baltic States consumption (Dailymail, 27.10.2014).

In Finland, active expansion of LNG is expected: until 2019 five terminals shall be constructed (Inkoo, Turku, Rauma, Pori and Tornio in the Bothnian Gulf). After long negotiations, Finland and Estonia have agreed on the details of a liquefied natural gas pipeline between Finland and Estonia (BalticConnector) and a terminal "Finngulf" in Inkoo, meaning the project can finally go ahead (launch target has been set for 2019) (Yle, 18.11.2014). In conjunction with this project, a small-scale terminal may be built in Muuga (Postimees, 20.11.2014). In Latvia, one terminal in Riga is under consideration.

In Russia, at least three LNG terminals are expected in the nearest future: a terminal for production and transshipment with capacity 660,000t annually in Vysotsk (Portnews, 26.11.14), a 10 mln t production facility in Ust-Luga (previously, it has been speculated that this facility will be built in Primorsk or even St. Petersburg), and a 2,4 mln t import facility in Kaliningrad.

5 DISCUSSION AND CONCLUSIONS

Two critical factors can be considered as the main drivers to develop LNG facilities in the eastern Baltic Sea: 1) willingness to diversify gas supply infrastructure; and 2) need to provide low emission (in particular, SECA-compliant) marine fuel. The geography of LNG facilities development in the eastern Baltic Sea region suggests that to date the first factor prevailed over the second one.

As LNG market is demand driven, LNG import/export and storage facilities in ports are built to meet existing or emerging demand rather than to create it. Current demand for LNG in eastern Baltic has been related to energy policy goals of supply diversification. In particular the 2014 Ukrainian events resulting in mutual sanctions between the EU and Russia have created even stronger motivations to lessen the interdependences between parties involved in gas trade. Whereas LNG terminals will allow importing countries (Estonia, Finland, Latvia and Lithuania) to be less dependent on Russian supply, Russia expects to find new customers by shipping its gas to alternative destinations. As a result, LNG terminals add flexibility both in terms of gas import and export. However, their potential to expand as bunkering facilities has not been utilized extensively.

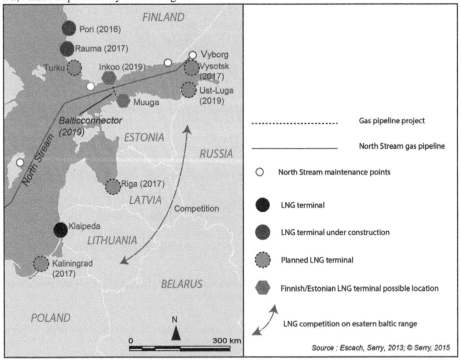

Figure 6. LNG facilities development in the eastern Baltic ports, as of 1.1.2015 (source: authors)

Market uptake of LNG as a bunker fuel in the Baltic Sea has been slow. It can be regarded as a typical "chicken and egg problem": absence of bunkering facilities undermined investment in LNG-powered ships, whereas lack of demand for LNG reduced incentives to develop LNG fuelling facilities. The uncertainties regarding rules, technical standards and future demand have further undermined private investment and limited the role for port authorities. Emerging LNG terminals in the eastern Baltic are mostly a matter of political decision-making. Public investment (including state-dominated energy companies and the EU funds) has been used to finance the projects. Yet, the ongoing development of LNG facilities incentivized by energy political motives can be seen as an interference with the "chicken and egg problem" of LNG bunkering from an adjacent policy area, which reduces the interdependence and increases the potential to use LNG as a SECA-complaint bunker fuel in the near future.

As the SECA regulation entered into force on 01.01.2015, most of shipping companies in the Baltic Sea switched to low-sulphur fuel. Due to the drop in oil price this change has had a relatively mild effect in comparison to earlier predicted severe increase of bunker price. Yet, the order books contain many orders for dual-fuel vessels to operate in the Baltic (Lloyd's Register EMEA, 2014). For instance, LNG-powered ships have been ordered for Helsinki-Tallinn Ro-pax traffic (Postimees, 11.12.2014). In particular, LNG can be expected to occupy a significant niche as a bunker fuel for shipping within the SECAs.

The structure of Baltic maritime transport which is characterized by high volume of intra-Baltic shipping (see Section 4.1) is supportive of LNG to become an attractive SECA compliance option. The idea of transition to clean shipping using the LNG as the most viable low emission fuel is actively promoted within maritime sector (DNV, 2012) and through public support for development of bunkering infrastructure (e.g., "LNG in Baltic Ports" project financed by the EU). Additionally, the large availability of natural gas resulting in possibility of economies of scale and relatively low price can become stronger driver in the future, once the initial infrastructure has been built.

This investigation concludes that in the eastern Baltic range ports do not actively compete to attract LNG projects. Rather, impulses for LNG terminal construction and subsequent investment come from energy companies and public authorities. This could explain a mismatch between those ports where LNG can be expected to be needed first (ports involved into the intra-Baltic trade such as Riga, Helsinki, Hamina-Kotka, Saint-Petersburg, and Tallinn serving major Ro-ro/Ro-pax and feeder/container lines) and those ports where terminals are being built

(Rauma, Pori, Inkoo, Ust-Luga), exemplified by comparing Figures 4, 5, and 6.

One illustration can be derived by looking at Riga and Klaipeda. The two "mid-range" ports are rather similar in terms of international trade volume, well-connected in the container network (yet not major nodes), their Ro-pax operations are stable, but not as large as in some other ports of the eastern range (Figures 4 and 5). Due to their involvement into intra-regional trade and transshipment, LNG bunkering demand can be expected to grow both in Riga and Klaipeda. However, whereas an LNG terminal in Klaipeda has become operational in fall 2014, finalization of decision to build an LNG terminal in Riga is pending since 2009. It could be suggested that Latvia remained reluctant to build an own LNG terminal as Klaipeda terminal in Lithuania allows to fulfill its energy diversification goal without additional investment. Whereas energy political goals do not require a prompt expansion of LNG in the port of Riga, the existence of an LNG terminal in Klaipeda may impact its future development as a multi-purpose port and the newly acquired LNG infrastructure may give it a competitive advantage, among others over the port of Riga.

Even though to date the effect of LNG facilities development on port competition in the eastern Baltic range cannot be empirically proven, changes may be expected once a critical amount of vessels have switched to LNG bunkering. For instance in the North Sea, especially in Hamburg - Le Havre port range, port authorities are adopting incentive policies to promote investments in the LNG import/export, storage and bunkering facilities in order to improve own competitiveness and market contestability. Eventually, also Baltic ports may need to take more proactive position and try to acquire LNG infrastructure at an early stage of LNG maritime applications development. The absence or presence of bunkering facilities may affect competitiveness of ports located in SECA areas in the nearest future.

REFERENCES

Acciaro, M. & Gritsenko, D., 2014. LNG bunkering infrastructure development in the seaports. In Proceedings of IAME 2014 Conference, 15-18 July, Norfolk, USA.

Andersson, H., Christiansen, M., & Fagerholt, K., 2010. Transportation planning and inventory management in the LNG supply chain. In Energy, natural resources and environmental economics, 427-439. Springer Berlin Heidelberg.

Bengtsson, S., Andersson, K., Fridell, E., 2011, A comparative life cycle assessment of marine fuels liquefied natural gas and three other fossil fuels. In Journal of engineering for the maritime environment, 225: 97–110.

Burel, F., Taccani, R., & Zuliani, N., 2013. Improving sustainability of maritime transport through utilization of

Liquefied Natural Gas (LNG) for propulsion. In Energy, 57: 412-420.

Dailymail, 27.10.2014, Lithuania installs LNG terminal to end dependence on Russian gas. Available at: http://www.dailymail.co.uk/wires/reuters/article-2809932/Lithuania-installs-LNG-terminal-end-dependence-Russian-gas.html#ixzz3Q7YiuWlG.

DNV, 2010. Greener shipping in the Baltic Sea, Det Norske Veritas, June 2010, available at: http://cleantech.cnss.no/wp-content/uploads/2011/05/2010-DNV-Greener-Shipping-in-the-Baltic-Sea.pdf.

DNV, 2012. LNG - The new fuel for short sea shipping. Available at: http://www.dnv.pl/Binaries/LNG%20The%20new%20fuel%20for%20short%20sea%20shipping%20WEB_tcm144-520777.pdf.

Einang, P. M., 2011. LNG fuelling the future ships. In Marintek, SINTEF, Shanghai.

EMSA, 2010. The 0,1% sulphur in fuel requirement as from 1 January 2015 in SECAs. An assessment of available impact studies and alternative means of compliance. European Maritime Safety Agency, Technical report.

Escach, N. & Serry, A., 2013. Les ports de la Mer Baltique entre mondialisation des échanges et régionalisation réticulaire. Géoconfluence. Available at: http://geoconfluences.ens-lyon.fr/test/doc/transv/Mobil/MobilScient7.html

Eurogas, 2014. Eurogas Statistical report 2014. Available at: http://www.eurogas.org/uploads/media/Eurogas_Statistical_Report_2014.pdf.

Hayes, M. H., 2007. Flexible LNG supply and gas market integration: a simulation approach for valuing the market arbitrage option. Program on Energy and Sustainable Development at Stanford University, Working Paper.

IGU, 2013. World LNG Report 2013. International Gas Union, 2013. Available at: http://hcbcdn.hidrocarburosbol.netdna-cdn.com/downloads/online_version_world_lng_report_2013_edition_original.pdf.

Järvi, A. 2010. Methane slip reduction in Wärtsilä lean burn gas engines. 26th CIMAC World Congress on Combustion Engines, Bergen, Norway.

Kalli, J., Karvonen T. & Makkonen, T., 2009. Sulphur content in ships bunker fuel in 2015. A study on the impacts of the new IMO regulations and transportation costs. Publications of the Ministry of Transport and Communications, 31. Available at: http://www.lvm.fi/c/document_library/get_file?folderId=339549&name=DLFE-8042.pdf&title=Julkaisuja%2031-2009.

Liuhto, K., 2012. A liquefied natural gas terminal boom in the Baltic Sea region? Electronic Publications of Pan-European Institute 5/2012.

Liuhto, K., 2013. Liquefied Natural Gas in the Baltic Sea Region. Journal of East-West Business, 19(1-2): 33-46.

Masane-Ose J., 2013. Competitive Position of the Baltic States Ports. In Conference proceedings of Transbaltika 2013, available at: www.rms.lv/tb2013materials.

McGinnis, M.D., 2011. Networks of Adjacent Action Situations in Polycentric Governance. Policy Studies Journal 39: 51–78.

Merk, O., Hilmola O. & Dubarle P., 2012. The Competitiveness of Global Port-Cities: The Case of Helsinki, Finland, OECD Regional Development Working Papers, No. 2012/08, OECD Publishing.

Pan-European Institute 5/2012, available at: http://www.tse.utu.fi/pei.

PortNews, 26.11.2014. Construction of LNG terminal at Vysotsk seaport to commence in 2016. Available at: http://en.portnews.ru/news/191141/.

Postimees, 11.12.2014. Available from: http://news.postimees.ee/3023159/tallink-grupp-to-order-lng-powered-ship-for-tallinn-helsinki-route.

Postimees, 20.11.2014. Vopak LNG terminal alters Balticconnector course. Available at: http://news.postimees.ee/2999183/vopak-lng-terminal-alters-balticconnector-course.

Rozmarynowska, M., 2010. LNG in the Baltic Sea region opportunities for the ports. Zeszyty Naukowe Akademii Morskiej w Gdyni, 89-100.

Rozmarynowska, M., 2012. LNG import terminals in Baltic Sea Region–review of current projects. In Baltic Rim Economies 6, 19.12.2012, Expert article 1149.

Serry, A., 2014. Dynamics of maritime transport in the Baltic Sea: regionalization and multimodal integration. In Proceedings of the 6th International Conference on Maritime Transport, Barcelona, 25-27 June 2014.

Stevens, P., 2010. The 'Shale Gas Revolution': Hype and Reality. A Chatham House Report. London: The Royal Institute of International Affairs.

World Maritime News, 27.11.2014, IMO approves IFG Code. Available at: http://worldmaritimenews.com/archives/144628/imo-approves-igf-code/.

Yle, 18.11.2014. Gasum: Kaasuhankkeiden toteutuminen edellyttää riittäviä tukia. Available at: http://yle.fi/uutiset/gasum_kaasuhankkeiden_toteutuminen_edellyttaa_riittavia_tukia/7634688

Influence of "Suezmax" Tankers Size Increase on Mooring Ropes at Existing Terminals

R. Mohovic, M. Baric & D. Mohovic
Faculty of maritime studies, University of Rijeka, Croatia

ABSTRACT: Growth in maritime traffic results in larger vessels and larger main traffic channels. Due the latter reasons and in comparison with the cargo terminals, angle of mooring rope is not the same as when cargo terminals were built. The aim of the proposed research is to see how the increment of the "Suezmax" tankers affects mooring rope angles in such a way that the safety of the vessel is reduced. In order to analyse the mooring rope angles it was necessary to determine the average size of the "Suezmax" tanker. When the aver-age size was obtained, the level of safety at the berth was observed on tanker terminal which was built in 1980`s, when the size of "Suezmax" tankers was significantly different. The result showed that mooring ropes are reducing the safety of the vessel. Using OCIMF[2] calculation for mooring rope strength it was possible to calculate holding force of the ropes for lateral and longitudinal straining. Reduction of safety is noticeable, but it does not endanger the cargo operations. However, conducted research led to the conclusion that constant increase of "Suezmax" tankers will affect the angles of mooring ropes and will represent safety issue in the future. Further research will consist of continuous monitoring of "Suezmax" tanker size and recognising the moment when the size will impede the safety at the berth.

1 INTRODUCTION

Constant growth of maritime transportation over the years caused increment of vessels size. Crude oil tankers were the first type of the vessels which had recorded major increase in size and in deadweight. That increment was possible due to tanker fast cargo manipulation and large crude oil demand all over the world. However, that increment was limited with size of the world major canals and ports. One of the reasons was limited depth in ports and size of terminals. But increment of vessels size does not mean increment of terminal size. In Croatia the biggest crude oil terminal is located in Omisalj bay. The terminal started with cargo manipulation in 1980, with storage capacity of 760.000 m³. From the point of view of maritime safety at the berth, beside depth, a very important factor is mooring rope arrangement. By analysing the size of "Suezmax" tankers from 2008 to 2012 it was possible to determine average dimensions of the vessels, like length over all, breadth and draught. After determination of average dimensions, mooring ropes

arrangement was analysed in order to determine if increment of the vessels size caused change in mooring ropes angles. Results of the analysis results were compared with mooring arrangement of "Suezmax" tanker which has 140,000 tons of deadweight and represents typical "Suezmax" tanker from 1980 to 2001.

2 "SUEZMAX" TANKERS DIMENSIONS ANALYSIS

„Suezmax" tanker represents the vessel with maximum dimensions for passage through the Suez Canal. That vessel is limited with length, draught and breadth. In 2007 *MAN Diesel A/S* analysed "Suezmax" tankers´ deadweight and dimensions, and average deadweight was 150.000 tons, length over all 274 m, breadth 48 m and draught 16.1 m.

From 2008 to 2012 draught of "Suezmax" tankers was between 12 and 17.99 m, breadth between 32.51 and 55 m and length over all from 200 to 350 m.

Table 1 shows the number of „Suezmax" tankers by draught, and in 2012 91% of "Suezmax" tankers had draught between 16 and 17.99 m.

---

[2] OCIMF – Oil Companies International Marine Forum - Mooring equipment guidelines

Table 1. Number of „Suezmax" tankers by draught from 2008 to 2012

Draught(m)	2008	2009	2010	2011	2012
12 – 13,99	3	2	2	1	0
14 – 15,99	45	41	38	40	36
16 – 17,99	321	319	355	373	405
18 – 19,99	6	5	4	4	4
Total	376	368	400	418	445

Breadth of "Suezmax" tankers from 2008 to 2012 was between 32.51 m and 55 m. In 2008 89.89% of the analysed vessels had breadth between 40 and 50 m and in 2012 86.07%. However, in the analysed time period the number of "Suezmax" tankers with breadth between 50 and 55 m was increased from 9.04% to 13.93%. According to data in Table 2, the number of "Suezmax" tankers with breadth between 40 and 50 m is lower due to the increment of the "Suezmax" tankers with breadth between 50 and 55 m. In the analysed time period "Suezmax" fleet increased for 45 new ships, which represents an increment of 15%.

Table 2. Number of „Suezmax" tankers by breadth from 2008 to 2012

Breadth(m)	2008	2009	2010	2011	2012
32,51 - 40	4	4	4	2	0
40 - 50	338	326	353	360	383
50 - 55	34	38	43	56	62
Total	376	368	400	418	445

Length over all (LOA) of "Suezmax" tankers from 2008 to 2012 was between 200 m and 350 m. In 2012 the 99.33% of "Suezmax" tankers have LOA between 250 and 300 m. "Suezmax" tankers with LOA between 200 and 250 m and LOA between 300 and 350 m represent only 4% of analysed fleet in 2008 and by 2012 only 0.67%. The number of "Suezmax" tankers by LOA is presented in Table 3.

Table 3. Number of „Suezmax" tankers by length over all from 2008 to 2012

LOA(m)	2008	2009	2010	2011	2012
200 -250	4	4	3	2	2
250–300	368	362	396	415	442
300–350	4	2	1	1	1
Total	376	368	400	418	445

From the analysed data it is noticeable that in the time period from 2008 and 2012 average "Suezmax" tanker dimensions were, for draught between 16 and 17.99 m, for breadth between 40 and 50 m and for length over all between 250 and 300 m. In 2012 "Suezmax" tankers with those dimensions represent 90% of the vessels in total tanker "Suezmax" fleet.

In 2008 "Suezmax" fleet had 376 vessels with 57 million tons of deadweight. Decrement of 1 million

tons of deadweight was recorded in 2009, when total fleet was decreased by 8 vessels. From 2010 to 2012 total number of the vessels increased. In five analysed years number of "Suezmax" tankers, with draught between 12 and 13.99 m and draught between 18 and 19.99 m, was decreased. Also the number of "Suezmax tankers, with breadth between 23.51 and 40 m and with length over all between 200 and 250 and length over all between 300 and 350, also decreased. In 2012 average deadweight of "Suezmax" tanker was 155.000 tons. The trend of "Suezmax" tankers number and deadweight in the analysed time period is shown in Figure 1.

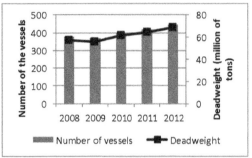

Figure 1. Number and deadweight of „Suezmax" tankers from 2008 to 2012

The analysed data shows that "Suezmax" tankers´ dimensions are increasing. One of the reasons is constant increment of Suez Canal. The draught of the passing vessels was increased to 20.1 m by 2010. Canal length is 193.30 km and cross-section area is 5200 m². Further step is to increase vessels maximum allowed draught to 21.9 m. The increment of the Suez Canal through history is shown on Figure 2.

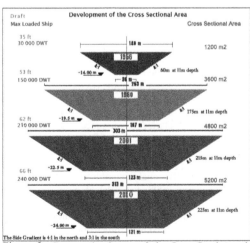

Figure 2. Cross-sectional area of the Suez Canal trough history, Suez Canal Authority ©

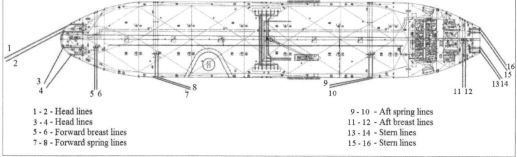

1 - 2 - Head lines	9 - 10 - Aft spring lines
3 - 4 - Head lines	11 - 12 - Aft breast lines
5 - 6 - Forward breast lines	13 - 14 - Stern lines
7 - 8 - Forward spring lines	15 - 16 - Stern lines

Figure 3. Vessel mooring ropes by location

3 VESSEL MOORING ROPES

Vessel mooring ropes are divided by location on: head lines, forward breast lines, forward spring lines, aft spring lines, aft breast lines and stern lines as shown in figure 3.

Mooring ropes should be positioned as much as possible symmetrically to the vessel centreline in order to ensure equal load on each mooring line. When planning vessel mooring arrangement, it is necessary to follow basic principles for every line. Head and stern lines should be on the vessel bow i.e. stern and horizontal angle should be close to 60°. Head and aft breast lines should be as much as possible perpendicular to the vessel centreline and positioned close to the vessel bow and stern. Horizontal angle of breast line should be around 90°. Forward and aft spring lines should be parallel to the vessel centreline and positioned on the ¼ of the vessel length from the bow and stern of the vessel. Horizontal angle of spring lines should not exceed 10°. Ideal vertical angle for all mooring lines is 0°, but should not exceed 25°, maximally 30°.

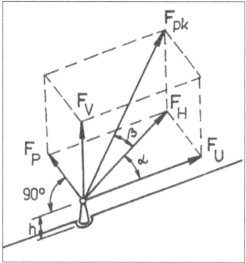

Figure 4. Components of mooring ropes angles

When analysing mooring ropes arrangements it is necessary to know the values of horizontal and vertical angle of mooring lines. Horizontal angle (α) is the angle which the mooring line makes with the longitudinal centreline of the vessel.

Horizontal component is divided on two components: lateral and longitudinal. Vertical angle (β) is the angle which the mooring line forms with the horizontal plain. Vertical angle can be used for calculating lateral and longitudinal mooring ropes holding force.

Tankers are moored by steel wires. Steel wire characteristics are small elasticity and large holding force. At the end of steel wire is the "tail". "Tail" is synthetic rope, its approximate length is around 11 meters and breaking force is 125% of breaking force of the attached steel wire. All mooring lines should have the same characteristics, like elasticity and if more mooring ropes are given from the same position their length should be equal. Mooring rope optimal length is between 35 and 50 meters, but mooring rope length depends on shore mooring arrangements. It is also necessary to take into consideration the lack of homogeneity of mooring ropes, which means that not all mooring ropes are equally loaded at the same time. In practice it is impossible to achieve homogeneity of mooring ropes, so in calculation it is necessary to take into consideration the safety factor. Forces acting on the vessel should not be larger than 55% of breaking load of the mooring ropes, which is equal to safety factor of 1.82.

The load of external forces on the vessel is considered through two components: force acting perpendicular to the vessel centreline and force acting parallel to the vessel centreline. Force acting perpendicular to the vessel centreline, lateral force, has the greatest effect on the breast lines. Force acting parallel to the vessel centreline, longitudinal force, has the greatest effect on the forward and stern spring lines. When calculating lateral and longitudinal components it is necessary to take into consideration the wind force, which acts on the vessel surface above sea level and has the greatest effect when the vessel is empty. Also it is necessary to take into consideration the sea current force and sea wave's force.

4 TANKER TERMINAL OMISALJ

Tanker terminal "Omisalj"[3] is located in Omisalj bay, and has two berths for mooring vessels from 30.000 to 350.000 tons of deadweight. The terminal was built in 1980.

Terminal consists of central part on pilots, 120 m in length. Mooring hooks are located at height of 3 meters above sea level.

On the vessel every mooring rope is attached to mooring winch, and from every winch two mooring ropes are given. This means that vessels has in total 8 mooring ropes at forward end and 8 mooring ropes at aft end of the vessel.

The analysed vessels were: tanker "Donat" built in 2007, dimensions length over all 280 m, breadth 48 m, summer draught 17 m and deadweight 166.000 tons; tanker "Jahre Target" built in 1990, dimensions length over all 269 m, breadth 44.5 m, summer draught 16.2 m and deadweight 140.000 tons.

5 MOORING ARRANGEMENTS GEOMETRY FOR "SUEZMAX" TANKERS AT "OMISALJ" TERMINAL AND COMPARATIVE ANALYSIS OF CALCULATED MOORING ROPES HOLDING FORCES

At the "Omisalj" terminal "Suezmax" tankers are berthed with the bow turned to bay exit, and all mooring ropes are of steel wire. On both analysed vessels there were 16 mooring lines, marked by numbers from bow to stern. At the bow there were four headlines, two breast lines and two spring lines. At the stern there were four stern lines, two breast lines and two spring lines. Horizontal and vertical angles of mooring ropes on both analysed vessels are shown in Table 4.

When holding force of mooring ropes was analysed, it was concluded that beside rope breaking forces, rope elasticity, mooring rope arrangements, horizontal and vertical angles for the analysed case, restrictions for mooring arrangement can arise from shore mooring hooks breaking load. In this analysis all calculations were made taking into account the breaking load of mooring hooks, which is 1250 kN for tanker terminal "Omisalj".

Table 4. Horizontal and vertical angles of mooring ropes

Name of the rope	Mooring line	Tanker „Donat"		Tanker „Jahre Target"	
		Horizontal angle(°)	Vertical angle(°)	Horizontal angle(°)	Vertical angle(°)
Head line	1	40	8,8	42	6,8
Head line	2	40	8,8	44	6,8
Head line	3	55	10,4	55	8,4
Head line	4	53	10,4	54	8,4
Fwd. Brest line	5	85	14,6	87	12,6
Fwd. Brest line	6	84	14,6	88	12,6
Fwd. Spring line	7	12	23,5	9	21,5
Fwd. Spring line	8	11	23,5	9	21,5
Aft Spring line	9	10	21,8	8	19,8
Aft Spring line	10	9	21,8	9	19,8
Aft Brest line	11	54	12,8	60	10,8
Aft Brest line	12	52	12,8	59	10,8
Stern line	13	70	12,7	65	10,7
Stern line	14	59	12,7	67	10,7
Stern line	15	48	9,9	42	7,9
Stern line	16	45	9,9	44	7,9

For calculating lateral holding force expression (1)

$$Fp_{pk} = F_{pk} * \sin \alpha * \cos \beta \qquad (1)$$

was used, and for calculating longitudinal holding force expression (2) was used.

$$F_{U\,pk} = F_{pk} \cdot \cos \alpha \cdot \cos \beta \qquad (2)$$

In both expressions the following figures were used: Fp_{pk} - is the lateral component of mooring rope holding force, Fu_{pk} – is the longitudinal component of mooring rope holding force, F_{pk} – the holding force of mooring rope, α – the horizontal angle of mooring rope and β – the vertical angle of mooring rope.

After calculating true lateral and longitudinal holding force of the mooring rope, due to lack of homogeneity of mooring ropes, mooring ropes holing force was corrected using the safety factor. In this case the safety factor is 1.82.

Longitudinal holding force, unlike lateral holding force, was calculated for two directions of action, from the bow and from the stern. Considering the direction of acting of longitudinal force it is necessary to determine which ropes are loaded in each case. For both analysed cases values of lateral and longitudinal forces were calculated and result is shown in Table 5.

---

[3] http://www.janaf.hr/sustav-janafa/naftni-terminal-luka-omisalj/ (5.01.2013)

Table 5. Results of analysed lateral and longitudinal forces for „Suezmax"

Mooring rope	Tanker „Donat"			Tanker „Jahre Target"		
	Lateral force (kN)	Longitudinal force (kN) Fwd	Aft	Lateral force (kN)	Longitudinal force (kN) Fwd	Aft
1	397,0	473,1	-	415,3	461,2	-
2	397,0	473,1	-	431,1	446,4	-
3	503,6	352,6	-	506,5	354,6	-
4	490,9	370,0	-	500,2	363,4	-
5	602,5	-	-	609,1	-	-
6	601,5	-	-	609,6	-	-
7	-	-	560,6	-	-	574,4
8	-	-	562,6	-	-	574,4
9	-	571,5	-	-	582,3	-
10	-	573,2	-	-	580,8	-
11	493,1	358,2	-	531,7	307,0	-
12	480,3	375,2	-	526,2	316,2	-
13	572,9	-	208,5	556,6	-	259,5
14	522,6	-	314,0	565,3	-	240,0
15	457,5	-	412,0	414,2	-	460,1
16	435,4	-	435,4	430,0	-	445,3
Total	5954,4	3547	2493,2	6095,9	3412	2553,6

Vessels` length and height have the largest influence on mooring arrangement. Difference between two analysed ships is 11 meters in length and 2 meters in height. That is not a large increase in size and in new terminals this will not have any effect. However, on terminal which was built in 1980 calculation showed some changes, but that change is still minimal. Mooring rope lateral holding force is reduced for 3% or 141.5 kN. Moring rope longitudinal holding force acting from the bow of the vessel is increased for 4% or 135 kN and mooring ropes longitudinal holding force acting from aft of the vessel is reduced for 3% or 60.4 kN. This small reduction in mooring ropes lateral and longitudinal holding force will not affect vessel safety, but vessel length is changed for only 11 meters or 4% so further analysis should keep track of "Suezmax" tankers dimensions.

From the point of safety at berth, external forces acting on the vessel have to be compared with the holding force of the mooring ropes. External lateral force acting on the vessels` will not be increased due to vessel small increment in length and height. However, external longitudinal forces will be increased because of increment in vessels` breadth and draught.

6 CONCLUSION

Analysing "Suezmax" dimensions, due to Suez Canal increment in size, it is noticeable that "Suezmax" tankers dimensions are increasing. Between 1980 and 2001 Suez Canal dimension did not change and average deadweight for "Suezmax" tanker was 140.000 tons. After the increase of the canal depth between 2001 and 2010 "Suezmax"

vessels` deadweight increased up to 240.000 tons. However, tankers did not follow that sudden increase in deadweight like container vessels, and increment of "Suezmax" tanker is noticeable but still very small. If we take "Suezmax" tanker which is berthed on "Omisalj" terminal, mooring arrangement can be analysed. Considering analysis and results, with current terminal limiting factors, the vessel safety at the berth is not significantly impaired. However, taking into consideration the trend of vessels` dimensions increment, current terminals at some point will not be able to ensure safe mooring arrangement. That means that ropes horizontal and vertical angles will be out of allowed limits and when external forces are taken into account, due to vessels` dimensions increase, vessels mooring ropes and terminal mooring hooks holding force will be questionable.

REFERENCES

Jadranski Naftovod, Joint Stock Co. (JANAF Plc.). Available at URL: http://www.janaf.hr/home/, accessed at 15.11.2014.
Mentes, A. & Helvacioglu, I. H. 2011. An application of fuzzy fault tree analysis for spread mooring systems. Ocean Engineering 38: 285-294.
Mohovic, R. & Mohovic, D. & Zorovic D. 2003. Possible measures in Maritime Safety of Manuevering and Navigation of Ships in the Omišalj oil Terminal. Rijeka: Technology transfer: 22-33.
Mohovic, R. & Mohovic, D. & Zorovic D. 2003. Possible measures of safety of navigation, maneuvering, and mooring of the ship's on Omišalj oil terminal. Centar za inovacije i transfer tehnologija Rijeka: 22-32.
Mohovic, R. & Mohovic, D. & Zorovic D. 2004. Characteristic of the Adriatic Sea waves in the function of safety and comfort of maritime traffic. Pomorstvo: journal of maritime studies (1332-0718) 18: 209-219
Mooring equipment guidelines, 3rd edition (MEG3), OCIMF, 2008.
Propulsion trends in tankers, MAN Diesel A/S, Copenhagen, 2007.
Review of maritime transport. UNCTAD Secretariat. 2012.
Shipping statistics and market review, Volume No. 3. Institute of shipping Economics and logistics. 2008-2012.
Suez Canal, Suez Canal Authority. Available at: http://www.suezcanal.gov.eg/sc.aspx?show=12, accessed at 10.10.2014.
Zorovic, D. & Mohovic, R. & Mohovic, D. 2002. Determination of the Relaion between the Length and Period of the Adriatic sea waves. Pomorstvo: journal of maritime studies (1332-0718) 16: 99-104
Zorovic, D. & Mohovic, R. & Mohovic, D. 2003. Towards determining the length of the wind waves of the Adriatic Sea. Naše more : znanstveni časopis za more i pomorstvo 50, 3-4:145-150.

Port Facilities
Safety of Marine Transport – Marine Navigation and Safety of Sea Transportation – A. Weintrit & T. Neumann (eds.)

The Analysis of Dredging Project's Effectiveness in the Port of Gdynia, Based on the Interference with Vessel Traffic

L. Smolarek & A. Kaizer
Gdynia Maritime University, Gdynia, Poland

ABSTRACT: The article presents modeling feasibility of dredging in the port of Gdynia, based on the available materials. Through increasing the technical depth, these measures will allow further development of the entire port. However, the effectiveness of this type of project depends to a great extent on the correct analysis of the factors hindering the implementation of the project, mainly the vessels traffic and uninterrupted operation of port terminals. The operation of the Port of Gdynia is problematic due to the strategic terminals being located in the western area and investment dredging covering the entire port canal. The vessel traffic surveyed and described in the article is based on the data available from the VTS system of the gulf of Gdansk, covering a period of two months of increased port operation activity.

1 INTRODUCTION

Seaports are very important elements of transportation networks, which strongly influence the efficiency of freights handling. Port productivity greatly depends on its proper operation and unhindered access to the terminals. Modern trends in constructing ever increasingly larger vessels of much greater parameters results in one of the main challenges being the development of sea ports in order to provide adequate maneuvering space and technical depth. That is why actually research of investment projects in ports, like a dredging are so important for an economic competitiveness.

What is more, fast development of containerization results in unit cargo freights becoming the dominant form of transport and significantly influences spatial layouts of both existing and newly designed seaports. Annual growth increase in shares of this branch in the global transport results in the number of containers handled being a significant determinant of the development of a specified port.

2 THE PORT OF GDYNIA

The port of Gdynia is a merchant seaport in the southern part of the Baltic Sea, on the Gulf of Gdansk. It is the third Polish seaport in the amount of transshipped cargo (after Gdańsk and Szczecin-Świnoujście port complex). The port of Gdynia is a universal port, specialized mainly in handling general cargo vessels. The Hel Peninsula is a natural protection for vessels anchored at the roadstead and the entrance to the port, which is 150 m wide and 14 m deep, makes the port easily accessible.

The port is non-tidal, accessible throughout the whole year because it never freezes up in winter. The overall premises of the object is 755,4 hectare, of which 508 ha is on land. The length of the wharf designated as transshipment operations area reaches about 11000 meters [6]. The greatest load capacity of the port of Gdynia is containerized cargo, handled mainly by two container terminals, located at the west port, that is Baltic Container Terminal and Gdynia Container Terminal. The port handles general cargo, as well as liquid and loose bulk cargo.

The ferry slip that operates vessels from Stena Line, which provides services on the route between Gdynia and Karlskrona, is located at the west port, at the Helsinki II wharf. According to the developmental strategy by the Port of Gdynia Authority, the emphasis is put on modernization of port infrastructure, which allows for handling even bigger vessels. The strong development of containerization and steadily growing volume of unit cargo freights provides ample developmental chances for general cargo terminals. However, capabilities to handle the largest ocean vessels are fairly limited at Gdynia terminals by the depth and spatial parameters.

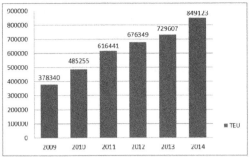

Figure 1. The container throughput development in Port of Gdynia [6].

3 THE CONCEPT OF INVESTMENT DREDING WORKS IN THE PORT OF GDYNIA

Modern trends in operating container terminals put strong emphasis on directly handling ocean vessels, which means giving up feeder shipping. Fulfilling the demands of ship owners is possible only by completing a number of development investments, which require a strong expansion into the area around the port and deepening of the port canals.

A significant developmental problem in the port of Gdynia is limitation to handle the biggest vessels, which is caused by the depth of the port canals. There are also sizeable difficulties with manoeuvring vessels over 350m in length. Due to spatial layout of the port of Gdynia, a great developmental chance is the feasibility to increase the parameters of the turning-basin, as well as dredging the port canal and the wharfs near container terminals up 15.5 m (baltimax size).

Following the analysis of equipment appropriate for the purpose of deepening the internal fairways of the port of Gdynia, authors propose to use a bucket dredger with auxiliary fleet in the form of two hopper barges (Fig.2). This choice is an optimal solution in terms of maneuverability and dredging efficiency. There is no possibility to use trailing suction hopper dredger because of lack of turning basin at the end of canal. Whilst the cutting suction dredger working with floating pipeline cannot be used due to lack of reclamation fields in the vicinity of the studied region.

However, a big technological impediment in case of the operation of this type of equipment is the anchorage (Fig. 6), which does not allow to bypass a working dredger during its operation. In this case, a model should be estimated prior to starting works, taking into account the eventuality of allowing to pass vessels calling at or leaving the port [3].

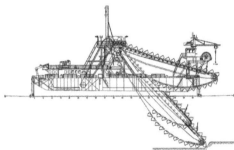

Figure 2. The draft of dredger working at canal in Port of Gdynia [8].

4 THE ANALYSIS OF VESSELS TRAFFIC IN PORT OF GDYNIA

Approach and port-channels are parts of the vessel movement concentration thus they are an area of high-priority traffic. Proper analysis of the stream of traffic flowing through aquatorium of dredging work is an essential element of the organization of this type of activity. Knowledge of the quantity and frequency of vessel traffic in the normal operation of port terminals allows you to plan the activities dredger which minimally interfere with the efficient handling terminals activities.

Port of Gdynia generates the most ship traffic on the canal port, which also requires the most capital dredging, because of location two container terminals in the western part of the channel.

4.1 Analysis of the amount of ships entering the Port of Gdynia

The Table 1 and Figure 3 shows summary statistics for Gdynia. It includes measures of central tendency, measures of variability, and measures of shape.

The standardized skewness and standardized kurtosis (parameter used in the choice of the probability distribution) can be used to determine whether the sample comes from a normal distribution. Values of these statistics outside the range of -2 to +2 indicate significant departures from normality, which would tend to invalidate any statistical test regarding the standard deviation. In this case, the standardized skewness value is not within the range expected for data from a normal distribution. The standardized kurtosis value is not within the range expected for data from a normal distribution.

Table 1. Summary Statistics for Gdynia [7].

Count	258
Average	199,19
Median	139,5
Standard deviation	198,314
Coeff. of variation	99,5603%
Minimum	0,0
Maximum	1116,0
Range	1116,0
Lower quartile	51,0
Upper quartile	294,0
Stnd. Skewness	10,1957
Stnd. Kurtosis	8,82776

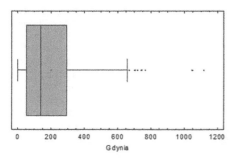

Figure 3. The. Box and Whisker plot of ship arrivals

Statistical analysis of the distribution of intervals between successive vessels for the stream of ships entering the port of Gdynia has demonstrated that it complies with the exponential distribution with density function of the formula

$$f(t) = \begin{cases} 0 & t < 0 \\ \lambda e^{-\lambda t} & t \geq 0 \end{cases} \quad (1)$$

This analysis shows the results of fitting an exponential distribution (mean=199,19) to the data on Gdynia, Table 2. We can test whether the exponential distribution fits the data adequately by selecting some goodness of fit tests.

The Table 2 shows the results of tests run to determine whether Gdynia can be adequately modeled by an exponential distribution. Since the smallest P-value amongst the tests performed is greater than or equal to 0,05, we cannot reject the idea that Gdynia comes from a exponential distribution (1) with 95% confidence, [5]. We can also assess visually how well the exponential distribution fits on graphs Figure 4 and Figure 5.

Table 2. Goodness of fit tests for Gdynia.

Kolmogorov-Smirnov Test	DPLUS	DMINUS	P-Value
Exponential	0,03877	0,04012	0,80043
Kuiper Exponential	V	Modified Form	
	0,07890	1,27535	>=0.10*
Cramer-Von Mises W^2 Exponential	W^2	Modified Form	
	0,00032	0,00032	0,99998*
Watson U^2 Exponential	U^2	Modified Form	
	-64,4997	-64,5397	1,0*

*Indicates that the P-Value has been compared to tables of critical values specially constructed for fitting the selected distribution. Other P-values are based on general tables and may be very conservative (except for the Chi-Squared Test).

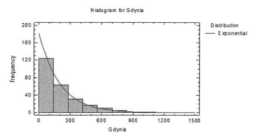

Figure 4. Frequency Histogram – displays a frequency tabulation of the data.

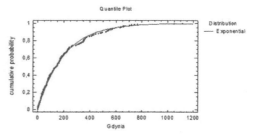

Figure 5. Quantile Plot – plots the empirical and fitted cumulative distributions.

The results of the statistical analysis determines that the traffic stream model is a Poisson point process with a parameter which is the inverse of the average interval between the ships. And the number of vessels at time T is described in the counting process. There are many ways to derive a Poisson process, [1]. Let X(t) (for t ≥ 0) be the number of ships arriving in the interval [0; t] (with X(t) = 0). Consider the following requirements:
− The number of ships arrivals in disjoint time periods are independent.
− X : representing the number of occurrences of ships in a continuous time interval, λ expected value of occurrences in this interval.
− For a small time period Δt the probability of a one ship arrival in the period is given by P [X(t + Δt) - X(t) = 1] = λΔt + o(Δt); where λ is known as the rate of the process and o(Δt) is some function

(which maybe negative or positive) which tends to zero in the limit as Δt tends to zero.
– The probability that two or more arrivals occur in the same small time period is negligible. More formally, P [X (t + Δt) - X (t) ≥2] = o(Δt):

Poisson Probability Distribution:
Let X be the random variable representing the number of occurrences of ships in some interval Δt. Then, the probability distribution function for X is

$$P(X=k) = \frac{\lambda^{k\Delta t}}{k!}e^{-\lambda\Delta t}, \quad k = 0,1,2,... \; E(X) = \lambda \quad (2)$$

$$D^2(X) = \lambda$$

where e = 2,7182... and λ is some parameter.
As the model of a time interval between consecutive ships we assume a random variable with an exponential distribution with parameter λ=1/199,19, Table 1.
The probability of the time interval (OCT) between consecutive vessels greater than T, according to formula (A), is given by

$$P(OCT > T) = e^{-\frac{T}{199,19}} \quad (3)$$

The probability that at the time interval [t1, t2] the number of vessels which want to enter the port will be equal k, can be calculated using formula (2) and is given by

$$P(X_{[t_1,t_2]} = k) = \frac{\left(\frac{t_2-t_1}{199,19}\right)^k}{k!}e^{-\frac{t_2-t_1}{199,19}} \quad (4)$$

Using the formula of total probability and formulas (3), (4) we can find the probability distribution of the ship's time which was lost due to dredging work.

5 THE MODELING OF DREDGING WORKS

Dredging of approach channels and the areas of increased flow of vessels inside the port is technologically and organizationally the most difficult area to implement this type of project. Modeling workflow strongly depends on the priorities established in the project. The analysis of the activity organization of dredging equipment shows that apart from digging efficiency, maneuvering characteristics of the dredger and its additional equipment is an important factor too. An operating dredger and its auxiliary ships (i.e. hopper barges, tugboats) pose navigational difficulties or block access to the port completely, as they have reduced maneuverability.
An additional difficulty for navigation derives from the anchorage of non-propelled vessels during

their work. During a non-propelled dredger avoidance maneuver, anchoring ropes must be either coiled or properly put on the seabed, so as to allow for a safe passage of other vessels (Fig. 6) [2].

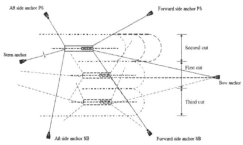

Figure 6. The draft of non-propelled dredger anchored during the work [4].

Where a priority is an unimpeded movement of vessels, works will often be paused during the dredging and lowering the dredging area, which prolongs the project and generates costs for the investors who commissioned the work. Whereas in a situation where the implementation of dredging is a priority, together with a high efficiency of dredging and a short duration of the project, vessels coming into the port must wait in queues. In this case, vessels waste time before entering the port, which generates significant costs for ship owners. Therefore, in order to obtain a balanced functioning of the port during dredging works, it is necessary to optimize the technical efficiency of dredging works against interferences with marine traffic (Fig. 7).

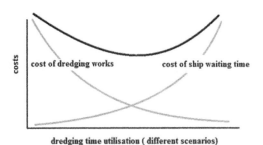

Figure 7. The idea of optimization task [2].

5.1 The Modeling of dredging works based on vessels traffic in western part of port in Gdynia

The modeling of dredging works in the Port of Gdynia is strongly dependent on the vessel traffic stream in a specific port area, digging performance and maneuverability of dredgers used for the project. The port of Gdynia has the most transshipment works done in its western part, where container terminals are located (Helskie I and Bulgarian wharfs). Therefore, these areas require special development in terms of accessibility for vessels with significant immersion. In order to implement this type of project, the whole port canal needs to be deepened. An additional difficulty in the implementation of dredging works in these areas are the Stena Line ferry traffic and the ships calling at the neighboring wharves and shipyards. Therefore, a sample model of workflow organization for dredging activity in the port channel should take into account the estimated time of inputs and outputs of vessels. In the case of regular ferry traffic and fixed services it is a relatively simple configuration. Other calls at the port may be, for the purpose of modeling, simulated by estimating the average time based on the observation of traffic stream. The analysis of the traffic in the port indicates that the average time between consecutive vessels passing through the port channel is approximately 3 hours. Assuming 1 hour pause in dredging activity as the time to let the vessels pass, the schedule of dredgers at the port canal may be modeled as 3 hours of operation and 1 hour of passing vessels, taking into consideration vessels with a fixed schedule (ferries).

The parameters established in the model :
- ferry service time at the port (fixed weekly timetable: 5^{00}-6^{00}; 7^{00}-10^{00}; 18^{00}-19^{00}; 21^{00}-22^{00}),
- average number of vessels observed navigating the port canal amounts to 7 vessels a day,
- estimated average time interval of individual vessel inputs equals about 3 - 4 hours,
- time assumed for passing vessels navigating through the canal is minimum 1 hour.

Table 3. The proposed dredging schedule (24 hours) at port canal in Port of Gdynia including modeling time parameters .

Time (hours)	1	2	3	4	5	6	7	8
Dredging schedule	0	1	1	1	0	1	0	0
Time (hours)	9	10	11	12	13	14	15	16
Dredging schedule	0	1	1	1	0	1	1	1
Time (hours)	17	18	19	20	21	22	23	24
Dredging schedule	1	0	0	1	0	1	1	1

0- no dredging activity, 1- dredging activity

On the basis of a model assumed for dredgers activity, according to the vessel traffic analyzed, dredging equipment will spend 105 working hours a week on deepening, which at the efficiency of dredging of 500 m3/h (dredger of inż. T. Wenda) makes for 52522 m³ of excavated material per week. According to the calculations of the model, at the assumed operation schedule dredgers will have to sacrifice 63 hours of dredging activity for vessels passing per week, which amounts to 31500 m³ of losses in dredging. An important issue, which influences both the vessel traffic and the port, is the matter of the so called „time window" (at least 1 hour) proposed each three or four hours, when vessels would be able to both enter and leave the port. In practice, vessels intending to call at the port or to leave the wharf at the moment of dredging

activity will have to wait in queue. Thus, estimating the volume of excavated material and verifying the vessel traffic at the port allow for planning the optimal timing for dredging equipment operation, which in turn allows for estimating the total duration of the dredging project.

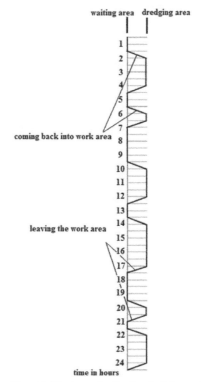

Figur 8. The intervals allowing dredging works defined by proposed model of ships' traffic.

6 CONCLUSIONS

The analysis of a vessels traffic and a proposed model of dredging works is introduced to describe the way how the assessment of mutual interdependence of the sea traffic and dredging activity in a port approaching channel could be made.

Created organization model enables to assess the ship losses caused by the time spent in the queue while the dredging works are performed. That could be used to calculate the total dredging works cost in different scenarios. It is shown how simulation performed for different scenarios could help to optimize the time schedule of dredging works by coordination them with a given schedule or random pattern of ship arrivals.

REFERENCES

[1] Gerlough Daniel L. Use of Poisson Distribution in Highway Traffic, The eno foundation for highway traffic controlsaugatuck. 1955 Connecticut, Printed in the United States of America by Columbia University Press.

[2] Gucma L., Modelowanie Czynników ryzyka zderzenia jednostek pływających z konstrukcjami portowymi i pełnomorskimi, Wydawnictwo Akademii Morskiej w Szczecinie, Szczecin 2005, p. 71-76.

[3] Kaizer A., Formela K. , The Concept of Modernization Works Related to the Capability of Handling E Class Container Vessels in the Port of Gdynia, "Marine Navigation and Safety of Sea Transportation", Edited by: Weintrit A., Neumann T., Taylor & Francis Group, Gdynia 2013 p. 201-205

[4] Vlasblom J. W., Designing dredging equipment. Lecture notes, published at: http://www.dredging.org, Central Dredging Association. Website dated on 29.12. 2014.

[5] Zhiming Ding, Guangyan Huang, Real-Time Traffic Flow Statistical Analysis Based on Network-Constrained Moving Object Trajectories, 20th International Conference, DEXA 2009, Linz, Austria, August 31 - September 4, 2009. Proceedings, Springer Berlin Heidelberg.

[6] Port of Gdynia Authority. Website dated on 19.01.2015:
http://www.port.gdynia.pl/en/about-port/basic-data
http://www.port.gdynia.pl/en/about-port/statistics

[7] Maritime Office in Gdynia. Website dated on 19.01.2015:
http://www.umgdy.gov.pl/ruchstatkow/

[8] Underwater works and dredging companies. Website dated on 19.01.2015:
http://www.prcip.pl/about-prcip.html
http://www.rohde-nielsen.dk/index.php?cID=125

Ship's Propulsion, Main Engine and Power Supply

Ship's Propulsion, Main Engine and Power Supply
Safety of Marine Transport – Marine Navigation and Safety of Sea Transportation – A. Weintrit & T. Neumann (eds.)

Reliability of Fuel Oil System Components Versus Main Propulsion Engine: An Impact Assessment Study

M. Anantharaman, F. Khan, V. Garaniya & B. Lewarn
Australian Maritime College (University of Tasmania), Australia

ABSTRACT: Reliability and safety are two vital factors of a large main propulsion engine which propels the vessel at high seas. The fuel oil system is a critical sub-system of the main propulsion engine. The essential components of this important system comprises of the fuel oil settling tanks, purifies, fuel oil service tanks and associated valves and pipings, filters, pumps, heaters and a temperature control valve. The Reliability of these individual components, will dictate the Reliability of the fuel oil system, which in turn determines the Reliability of the main propulsion system. This paper investigates the correlation between the reliability of the fuel oil system components and the impact of the same, on the Reliability of the main propulsion engine. Also methods to improve the reliability of the fuel oil system components will be studied and suggestions made to improve the reliability. This will lead to an improvement in the Reliability of the main propulsion engine.

1 INTRODUCTION

These main propulsion engine which propels the vessel sat sea have to be highly reliable and safe at all times, whilst sailing at high seas, transiting through canals and maneuvering in ports. It is imperative that the maintenance regime on board the vessels have to be very well structured, with utmost consideration to safety and reliability of the main propulsion system.

The Reliability of the main propulsion system is interdependent on the reliability of its subsystems, which are listed below.
1 Main Engine Lubricating oil system
2 Main Engine Jacket Cooling Water system
3 Main Engine Fuel Oil system
4 Main Engine Scavenge system
5 Main Engine Air Start system
6 Main Engine Safety System

This paper discusses the methodology adopted to quantify reliability of one of the vital sub-system viz. the fuel oil system, [6,10], and development of a model thereof.

Presently all large two stroke marine diesel engines, burn residual fuel oil the viscosity of which ranges between 380 to 700 cSt at 50 degs C. These fuel need to be purified, filtered and heated to a temperature as high as 150 degs C, to obtain a viscosity of 10-12 cSt at the inlet to the main engine fuel injection pump, for effective and efficient burning of the fuel. A fuel oil circuit for a large engine is shown in Figure 1. This comprises of a service tank where purified fuel oil is stored at a temperature of 90 degs C. The tank is provided with a quick closing valve, which can be operated remotely in case of any emergencies, to cut off fuel to the main engine. The oil from the tank is drawn by means of a supply pump, which is of the gear type, where a suction filter is provided at the downside of the pump to filter out any sediment. The supply pressure of the pump varies between 5 to 7 bars, The pump discharges the fuel oil to a discharge filter which is of the automatic back flush type and has a fine mesh where filtration of the fuel oil can be done up to a sediment size of 25 microns. In case of any problem with the automatic back flush filter, it is possible to isolate and bypass this automatic back flush filter by means of a bypass filter, until the automatic back flush filter is decommissioned. The fuel is then led to a flow meter, where the fuel consumption is monitored. In case of any malfunction of the flowmeter it is possible to bypass this flowmeter, by means of a bypass valve. The next component in the system is a buffer tank, where the fuel mixes with fuel returned from the return rail of the engine. A booster pump on the upstream of the buffer tank boosts the system pressure up to 10 bars and discharges it to steam heaters, where the fuel is heated to the requisite temperature. The temperature is controlled by a viscometer, located at the upstream of the heaters. Normally one heater is in use, the other being a standby. It is to be noted that the fuel supply pumps and booster pumps are identical. Normally one pump is in use, the other being on standby, [9]. The fuel oil at the correct viscosity between 10 to 12 cSt is supplied to the inlet rail, from where it is led to the individual fuel injection pump. The fuel injection pump, one for each cylinder, supplies the fuel at high pressures up to 600 bars, to fuel injectors on the cylinders. The return oil from the injectors comes to a return rail and led back to the buffer tank. A pressure regulator valve on the return line is provided, to maintain a back pressure of 8 to 9 bars.

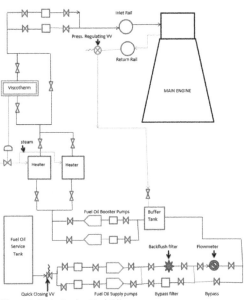

Figure 1. Main Engine Fuel Oil system

To evaluate the reliability of the fuel oil system, a systematic approach involving a Fault Tree Analysis (FTA), for the system will be considered,[5]. This will be followed by a critical component identification (CCI) and then a Reliability Block Diagram shall be developed (RBD). A model to evaluate the reliability for each of the system component is developed and the overall reliability of the system can be determined, [9]. In order to estimate the reliability of the Main Engine Fuel Oil system, we need to evaluate the reliability of the individual components of the fuel oil system. The first step in the evaluation will be to draw a RBD (reliability block diagram), for the Main Engine Fuel Oil system. This is shown in Figure 2 below. The reliability will be evaluated using Markov principle.

We shall now look into the various components of the Main Engine Fuel Oil system,[4], and determine the reliability of the system,[11]. The following steps are followed:

1 The Fault Tree Analysis (FTA) for the Main EngineFuel Oil system, [14].
2 Develop a Reliability Block Diagram (RBD) for the Main Engine Fuel Oil system, [2].
3 Look at the individual components in the Main Engine Fuel Oil system and draw the state diagram for these components
4 Carry out a Markov Analysis for these components ,[8].
5 Carry out a relaibility analysis.
6 Consider measures for improving the system reliability.
7 Draw comclussions based on the analysis.

2 THE FAULT TREE DIAGRAM FOR THE MAIN ENGINE FUEL OIL SYSTEM

There are eight (8) main components of the Main Engine Fuel Oil system, failure of which will lead to the failure of the main propulsion engine.

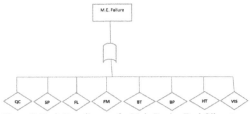

Figure 2. Fault Tree diagram for Main Engine Fuel Oil system

In the above figure 2, QC represents the Quick closing valve, SP represents the Fuel Supply pumps, FL is the discharge filters, FM is the flow meter, BT is the Buffer tank, BP represents the booster pumps, HT represents the steam heater and VIS the Viscotherm. The next step in the analysis of evaluating the Reliability of the Main engine Fuel Oil system is as shown below.Each of the above component will be analysed.

3 RELIABILITY BLOCK DIAGRAM FOR THE MAIN ENGINE FUEL OIL SYSTEM

Tthe following points are taken into consideration.
1 Each block represents the maximum number of components in order to simplify the diagram.
2 The function of each block is easily identified
3 Blocks are mutually independent in that failure of one should not affect the probability of failure of another, [1,3].

Figure 3RBD for Main Engine Fuel Oil system

176

4 RELIABILITY OF THE QUICK CLOSING VALVE

The quick closing is the main tank outlet valve, which can be operated remotely in case of an emergency. If we assume a constant failure rate λ,[12], then the reliability of this component may be expressed as $R_{QC}(t) = e^{-\lambda t}$, where the mean time to failure $MTTF = 1/\lambda$.

5 RELIABILITY OF THE FUEL OIL SUPPLY PUMP

The fuel oil supply pumps FS are of the gear type and identical in design and construction. The state diagram for the pumps is shown in figure 4 below. The reliability function is an exponential function of time t and the failure rate λ expressed as number of failures per running hours.

Figure 4. Markov Model analysis for the fuel oil supply pump FS

Table 1. Fuel oil supply pump

State of Fuel oil supply pump	Pump 1	Pump 2
1	Operating	Standby
2	Failed	Operating
3	Failed	Failed

From the Table 1 above we see that there are 3 states. In this case the two Fuel oil supply pumps are identical standby units, on of which is on line and the other standby.The reliability of the two identical systems isderived as,

$$R_s(t) = e^{-\lambda t} \sum_{i=0}^{1} \frac{(\lambda t)^i}{i!} \qquad (1)$$

In this case $R_s(t) = e^{-\lambda t}(1 + \lambda t)$ and MTTF (Mean time to failure) $MTTF = 2/\lambda$.

6 RELIABILITY OF THE FUEL OIL DISCHARGE FILTER

In case of failure of the backflush discharge filter, the standby filter comes on line. The failure rate of this bypass filter will be different than that of the main discharge filter. The failure rate of this standby filter depends upon the state of the main backflush filter. Figure 5, below shows the Markov Model for the fuel oil discharge filter FL. An analysis is carried out as shown below.

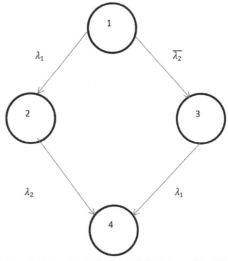

Figure 5. Markov Mode for the fuel oil discharge filter FL

$$R_p(t) = e^{(-\lambda_1 t)} + \frac{\lambda_1}{\lambda_1 + \overline{\lambda_2} - \lambda_2}\left\{ e^{(-\lambda_2)t} - e^{(\lambda_1 + \overline{\lambda_2})t}\right\} \qquad (2)$$

$$MTTF = \frac{1}{\lambda_1} + \frac{\lambda_1}{\lambda_2\left(\lambda_1 + \overline{\lambda_2}\right)} \qquad (3)$$

In case $\lambda_1 = \lambda_2 = \lambda$ and $\overline{\lambda_2} = \overline{\lambda}$, then

$$R_p(t) = e^{(-\lambda t)} + \frac{\lambda}{\overline{\lambda}}\left\{ e^{(-\lambda)t} - e^{(\lambda + \overline{\lambda})t}\right\} \qquad (4)$$

$$MTTF = \frac{1}{\lambda} + \frac{1}{\left(\lambda + \overline{\lambda}\right)} \qquad (5)$$

7 RELIABILITY OF OTHER COMPONENTS OF FUEL OIL SYSTEM

On similar lines Markov model analysis for the other components of the fuel oil system could be evaluated. Having done that the Reliability of the Main Engine Fuel Oil system could be evaluated.

8 FAILURE RATE OF FUEL OIL SYSYTEM COMPONENTS TO DETERMINE RELIABILITY

Research on reliability studies of the Main Engine Fuel Oil system has been carried out by Prof. Alfred Brandowski at the Polish Maritime Research. Failure rates for various components of the Main Engine Fuel Oil system has been obtained from his research,[3]. This is seen in the following Table 2.

Table 2. Failure rate of fuel oil system components

Component	Abbreviation	Failure rate $\lambda \times 10^{-6}$
Fuel oil service tank Quick Closing Valve	**QC**	2.2062
Fuel Oil Supply pump	**FS**	7.5572
Fuel Oil Filter	**FL**	6.9599
Fuel Oil Flowmeter	**FM**	5.2965
Fuel Oil Buffer Tank	**BT**	5.2965
Fuel Oil booster pump	**BP**	8.2840
Fuel Oil Heater	**HT**	4.2666
Fuel Oil Viscotherm	**VIS**	10.6323
Piping/heating/miscellaneous		5.2965

9 THE RELIABILITY OF MAIN ENGINE FUEL OIL SYSTEM

$$R_{F.O}(t) = R_{QC}(t) R_{FS}(t) R_{FL}(t) R_{FM}(t) R_{BT}(t) R_{BP}(t) R_{HT}(t) R_{VIS}(t) \quad (6)$$

where
$R_{QC}(t)$ is the reliability of the Quick Closing valve
$R_{FS}(t)$ is the relaibility of the Fuel oil supply pump
$R_{FL}(t)$ is the reliability of the discharge filter
$R_{FM}(t)$ is the reliability of the flowmeter
$R_{BT}(t)$ is the reliability of the buffer tank.
$R_{BP}(t)$ is the reliability of the booster pump
$R_{HT}(t)$ is the reliability of the bsteam heater
$R_{VIS}(t)$ is the reliability of the Viscotherm

The Reliability the fuel oil system components versus the running hours of the main engine, and the Reliability of the Main Engine Fuel Oil system as a whole versus the running hours of the main engine is shown in the graphs below. These graphs are based on the failure rates obtained rom Table 2. Figure 6 shows the Reliability of Quick Closing valve QC versus running hours. Likewise Figure 7 shows the Reliability of fuel oil supply pump FS, Figure 8 the Reliability of the discarge filter FL, Figure 9 the Reliability of the flowmeter FM,Figure 10 the Reliability of buffer tank FT, Figure 11 the Reliability of booster pump BP, Figure 12 the Reliability of heater HT, Figure 13 the Reliability of Viscotherm vis and finally Figure 14 depicts the Relibility of the fuel oil system.It can be seen from the equation 6 above that the Reliability of the fuel oil system could be improved by improving the Reliability of the individual fuel oil system component. But it is essential to take into consideration the cost of the Reliability improvement.

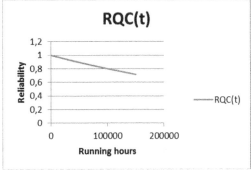

Figure 6. Reliability of Quick Closing valve QC

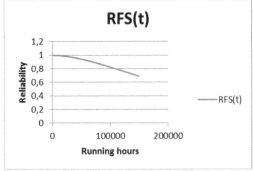

Figure 7. Reliability of fuel oil suppply pump FS

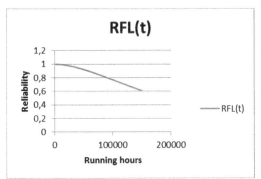

Figure 8. Reliability of discharge filter FL

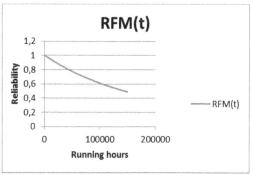

Figure 9. Reliability of flowmeter FM

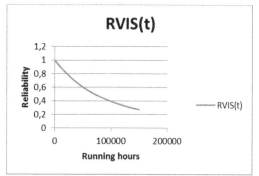

Figure 13. Reliability of viscotherm VIS

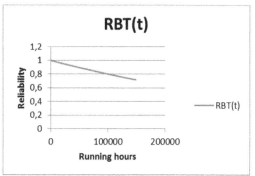

Figure 10. Reliability of buffer tank BT

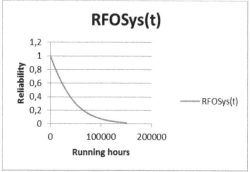

Figure 14. Reliability of fuel oil system FOS

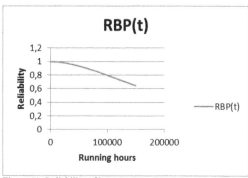

Figure 11. Reliability of booster pump BP

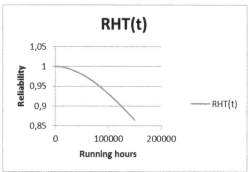

Figure 12. Reliability of heater HT

10 IMPROVING RELIABILITY

Reliability of the system can be improved by improving the componnet reliability as seen in the above equation. For instance in the case of the fuel oil supply pump FS shown in Figure 7 above, physical introduction of an additional pump will increase the releiability.This cost for improvement of reliability need to be assessed and the cost benefit for the incremental reliability to be determined.If the original value of Reliability R_O. at cost x is improved to Reliability R_I at cost y, then the incremntal reliability for the differential cost $\frac{R_I - R_O}{y - x}$. should be compared with the base relaibility to cost ratio which in this case is $\frac{R_O}{x}$. For cost benefit $\frac{R_I - R_O}{y - x} > \frac{R_O}{x}$.

This could be a feasible proposition for some components, but not for all components. Similar study needs to be done for all other components and a cost beneficial CBM model could be developed.

The failure rates shown in Table 2 above, has been utilised in the mathematical model to determine the Relaibility of the fuel oil system R(FOS). This is shown in figure above. Further the author has slected an identical value of the failure rate for all components of the fuel oil system. A failure rate of 2.2062*10^-6 or the Quick Closing valve QC, was selected, being the lowest failure rate for all components in the mathematical model. It can be shown that there is a 30% improvement in the overall reliability as can be seen in the Figure 15 below.

Further feasibility studies need to be conducted to look into the cost benefit for the improved reliability.

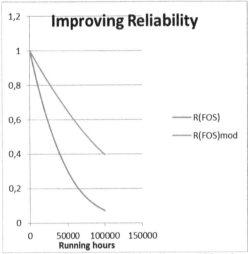

Figure 15. Reliability of improved fuel oil system FOS mod.

11 CONCLUSION

In this paper the Main Engine Fuel Oil system, which is a very vital part of the Main propulsion system was analyzed Failure of the Main Engine fuel system may result in serious damage to the engine components and failure of the Main Engine. A step by step approach for evaluating the reliability of the Main Engine Fuel Oil system was presented. Also it was shown that utilizing the least failure rate of the fuel oil system component, as an identical value of failure rate for all components in the fuel oil system, the overall reliability of the Main Engine Fuel Oil system, could be improved considerably. Similar approach could be looked at to evaluate the reliability of other sub systems of the main propulsion engine. Thus we can improve the overall reliability of the main propulsion engine.

REFERENCES

[1] Anantharaman, M 2013, 'Using Reliability Block Diagrams And Fault Tree Circuits, To Develop A Condition Based Maintenance Model For A Vessel's Main Propulsion System And Related Subsystems', *Transnav: International Journal On Marine Navigation And Safety Of Sea Transportation*, Vol. 7, No. 3.[2] Bhattacharya, D & Deleris, La 2012, 'From Reliability Block Diagrams To Fault Tree Circuits', *Decision Analysis*, Vol. 9, No. 2, Pp. 128-137.

[3] Brandowski, A 2009, 'Estimation Of The Probability Of Propulsion Loss By A Sea Going Ship Based On Expert Opinion.', *Polish Maritime Research(1)*, Vol. 16, Pp. 73-77.

[4] Cicek, K & Celik, M 2013, 'Application Of Failure Modes And Effects Analysis To Main Engine Crankcase Explosion Failure On-Board Ship', *Safety Science*, Vol. 51, No. 1, Pp. 6-10.

[5] Conglin Dong, Cy, Zhenglin Liu, Xinping Yan 2013, 'Marine Propulsion Sysytem Reliability Research Based On Fault Tree Analysis', *Advanced Shipping And Ocean Engineering*, Vol. 2, No. 1, P. 7.

[6] Epsma 2005, 'Guidelines To Understanding Reliability Prediction', No. 24 June 2005, P. 29.

[7] Erick M. Portugal Hidalgo, Eub, Dennis W. Roldán Silva, Dsub & Gilberto F. Martha De Souza, Gub 'Fmea And Fta Analysis Applied To The Steering System Of Lng Carriers For The Selection Of Maintenance Policies', In *21st Brazilian Congress Of Mechanical Engineering*, Natal, Rn, Brazil.

[8] Gowid, S, Dixon, R & Ghani, S 2014, 'Optimization Of Reliability And Maintenance Of Liquefaction System On Flng Terminals Using Markov Modelling', *International Journal Of Quality & Reliability Management*, Vol. 31, No. 3, Pp. 293-310.

[9] Liberacki, R 'Influence Of Redundancy And Ship Machinery Crew Manning On Reliability Of Lubricating Oil System For The Mc-Type Diesel Engine', In Gdansk University Of Technology Ul. Narutowicza 11/12, 80-950 Gdańsk, Poland.

[10] Mollenhauer, K & Tschöke, H 2010, 'Handbook Of Diesel Engines', *Handbook Of Diesel Engines, Edited By K. Mollenhauer And H. Tschöke. Berlin: Springer, 2010.*, Vol. 1.

[11] Navy, Us 1994, *Handbook Of Reliability Prediction Procedures For Mechanical Equipment.*

[12] Pcag 2012, 'Failute Rate And Event Data For Use Within Risk Assesments.'.

[13] Xu, H 2008, 'Drbd: Dynamic Reliability Block Diagrams For System Reliability Modelling', University Of Massachusetts Dartmouth.

[14] Zhu, Jf 2011, 'Fault Tree Analysis Of Centrifugal Compressor', *Advanced Materials And Computer Science, Pts 1-3*, Vol. 474-476, Pp. 1587-15

Ship's Propulsion, Main Engine and Power Supply
Safety of Marine Transport – Marine Navigation and Safety of Sea Transportation – A. Weintrit & T. Neumann (eds.)

Impact of Electricity Generator on a Small-Bore Internal Combustion Engine at Low and Medium Loads

P. Olszowiec, M. Luft & E. Szychta
University of Technology and Humanities in Radom, Poland

ABSTRACT: The paper outlines results of steady-state tests carried out on a Renault 1.2 TCe (100hp) engine at varying electric loads of the nominal power rating AC electric generator. These tests are the first stage in the study dealing with the impact of classical sources of energy in a passenger vehicle on an internal combustion unit. The aim of the above mentioned research is to obtain information on the possibilities of implementation of innovative sources of electric energy based on the combustion unit loss in performance of a passenger car. During the study suitability of innovative electric energy sources for dual mode (hybrid) vehicles was analysed.

1 INTRODUCTION

The paper is within the framework of the **R**ecovery **E**nergy **S**ystem with **T**urbogenerator – **REST** – project implemented at the University of Technology and Humanities in Radom being an attempt to develop a system of the electric energy recovery from the combustion engine losses resulting from the gas flows in the exhaust system. Examination of the impact of the nominal power rating AC electric generator (alternator) at varying electric loads on the wear of the power supply medium was carried out. It was decided that during the examinations the combustion engine load would change every 15 Nm for a specific value of the rotational speed. The scope of the examinations was defined to be in the range of rotational speeds from 1,500 rpm to 3,500 rpm, and loads from 0 to 90 Nm. The above load and speed ranges were based on the average values in use noted for vehicles used in urban agglomerations. (M. Luft, 2011, Application of a turbogenerator ...) The examinations aimed to establish the rotational speed and load ranges within which there is a possibility of using electric energy obtained from the turbogenerator.

2 SELECTION OF THE COMBUSTION ENGINE

While choosing the combustion unit for the REST project and examinations accompanying it, one of the criteria for the choice were global and European market trends. The research conducted by ACEA – European Automobile Manufacturers' Association – shows that in 2013 reduction of combustion unit's capacity was still the major trend. In the EU the top five places among the best selling cars belonged to urban cars (45%). As the ACEA analyses indicate, the average engine capacity for new cars purchased by European customers declined from 1,706 cm^3 to 1,635 cm^3, whereas the average power from 86 kW/117 hp to 82 kW/111.5 hp. Moreover, attention needs to be paid to a recent tendency in the modern combustion unit construction which is subject to ecological regulations. The result of this is a trend known as downsizing which aims to reduce the unit capacity to the benefit of the implementation of the direct injection of fuel, turbocharger and reduced number of cylinders. One should not mix up engine "downsizing" with engine "tuning" because it is not about increasing the maximum power, permitted rotations, but about causing that the performance of a smaller piston engine (sometimes with reduced number of cylinders) is practically the same as that of a larger but less efficient unit. For this reason, for the purposes of this research, we have chosen a turbocharged unit with spark ignition made by Renault, with the capacity of 1,200 cm^3 and power of 100 hp (Fig.1) (1.2 TCE), code: D4F78 with the SIM32 CAN controller. This unit fits into the downsizing trend and represents the group of engines with spark ignition which are frequently used in vehicles of the Renault team. A confirmation that our choice of the unit was right is also the fact that Sławomir Dziubański, Ph.D., Eng., and Jerzy Jantos, Ph.D., Eng., from the Opole University of Technology decided to continue and develop their research into a turbogenerator solutions (Dziubański S, 2008, The use of exhaust …).

The experiments of the above mentioned researchers were carried out on the unit with spark ignition, naturally aspirated, with the capacity of 1,400 cm^3 made by the Fiat. After the research team from the Opole University of Technology achieved promising results, in the implementation of the REST project it was decided to move a step further and implement a turbogenerator in the supercharged engine and create a logical system managing the work of the electric energy sources.

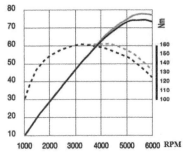

Figure 1. Graph of Renault 1.2 TCe engine power [5]

3 ELECTRIC ENERGY DEMAND

A tendency for the by-wire system development, a steadily growing active safety level required by the European Commission and continuous progress in providing vehicles, even those of a lower standard, with additional equipment, increase appetite and enforce more demand for electric energy in a vehicle. The necessity to provide urban cars with such solutions as an ABS (anti-lock braking) system, ESP (electronic stability programme) system, air-conditioning or navigation which so far have been limited to large cars of the premium class, caused that the power of alternators (3-phase synchronous generators) in vehicles with combustion units having the capacity of up to 1,600 cm^3 increased drastically. The combustion unit used in our investigations was applied in five models of the Renault brand. The manufacturer offered the top additional equipment in the Megane and Clio Spor models. Analysis of electric energy demand by a vehicle with the best equipment indicated that the value approaches 700 W. The Valeo company alternator characterised by such parameters as 12 V, 90 A and power of 1,080 W was mounted in these vehicles (Fig.2).

A tendency for the by-wire system development, a steadily growing active safety level required by the European Commission and continuous progress in

providing vehicles, even those of a lower standard, with additional equipment, increase appetite and enforce more demand for electric energy in a vehicle. The necessity to provide urban cars with such solutions as an ABS (anti-lock braking) system, ESP (electronic stability programme) system, air-conditioning or navigation which so far have been limited to large cars of the premium class, caused that the power of alternators (3-phase synchronous generators) in vehicles with combustion units having the capacity of up to 1,600 cm^3 increased drastically. The combustion unit used in our investigations was applied in five models of the Renault brand. The manufacturer offered the top additional equipment in the Megane and Clio Spor models. Analysis of electric energy demand by a vehicle with the best equipment indicated that the value approaches 700 W. The Valeo company alternator characterised by such parameters as 12 V, 90 A and power of 1080W was mounted in these vehicles.

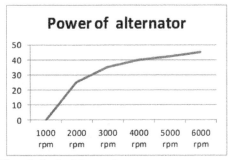

Figure 2. Graph of alternator's power

4 ENGINE TEST STAND

In order to conduct tests on the engine bed of the Institute of Vehicle and Machine Operation of the University of Technology and Humanities in Radom, a test stand was created based on the above mentioned combustion unit combined with a 150Nm eddy current brake. To ensure full control of the tested object, the factory ignition map was reproduced from the original Siemens controller and transferred to a fully programmable Perfect Power controller. This controller enables programming the combustion engine operation parameters and data recording during its operation. A number of sensors serving and supervising the combustion engine operation were connected to the controller. The sensor-controller connection was accomplished by means of four connectors. Each pin was described with regard to managing an appropriate sensor. In order to connect the sensor to an appropriate pin of a given plug, the cables were led out in accordance with the producer's description. The cables were numbered so as to avoid any mistake during the

assembly. After all pins had been connected, the program operating the universal controller instead of the nominal one was started off. The settings of the performance parameters of the power unit were verified for the Renault engine. This verification consisted in the determination of:
– the number of cylinders,
– the number of ignition coils,
– maximum rotational speed,
– calibration of the vacuum pressure sensors in the intake manifold,
– calibration of the temperature sensor in the intake manifold,
– calibration of the coolant temperature sensor and determination of the temperature at which the fan is to be switched on,
– selection of the appropriate sensor for a given model of the combustion engine from the list,
– mapping of the ignition advance angle,
– mapping of the fuel injection timing,
The Perfect Power enables the process of controlling and recording the parameters read off the sensors as well as the controlling signals with the frequency of 1, 10, 100 Hz (Fig.3). The controller was connected by means of an interface. The controller was programmed using an appropriate module which allowed for any configuration of the testing system and controller's algorithms.
(Service Organisation, 2002, Board network, Seat Barcelona)

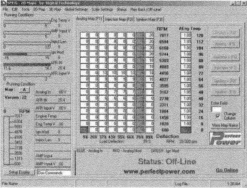

Figure 3. Perfect Power program window [own screenshot]

5 RESEARCH RESULTS

Considering statistics of modern vehicle operation conditions and the range of the unit's nominal power it was assumed that the load drift of the combustion engine by the eddy current brake would be every 15 Nm throughout the range of up to 90 Nm (Fig.4,5,6). The range of rotations tested in the project started from 1,500 rpm because the pressure values in the exhaust system measured at neutral gear are too low to significantly speed up the turbogenerator and thus let us achieve credible results.

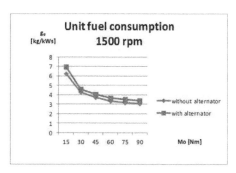

Figure 4 Specific fuel consumption with and without an alternator at 1,500 rpm

During the tests of specific fuel consumption in the measurement variant with an alternator, the generator was under the load of the power of 700W corresponding to maximum electric energy demand of vehicles in which it is used.

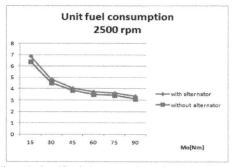

Figure 5. Specific fuel consumption with and without an alternator at 2,500 rpm

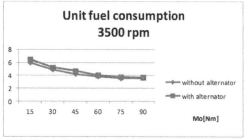

Figure 6. Specific fuel consumption with and without an alternator at 3,500 rpm

The tests were carried out using unleaded petrol Pb 95. During the tests, the exhaust emission was also under control including such factors as carbon oxides, hydrocarbons, nitrogen oxides and the lambda settings. The exhaust analyser used for tests was AVL model DiCOM 4000.

183

6 CONCLUSIONS

The conducted test indicated, that the usefulness of the electric energy recovered from the turbogenerator throughout the range of rotations of up to 3,500 rpm and loads not exceeding 90 Nm is worth considering in a supercharged spark ignition unit of small capacity. The reason for this is that the specific consumption measured for the supply medium is almost identical for loading with an AC generator and without it and possible differences are on the border of the measurement error.

BIBLIOGRAPHY

[1] Postrzednik S., Żmudka Z.: Termodynamiczne oraz ekologiczne uwarunkowania eksploatacji tłokowych silników spalinowych,(Thermodynamic and ecological conditions of operation of internal combustion piston engines) Gliwice 2007. (Published in Polish)

[2] Kozaczewski W.: Konstrukcja grupy tłokowo cylindrowej silników spalinowych (The design of the cylinder piston internal combustion engines). WKiŁ 2010. (Published in Polish)

[3] Dziubański S., Jantos J., Mamala J.: Wykorzystanie energii spalin do napędu turbogeneratora w silniku ZI, (The use of exhaust gas energy to drive the turbo-generator in SI engine) Czasopismo Techniczne Politechnika Krakowska 2008. (Published in Polish)

[4] Teodorczyk A., Rychter T.: Teoria silników tłokowych, (Theory piston engines) WKiŁ 2006. (Published in Polish)

[5] Luft. M, Olszowiec. P: Application of a turbogenerator in the I.C. engine exhaust system, Cesds Słowacja 2011.

[6] Service Organisation: Sieć pokładowa, (Board network) SEAT S.A Sdad. Barcelona. Tomo styczeń 2002. (Published in Polish)

A Comparative Approach of Electrical Diesel Propulsion Systems

A. Arsenie, R. Hanzu-Pazara, A. Varsami, R. Tromiadis & D. Lamba
Constanta Maritime University, Romania

ABSTRACT: In this paper we intend to discuss different types of propulsion systems using engines with internal combustion which involves electrical generators for producing electrical energy. This electrical energy is used further on for powering electrical engines and providing the necessary revolution and power for the propellant. Further on we are going to present the advantages of this installation such as: eliminating the speed reduction unit and the axial line which impose a more difficult maintenance and surveillance; considerable reduction of vibrations and noise allowing their use onboard submarines; conceiving a more flexible system regarding the commutation of power from one propulsion system to another; increasing space in the engine department by eliminating the speed reduction unit and axial line; the possibility of using a better performing automation and lowering the number of human operators; better possibilities for rationally placing onboard the engine department and component elements of the propulsion system. Besides these advantages, there is also one main disadvantage of such installations which shall be discussed being represented by the increased complexity of the electrical system which determines a higher probability of failure in exploitation. In order to emphasise the comparative approach of our analysis we also intend to analyse in detail the main component elements of an electrical Diesel propulsion system which are: the electrical engine (EE) which provides the transformation of the electrical energy in rotation mechanics of the propeller shaft; electrical conductors (EC) providing the transfer of electrical energy; electrical breaker (EB) which allows disruption of the connection between the generator and electrical engine meaning between the thermal engine and the propeller.

1 INTRODUCTION

In the last twenty years, different technologies have been developed for the reduction of the temperature and consequently reduction of NO_x emissions. Some of these technologies are: water injection, steam injection and dry combustion room with a high coefficient of combustion excess.

This type of installations uses internal combustion 4-stroke engines, smaller in dimension and with lower revolutions (than 2-stroke engines) which are reduced by the revolution reducer in order to provide an optimal functioning of the propeller.

We will further on describe the main components of these systems and we will present the advantages and disadvantages presented during our research.

2 ADVANTAGES AND DISADVANTAGES

Advantages of this installation are: the weight of the aggregates and the volume of the engine room, per power unit are reduced by 40-50% as compared with the installations without a revolution reducer; the cost of the installation is lower by 10-15% than the cost of the installations without a revolution reducer; There is a possibility for powering electrical generators by the main engines, which reduces both the cost of electrical power and the number of diesel generators onboard; propulsion systems may be designed in larger range of power with only one type of engine; ageing of pistons, cylinders and segments is much more reduced; better possibilities for rational placing onboard of the engine department; simpler introduction and extraction of engines in/from the engine department; higher safety of the propulsion system; better organization of maintenance especially in case of automatic systems with two or more engines; lower specific effective consumption of fuel.

Disadvantages of the installation are: more difficult maintenance of the distribution system of engines (especially evacuation valves); higher number of cylinders per power unit leads to an increase of the possibility of malfunctioning and necessity of providing a higher number of spare parts; more rigorous requirements concerning fuel quality.

3 ELECTRICAL DIESEL PROPULSION SYSTEMS

These types of propulsion systems use internal combustion engines powering electrical generators, by producing electrical power. This electrical power is further used to power electrical engines and provide the necessary revolution and power for the propellant.

Advantages of this installation are:
– eliminating revolution reducers and the axial line which imposed more difficult maintenance and surveillance;
– considerable reduction of vibrations and noise allowing their use onboard submarines;
– designing a much more flexible system concerning power exchange from one propulsion system to the other;
– increase of space in the engine department by elimination the revolution reducer and the axial line;
– possibility of using a better automation and decreasing the number of human operators;
– better possibilities for a rational placing onboard of the engine department and component elements of the propulsion system.

Disadvantages of the installation are:
– higher complexity of the electrical system determining a higher probability of failure in exploitation.

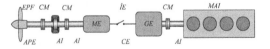

Figure 1. General Scheme of the electrical diesel propulsion system

Main component elements of a propulsion installation electrical diesel type are: the electrical generator (GE) providing mechanical energy transformation of the engine into electrical power; electrical engines (ME) providing electrical energy transformation in revolution mechanical energy of the propeller shaft; electrical conductors (CE) providing electrical power transfer; electrical

breaker (IE) allowing the interruption of the connection between the generator and electrical engine meaning between thermal engine and propeller.

Distribution of energy forms and revolutions on the electrical diesel propulsion system are presented in the next figure.

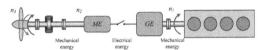

Figure 2. Distribution of energy forms and revolutions

In case of electrical diesel propulsion, the internal combustion engines revolution is the same with the generator's revolution and the propeller's revolution is the same with the powering electrical engine's revolution.

In figure 3 the scheme of an electrical diesel propulsion installation is presented using three internal combustion engine for powering three electrical generators.

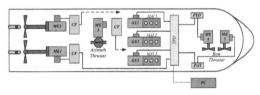

Figure 3. Electrical diesel propulsion installations with fast engines

Notations in the figure are: GE1, GE2 and GE3 – electrical generators; ME1 and ME2 –electrical engines for the main propulsion; ME3 – electrical engine for Azimuth Thrusters; ME4 and ME5 – electrical engines for Bow Thrusters; CF – frequency convertor; PYD – star-triangle power provider; PAT – auto-transformation power provider; PC – power switch board; TPD – main distribution panel.

In figure 4 a propulsion system with three main diesel generators is presented and an emergency diesel generator, and two electrical engines for each propeller shaft. Notations in the figure are: DG1, DG2 and DG3 – main diesel generators; ME1, ME2, ME3 and ME4 – main propulsion electrical engines; ME5 and ME6 – electrical engines for Bow Thrusters; PP –main propulsion; DGA – emergency Diesel-generator; TDA – emergency distribution panel.

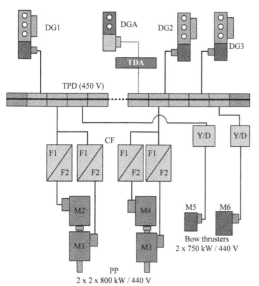

Figure 4. Scheme of an electrical diesel propulsion installation

4 COMBINED PROPULSION INSTALLATIONS WITH INTERNAL COMBUSTION ENGINES

Naval propulsion combined installations provide a mixture between the advantages of the different types of naval thermal machines as well as a greater adaptability of the propulsion system to the requirements imposed by maritime and river ships according to superior thermo-dynamic capacities.

Further on the most well-known types of mixed installations of the naval propulsion are presented.

4.1 Combined propulsion installation type CODAD

CODAD (COmbined Diesel And Diesel) is a very well-known propulsion system, characteristic for ships using two propulsion engines and only one axial line with one propeller.

In the figure below, along with notations used for propulsion installations previously presented, with DCD on-off device was noted (the clutch).

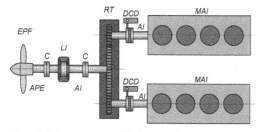

Figure 5. Scheme of a combined propulsion installation type CODAD

4.2 Combined propulsion installation type CODOG

Installation type CODOG (COmbined Diesel Or Gas) is designated generally for ships operating at full speed for a long time, as in the case of war ships such as corvettes and frigates.

In figure 6, TRD represents the transmission on carriers for coupling the gas turbine in the propulsion system.

For each propeller shaft there is an internal combustion engine designated in order to obtain cruise speed and a gas turbine in order to obtain maximum speed. Both are connected to the propeller shaft by means of couplings.

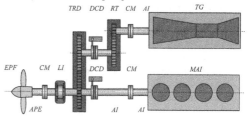

Figure 6. Scheme of a combined propulsion installation type CODOG

Unlike CODAG system in this case only one propulsion machine drives the propeller shaft while the second machine is in standby. The advantage of the CODOG system is that the transmission system of movement and revolution reduction is much simpler. Disadvantages are: it necessitates propulsion machines much larger in order to obtain the same power as in the case of the propulsion system CODAG and fuel consumption at high speed is much higher in the case of the CODOG system. One of the first ships using this type of propulsion was Corvette FNS Turunma belonging to Finland's Marine. She had a tonnage of 1330 [t] and a propulsion made up of three engines of 2200 [kW] made by Wärtsilä each driving a propeller and a gas turbine Rolls Royce Olympus TM1 of 16000 [kW] driving a jet pump. Cruise speed of the ship was 17 knots and full speed obtained with the gas turbine was 37 knots. The ship was launched in 1963 and functioned up to 2002 when it was withdrawn.

4.3 Combined propulsion installation type CODAG

The installation type CODAG (COmbined Diesel And Gas) is also designated to ships operating at full speed for a long time, a speed which is considerably higher than cruise speed. It's the case of corvettes and modern frigates.

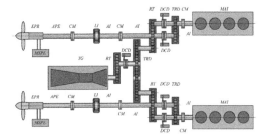

Figure 7. Scheme of a combined propulsion installation type CODAG

The installation is composed of one or more internal combustion engines, according to the number of shaft lines, and a gas turbine which may be coupled when it goes into the high speed regime. In most situation the difference between power given by engines when they function alone and the situation when they are coupled with the turbine, is too high for set screws in order to limit revolution without changing the transmission report of the revolution redactor in the engines system. Due to this fact special revolution redactors are necessary allowing the change in the transmission report.

This represents a disadvantage and an essential difference from the CODOG system which necessitated a revolution reducer much simpler with a fixed report. For example for the CODAG propulsion system with equips frigates of Fridtj of Nansen class in the Norwegian Royal Navy, transmission report of the revolution reducer for the propulsion engine, may be changed from 1:7.7 (for the situation engine – propeller) to 1:5.3 (for the situation engine and turbine - propeller). Other ships may even have three different transmission reports, one for the case when only one engine is used, the second for two engines and the third in case the turbine is used.

These propulsion systems occupy a much smaller space in the engine department, than those which are using only internal combustion engines for the same power of propulsion installations because much smaller engines may be used and the gas turbine and revolution reducer do not need a lot of space.

The system has much smaller fuel consumption on the functioning in cruise regime comparative with ships having only gas turbine propulsion systems. On the other hand, a disadvantage is represented by the fact that it necessitates a transmission system with revolution reducers with a much higher mass and much more complex from a constructive point of view. Regular cruise speed is 20 knots (using only engines) and maximum speed (obtained by coupling the turbine) is 30 knots.

CODAG system was used for the first time on the German frigate Köln, which had a displacement of 3000 [t]. The propulsion system contains two gas

turbines Brown Boveri&Cie of 8832 [kW] each, four-stroke Diesel engines MAN of 2208 [kW] each and two axial lines ending each with an adjustable blades propeller, with three blades of 2.95 [m.] in diameter. Maximum speed with Diesel engines was 24 knots, with turbines and two engines was 30 knots and with turbines and engines was 34 knots. The ship was operational between 1961 and 1982.

5 CODLAG TYPE COMBINED PROPULSION INSTALLATION

CODLAG propulsion system (COmbined Diesel-eLectricAnd Gas) is a particular case of the CODAG system. It contains electrical engines coupled with the propeller shaft (usually two). Electrical engines are fuelled by Diesel generators and used for low speed. When maximum speed is desired, a gas turbine is used by means of a revolution reducers system.

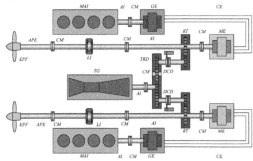

Figure 8. CODLAG type combined propulsion installation scheme

This system combines Diesel engines used for propulsion with those used for generating electrical current, which leads to reducing the cost of usage and construction, and reducing the number of Diesel engines used onboard. The system also implies a much easier maintenance. Electrical engines function very well for long time and may be directly connected with propellers (through transmission shafts) eliminating revolution reducers and necessitating only one reducer to activate the turbine in the propulsion system.

Another advantage of the electrical Diesel system, generally, is represented by the fact that, eliminating the mechanical transmission through the axial line, Diesel engines may be much better isolated, especially acoustically, by the ship's hull making propulsion much more silent. This made the system to be often used onboard submarines, but it may also be used for some surface ships such as submarines' hunters. Usually ships and submarines having a CODLAG propulsion system are equipped with rechargeable batteries allowing silent manoeuvrability without using any type of Diesel engine.

A particular case of the CODLAG system is the IEP system (Integrated Electric Propulsion) or IFEP (Integrated Full Electric Propulsion). For this system both diesel and gas turbine engines are used for the electrical generators. Such systems are used on passenger ships such as RMS Queen Mary 2.

6 CONCLUSION

As a conclusion for our research we should first emphasize that equilibrium between the advantages and disadvantages of using such an installation is difficult to reach and it could prove subjective according to the needs of individual users. Weight and costs as advantages opposed to maintenance and risk of malfunction as disadvantages could prove that the system is not a perfect one. Therefore, it only depends on whether the user would prefer to invest more money from the beginning without the risk of particular malfunctions later on or less money in the beginning assuming the risk that malfunctions could arise at some point.

Further on, the paper presents the electrical diesel propulsion system from a technical point of view. Here again our research is divided into the analysis of the advantages and disadvantages of such a system. In opposing advantages such as a considerable reduction of vibrations and noise and the possibility of using a better automation decreasing therefore the number of human operators to the disadvantage of higher probability of failure in exploitation, the need for an alternative solution arises.

This is why in the final chapter the paper presents the alternative of combined systems meant to satisfy the particular requirements of their users. Finally the most well known types of mixed installations are presented in order to prove that each one of them provides a positive combination between the advantages of the different types of naval thermal machines.

ACKNOWLEDGEMENTS

This article is a result of the project "Increasing quality in marine higher education institutions by improving the teaching syllabus according to International Convention STCW (Standards of Training, Certification and Watchkeeping for Seafarers) with Manila amendments". This project in co funded by European Social Fund through The Sectorial Operational Programme for Human Resources Development 2007-2013, coordinated by Constanta Maritime University.

The content of this scientific article does not reflect the official opinion of the European Union. Responsibility for the information and views expressed in the article lies entirely with the authors.

BIBLIOGRAPHY

Takagi M., Atsuto O., Masaru I, Katsuhide H., NOx Emission Value - Study of On-board Measurement, Journal of the Japan Institute of Marine Engineering, 2006

Trifan A. Instalaţii energetice navale cu motoare navale. Emisii poluante, PhD research paper, 2009

U.S. Department of Energy, Progress report for advanced combustion engine technologies, Advanced Combustion Engine R&D Office of Vehicle Tech. 2008

Warnatz, J., Mass, U., Combustion, Springer, Varlag, 1996

Wärtsilä, Wärtsilä 38 and 64 - Technology Review, 2006

Wärtsilä, Sulzer RT Flex 50, Engine documentation, 2005

Wärtsilä, Sulzer RTA 72 U-B, Engine documentation, 2005

Woodyard D., Pounder's Marine Diesel Engines and Gas Turbines, Elsevier, 2004

Method of Determining Operation Region of Single-transistor ZVS DC/DC Converters

E. Szychta, M. Luft & D. Pietruszczak
Kazimierz Pułaski University of Technology and Humanities in Radom, Poland

L. Szychta
University of Technology, Lodz, Poland

ABSTRACT: This paper introduces a method of determining ZVS operation regions for selected multiresonant DC/DC converters in the state of stable operation. These regions are determinded on the basis of simulation results by means of Simplorer software. Such variation ranges of the control frequency and the corresponding duty cycle of the transistor control signal are adopted for purposes of the testing that semiconductor elements are switched at zero voltage.

1 INTRODUCTION

Multiresonant DC/DC converters convert electricity by using resonance when switching semiconductor elements at high frequencies. Resonant circuits of these systems oscillate at a minimum of two free vibration frequencies in a full switching cycle. Attempts at obtaining high switching frequencies are combined with great energy efficiencies.
(Citko, 2001. Resonant Circuits in ...),
(Tabisz, 1989. Zero-voltage-switching ...),
(Tunia, 2003. Theory of Converters)

Efficiency of resonant converters is largely dependent on switching processes of semiconductor components. Power losses, the current times voltage across switched semiconductor components, occur at the time of turn on and off. In order to minimise switching losses, techniques of the so-called soft switching of semiconductor are applied, i.e. at zero voltage (ZVS) or zero current (ZCS). Energy efficiency coefficient of the systems is then mainly dependent on losses in the state of conductance of semiconductor components.

At high operation frequencies of the converters, parasitic capacitances of transistors and diodes (as well as parasitic inductances of switches) begin to have adverse impact on electromagnetic effects which arise in the circuit as parasitic oscillations of electricity. As part of multiresonant converter structures, parasitic capacitances of transistors and diodes and parasitic inductances of switches become parts of the resonant circuit. This enables elimination of parasitic vibrations.

Multiresonant DC/DC converters, depending on design, have the following properties of voltage conversion: buck, boost or buck of the output voltage relative to the input voltage at an appropriate gain. These converters are applied, for instance, to powering of control systems for power electronic equipment, to portable power supply systems of electronic or telecommunications equipment, etc.

The aim of this paper is to introduce a method of determining ZVS operation region of selected single-transistor DC/DC multiresonant converters by means of simulation testing.

2 SIMULATION MODELS OF CONVERTERS

Figure 1 shows some single-transistor multiresonant converters modelled using Simplorer (Szychta, 2007. Control characteristics ...). Their designs include:

a MOSFET T of an output capacitance C_{OS} and integrated diode D_S, a turn diode D of junction capacitance C_{OD} and resonant circuit elements: resonant inductance L, resonant capacitance C_S in parallel with the transistor and resonant capacitance C_D in parallel with D. L is assumed to comprise parasitic inductances of the circuit. C_S and C_D in parallel with C_{OS} and C_{OD} constitute equivalent resonant capacitances of the circuit. The resonant circuit of a variable configuration operates at two frequencies of free vibrations. C_T and L_T store energy in the circuit when the transistor is conducting. C_F and L_F are parts of a downpass filter.

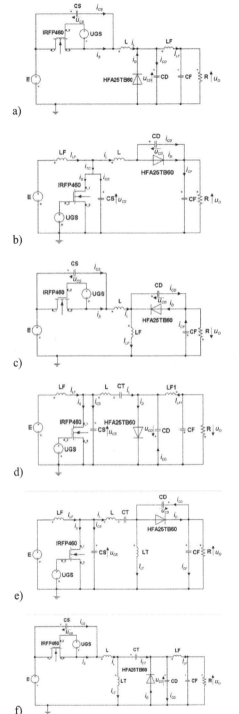

a)

b)

c)

d)

e)

f)

Figure 1. Diagrams of single-transistor ZVS multiresonant converters modelled using Simplorer: a) buck system, b) a boost system, c) a buck-boost system, d) Cuk-type system, e) Sepic-type system, f) Zeta-type system.

The simulation systems shown in Figure 1 include a MOSFET IRFP460, a diode HFA25TB60 and models of the following passive elements: L=7µH, C_S=7nF, C_D=23nF, L_F=L_{F1}=L_T=600µH, C_F=10µF, C_T=0.4µF and R_N=1. The resonant frequencies are: f_S=678kHz and f_D=396kHz. The supply voltage is E=50V DC.

3 REGION OF ZVS OPERATION OF CONVERTERS

Current and voltage changes derived from simulation testing of a boost converter (Figure 1.b) are displayed in Figure 2:

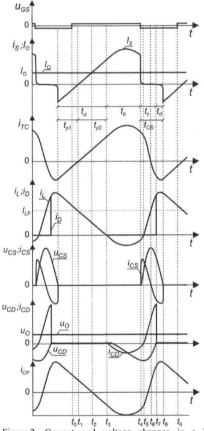

Figure 2. Current and voltage changes in a ZVS boost converter.

The following quantities are used to determine operation regions of the individual structures:
load resistance RN in relative units:

$$R_N = \frac{R}{Z_S} \qquad (1)$$

wave impedance Z_S:

$$Z_S = \sqrt{\frac{L}{(C_S + C_{OS})}} \qquad (2)$$

switching frequency of T in relative units f_N:

$$f_N = \frac{f}{f_S} \qquad (3)$$

free vibration frequency f_S of the resonant circuit L, $(C_S + C_{OS})$:

$$f_S = \frac{1}{2\pi\sqrt{L(C_S + C_{OS})}} \qquad (4)$$

free vibration frequency f_D of the resonant circuit L, $(C_D + C_{OD})$:

$$f_D = \frac{1}{2\pi\sqrt{L(C_D + C_{OD})}} \qquad (5)$$

capacitance CN:

$$C_N = \frac{C_D + C_{OD}}{C_S + C_{OS}} \qquad (6)$$

The multiresonant ZVS converters in Figure 1 are controlled by means of frequency regulation at a constant time of transistor turn-off toff. (Szychta, 2008. Zero Voltage ...) (Szychta, 2007. ZVS operation region ...)
Another method of control is possible, based on frequency regulation at a variable time of transistor turn-off toff. In both cases, the transistor's control modulation ratio β is variable, while it should be of such values that the converter's semiconductor elements are switched at zero voltage. β is expressed as:

$$\beta = \frac{t_{on}}{T} = \frac{T - t_{off}}{T} = 1 - t_{off} \cdot f \qquad (7)$$

where: $f = \dfrac{1}{T} = \dfrac{1}{t_{on} + t_{off}}$

Based on simulation results, the converters' ZVS operation regions are defined as $\beta = f(f_N)$ (Figure 3). ZVS regions were determined by observation, which clearly indicates the instant at which the boundary of ZVS region is crossed. It was manifested as a momentary excess current across the transistor. ZVS regions are delimited by curves plotted for minimum β_{min} and maximum β_{max} within the acceptable range of frequency variations f_N. β_{min} corresponds to maximum times of transistor turn-off $t_{off\,max}$, whereas β_{max} corresponds to minimum times of transistor turn-off $t_{off\,min}$.
In the ZVS operation regions of each system under discussion, a transistor control pulse is supplied within the time interval when the diode D_S

integrated into T, conducts at zero voltage of D. The transistor begins conducting at $t = t_1$, when the diode current D_S is zero for $u_{CS} = 0$ (Figure 2). The transistor is always turned off at an approximately zero voltage u_{CS}. Turn-on of D involves interception of the current i_{CD} at $u_{CD} = 0$. The diode starts to be turned off when $i_{CD} = 0$ and $u_{CD} = 0$. The capacitor C_S should overload within the time interval of transistor turn-off t_{off}. Where t_{off} equals t_{CS} of C_S overloading, $t_{off} = t_{off\,min} = t_{CS}$, (which corresponds to the curve for $\beta = \beta_{max}$) at a variable f_N in the ZVS operation regions, and then the converter operates on the boundaries of the ZVS region. Going beyond the region at $\beta > \beta_{max}$ $(t_{off} < t_{off\,min})$ will cause the transistor to turn off at an unloaded C_S, with the risk of hard commutation and greater losses of switching power. Where t_{off} is excessive, $t_{off} > t_{off\,max}$ $(\beta < \beta_{min})$, the direction of i_{TC} may change (Fig. 2.) and C_S will load. As a consequence, the transistor will be turned on at other than zero voltage. This implies the ZVS region of the converter's operation is determined by such values of β for which the following condition obtains:

$$\beta_{max} \geq \beta \geq \beta_{min} \qquad (8)$$

where: $\beta_{min} = \dfrac{t_{p2} + t_b}{T}$, $\beta_{max} = \dfrac{t_a + t_b}{T}$

for which $t_{off\,min}$ and $t_{off\,max}$ meet the following condition across the entire range of control frequency variations f_N (Szychta, 2008. Zero Voltage ...):

$$t_{off\,min} \leq t_{off} \leq t_{off\,max} \qquad (9)$$

where:
$t_{off\,max} = t_{p1} + t_c + t_d$, $t_{off\,min} = t_{CS}$,
$t_{p1}, t_{p2}, t_a, t_b, t_c, t_d, t_{CS}$ - the time intervals shown in Figure 2.

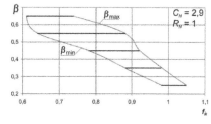

a)

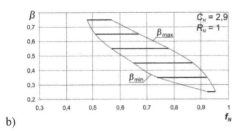

b)

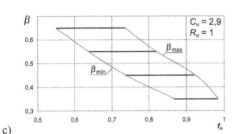

c)

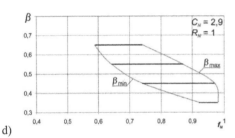

d)

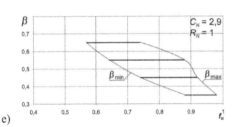

e)

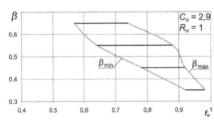

f)

Figure 3. Operation regions of single-transistor ZVS multiresonant converters: a) buck system, b) a boost system, c) a buck-boost system, d) Cuk-type system, e) Sepic-type system, f) Zeta-type system (simulation results).

All geometrical interpretations of simulation results which describe the operation regions of the convertors in operation are presented as continuous functions, results of approximations to calculation points. The latter are generated by simulation testing at a resolution of $f_N=0.01$.

4 CONCLUSION

ZVS operation regions of single-transistor multiresonant converters are determined by recommended operation ranges of the transistor and help to define parameters of the control system.

The boost converter is characterised by a maximum range of frequency variations Δf_N in the ZVS operation region among systems with voltage boosting capabilities.

In the ZVS operation region determined by variations of the maximum $\beta_{max}(f_N)$ and minimum $\beta_{min}(f_N)$ control modulation ratio of the transistor controlling pulse as a function of frequency f_N, the controlling pulse occurs within control ranges of the diode D_S integrated into the transistor. The transistor is turned off at zero voltage. Switching of semiconductor elements at zero voltage provides for elimination of switching power losses and generates high efficiencies of this equipment.

REFERENCES

Citko, T., Tunia, H., Winiarski B. 2001. Układy rezonansowe w energoelektronice. (Resonant Circuits in Power Electronics). Wydawnictwa Politechniki Białostockiej Białystok. (Published in Polish)

Szychta, E. 2008. Zero Voltage Switched Multiresonant Converters - Analysis and Design. Monograph University of Zilina Sk, 126 pages ISBN 978-80-8070-889-4.

Szychta, E. 2007. ZVS operation region of multiresonant DC/DC boost converter. Journal of Advaces in Electrical and Electronic Engineering Faculty of Electrical Engineering Vol.6 No.2 p.60-62 Zilina University Sk.

Szychta, E. 2007. Control characteristics of multiresonant ZVS boost converter. Przegląd Elektrotechniczny No. 2/2007, p.127–132.

Tabisz, W.A., Lee, F.C.Y. 1989. Zero-voltage-switching multiresonant technique-a novel approach to improve performance of high-frequency quasi-resonant converters. IEEE Transactions on Power Electronics vol.4 No.4 p.450–458.

Tunia, H., Barlik, R. 2003. Teoria przekształtników. (Theory of Converters). Oficyna Wydawnicza Politechniki Warszawskiej Warsaw. (Published in Polish)

Maritime Law and Policy

Maritime Law and Policy
Safety of Marine Transport – Marine Navigation and Safety of Sea Transportation – A. Weintrit & T. Neumann (eds.)

The Implementation of a New Maritime Labour Policy: the Maritime Labour Convention (MLC, 2006)

F. Piniella
Department of Maritime Studies, Universidad de Cádiz, Spain

J. González-Gil
European Maritime Safety Agency

F. Bernal
Department of Labour Law and Social Security, Universidad de Cádiz, Spain

ABSTRACT: This paper explores the implementation of a new Maritime Labour policy based on the Maritime Labour Convention (MLC, 2006), which was adopted in 2006 and came into force in 20 August 2013. It aims to investigate its impact on the development of regulations among Member States and on vessel inspections under the Port State Control (PSC) system. It also focuses on the most important contents of the Convention and its implementation by Flag and Port States. Last, the States's response to the MLC, 2006 is also discussed.

1 INTRODUCTION

1.1 *Aim and scope*

This paper explores the implementation of a new Maritime Labour policy based on the Maritime Labour Convention (MLC, 2006).

20 August 2014 was the first anniversary of the entry into force of the MLC, 2006. During these first 12 months, and as a consequence of MLC-related deficiencies, about a hundred ships have been detained by one of the ratifying Authorities within the framework of the Memorandum of Understanding of Paris (Paris MoU).

This regional Port State Control Agreement (PSC) among Maritime Administrations was signed in 1982 by fourteen European countries. At present, twenty-seven countries are parties to the Paris MoU. The European Commission, although not a signatory to the Paris MoU, is nowadays represented in the Paris MoU bodies.

With this background, this study focuses on a relevant aspect of the Convention: its implementation by those parties in the shipping industry. The response of the States to the MLC, 2006 is examined and the outcome of the first year of implementation is discussed in relation to the publicly available data of the detained vessels within the framework of the Paris MoU inspections.

1.2 *Literature review*

In the past decades maritime transportation safety has faced many changes. Yang, Wang & Li (2013)

reviews the challenges of maritime safety analyses and the different approaches used to quantify risks in maritime transportation and the importance of risk quantification analysis to facilitate the transformation of a maritime safety culture. Almost twenty years ago, Brooks (1996) alerted about the phenomenon of delegation of ship safety surveys.

Previously, authors like Li & Wonham (2001) or Li & Zheng (2008) revealed areas where IMO regulations pertaining to the safety of life at sea could be enhanced. Even, more recently, Schröder-Hinrichs et al. (2013) studied the possibilities of IMO in a proactive policy of responses to maritime accidents.

Taking a quantitative approach to shipping registry selection, Alderton & Winchester (2002a and 2002b) established what they called the Flag State Conformance Index (FLASCI). More recently, M. Perepelkin et al. (2010) put forward a new framework, of a theoretical nature as well, but with a considerable improvement of the system. Those scholars showed that the assessment system of the Paris MoU had three important flaws: although a system for Flag States or ROs conducting a large number of inspections, it did not include all those vessels that had not been inspected (47% of the total) and was merely based on ship detentions and not deficiencies. For this reason they formulated the concept of Quality of Flag (Q) based on other considerations different from the number of detentions so that the deficiencies and accidents that occurred were also taken into account. These concerns had already been expressed in the econometrics studies conducted by Cariou et al. (2007-2009) and Knapp & Franses (2007).

The relevance of MLC, 2006 has been reviewed in (2009) or Ruano-Albertos & Vicente-Palacio (2013). Also, Gonzalez-Gil (2013) and Piniella et al. (2013) have recently revealed the "dramatic" relationship between PSC and MLC, 2006: "Who will give effect to the ILO's Maritime Labour Convention 2006?"

2 THE PROTECTION OF SEAFARERS: THE EMERGING NEW INTERNATIONAL REGIME.

2.1 *Standards of maritime labour*

ILO has a specific maritime programme, which aims at promoting the social and economic progress of maritime employees. Since its foundation, the ILO has a Joint Maritime Commission that advises the Board on maritime matters; it also organizes special meetings of the International Labour Conference with the sole responsibility of drawing up and adopting standards pertaining to work at sea. The ILO cooperates with other bodies of the United Nations organization that are competent in the maritime sector, particularly the International Maritime Organization (IMO) and the World Health Organization (WHO).

2.2 *The Maritime Labour Convention (MLC, 2006)*

The MLC, 2006 came into force on 20 August 2013 – one year after being ratified by at least 30 ILO member States accounting for one-third of the world's gross shipping tonnage. Its aim is to improve the existing framework established under the Merchant Shipping (Minimum Standards) Convention, 1976 (No. 147) by means of a single, comprehensive instrument that would, as far as possible, consolidate all the standards currently in force under the ILO's maritime Conventions and Recommendations, and include the main principles featuring in other international labour Conventions. It updates some 65 of the ILO's maritime labour standards and is expected to extend comprehensive protection to more than 1.2 million seafarers around the world (Bauer, 2007; McConnell, 2011). It prescribes minimum standards for their conditions of employment, accommodation, recreational facilities, food and catering, health, medical care, welfare and social security protection (Chaumette, 2009-2010).

Two important issues stand out: first, the Convention provides mechanisms for effective implementation and real protection of seafarers' rights through inspection and certification procedures as well as reporting to the ILO's supervisory machinery; second, and as already noted above, the Convention supplements the existing body of international maritime law contained in the IMO Conventions and governing commercial aspects (shipping), maritime safety and marine pollution control.

The Convention is meant to be the fourth pillar of the international maritime regulatory regime alongside SOLAS, MARPOL and STCW. Whether the new Convention may be a success or a failure, however, it will heavily depend on its implementation by the flag States and within the framework of Port State Control. At the outset, the main responsibility for compliance is clearly assigned to the flag State's survey and certification system, in particular: the issuance of the Maritime Labour Certificate and the Declaration of Maritime Labour Compliance. In practice, therefore, much of the responsibility of inspection is likely to fall on PSC officers – hence the role explicitly given to PSC (Article V (4) and Regulation 5.2) – albeit on a discretionary basis.

A final option that remains to be considered in strengthening the application of the MLC, 2006 is the possibility of addressing these issues by referring to the social and corporate responsibility of the shipowners, at least when they can be identified. While it is difficult to say which, if any, mechanisms might be effective, some shipowning companies might have an interest in reaching agreements with trade unions so as to further the goals of the MLC, 2006 (Piniella et al. 2013).

3 EU REGULATIONS

3.1 *First Council Decisions*

Within the EU, the Council Decision 2007/431/EC of 7 June 2007 authorised Member States to ratify the MLC, 2006, of the International Labour Organisation (EC 2007) in the interest of the European Community. Some Member States like Spain, Bulgaria, Luxembourg, Denmark, Latvia, or the Netherlands had already ratified it some years before. An extensive and time-consuming gap analysis of the national legislations was the prerequisite for international and national standards to be consistent and in order to ratify the Convention. This was stated in the Proposal of a Directive concerning flag States responsibilities (EC 2012).

The EU also adopted the Council Directive 2009/13/EC of 16 February 2009, amending Directive 1999/63/EC2 (EC 2009a). Firstly the EU had set up a legal framework to increase maritime safety by adopting three maritime safety packages, the latest dating from 2009: the mentioned Directive 2009/13/EC and the Directive 2009/21/EC on compliance with flag State requirements to ensure that EU States effectively and consistently discharge their obligations as flag States, to enhance safety and prevent pollution from ships flying their flag. To this effect, IMO standards, in particular the mandatory audit plan of national maritime administrations and the IMO Flag State Code apply (EC, 2009b).

As early as 1994, the EU Council issued the Council Directive 94/57/EC on common rules and standards for ship inspection and survey organizations and for the relevant activities of maritime administrations, within the framework of its common policy on Maritime Safety (EC, 1994).

Later, in 2001, this earlier Directive was modified by that of 2001/105/CE; this increased, in particular, the requirements for the ROs to tighten even further the existing framework and meet the IACS standards. Even more recently, in 2009, with the Directive 2009/15/CE and the Ruling (EC) 391/2009, more severe changes were made in the regulation of the ROs. The creation of a certification body intended to be independent and the reform of the system of sanctions contributed to such changes. There have also been significant advances in the mutual recognition of the certificates among the better ROs, provided the certificate has been issued on the basis of equivalent technical standards (EC, 1994-2009c,d).

3.2 *The new Directive 2013/54/EU*

Different sections of the MLC, 2006 have been inserted into different Union instruments both as regards flag State and port State obligations. The aim of the new Directive 2013/54/EU (EC 2013) was to introduce some degree of compliance and enforcement provisions, envisaged in Title 5 of MLC, 2006. These are: ensuring the effective discharge of their obligations as flag States with respect to the implementation, by ships flying their flag, of the relevant parts of MLC, 2006 and establishing an effective system for monitoring mechanisms, including inspections.

The new Directive of 2013 requires each Member State to provide the ILO with a current list of any recognised organisations authorised to act on its behalf, and to keep this list up to date. The list shall specify the functions that the ROs have been authorised to carry out.

4 PARIS MOU: RESULTS OF THE FIRST YEAR OF MLC, 2006

4.1 *Happy birthday MLC, 2006*

As mentioned earlier in this paper, the MLC, 2006 came into force on 20 August 2013 – one year after fulfilment of the requirement that it be ratified by at least 30 ILO member States accounting for one-third

of the world's gross shipping tonnage. Therefore 20 August 2014 marked the first anniversary of the entry into force of the Maritime Labour Convention (MLC, 2006).

4.2 *Methodological aspects*

In this paper we have used inspection information as publicly available through the Paris MoU website. The data include the time period from 20 August 2013 to 20 August 2014.

The main aim of this paper is to analyse the first 12 months of existence of the Convention, focusing on those ships which have been detained by ratifying Paris MoU Authorities as a result of MLC-related deficiencies. In this study we have considered: those deficiencies directly related to the MLC, 2006 and those deficiencies related to the documentation required by the Convention (01139, 01140, 01212, 01218, 01219, 01220, 01221, 01306, 01307, 01308, 01329, 01330, and 01331), according to the list of codes published in Paris MoU website. (Source: https://www.parismou.org/sites/default/ files/List%20of%20Paris%20MoU%20deficiency% 20codes%20on%20public%20website_0.pdf). The deficiencies related to ILO147 were not taken into account.

4.3 *Results*

We first analyse the deficiencies and next the detentions of the vessels. In total: 81,837 deficiencies were detected in the period under study.

In accordance with the MLC-related deficiencies, 5.43% (4,441) deficiencies were found (see Table 1). These may be classified as:
- 181** Minimum requirements to work on board ship 1.60%
- 182** Employment conditions 7.30%
- 183** Accommodation, recreational facilities, food and catering 34.22%
- and 184** Health protection, medical care, social security 56.88%

Table 1. Number of deficiencies found in the MLC groups.

Deficiency Code	No. of deficiencies
*181***	71
*182***	324
*183***	1,520
*184***	2,526
Total	4,441

It is deemed necessary to add the codes related to the required documentation by the Convention, which shows a total of 1,530 deficiencies (see Table 2):

Table 2. Number of deficiencies in documentation groups.

Deficiency Code	No. of deficiencies
01139	4
01140	50
01212	14
01218	200
01219	10
01220	213
01221	8
01306	130
01307	137
01308	578
01329	13
01330	156
01331	17
Total	1,530

The most important groups are:
- 01308 Records of seafarers' daily hours of work or rest.
- 01220 Seafarer' employment agreement SEA.
- 01218 Medical certificate.

If we take into account both groups, the total number of deficiencies is 5,971 and the corresponding percentage rises to 7.30% of the total deficiencies of the year under study.

We then examined the number of deficiencies that lead to the detention of the ship. This is very probably the most important aspect of the analysis.

Table 3. Number of deficiencies and detentions.

Total deficiencies	Deficiencies are ground for detention	Related to MLC
81,837	6,848	238

Broadly speaking, 8.37% correspond to deficiencies for detentions but only 238 of these were related to MLC.

If we analyse the data by groups the results are as shown in Table 4.

Table 4. Number of deficiencies for detention according to the 18 Deficiency Code groups (MLC, 2006).

181**	Def.	182**	Def.	183**	Def.	184**	Def.
18103	1	18201	15	18302	24	18401	7
		18203	79	18306	5	18406	3
		18204	19	18312	1	18407	1
				18313	4	18408	9
				18314	16	18410	4
				18316	1	18412	5
				18321	6	18415	2
				18324	10	18417	3
				18325	4	18418	2
						18420	8
						18424	4
						18425	3
						18428	2
1		113		71		53	
0.42%		47.48%		29.83%		22.27%	

Table 5. Port State and number of MLC-related detentions.

Port State	No. of detentions
Bulgaria	10
Canada	6
Croatia	1
Cyprus	4
Denmark	2
France	5
Germany	1
Greece	11
Italy	2
Malta	2
Netherlands	5
Poland	2
Romania	1
Russia	9
Spain	25
Sweden	4
United Kingdom	2

There is no correlation between the groups (Table 1 and 4): whereas the group "Condition of employment" shows 7.30% of the deficiencies, these involve 47.48% of deficiencies with detentions. It's important to underscore the 18203 Code: "Wages".

The number of MLC-related detentions were 92, including the groups in Table 2 (some vessels were detained in more than one inspection, on occasions even three times). Tables 5 and 6 show the Maritime Administration of the detention ports and the Flag State of the vessels.

Table 6. Flag State and number of MLC-related detentions.

Port State	No. of detentions
Algeria	1
Antigua Barbuda	2
Bahamas	1
Belize	4
Cambodia	4
Cook Islands	1
Curaçao	1
Cyprus	4
Dominica	1
Finland	1
Gibraltar UK	2
Italy	1
Liberia	5
Malta	5
Marshall Islands	3
Moldova	7
Netherlands	2
Norway	1
Panama	17
Poland	1
Russia	2
Saint Kitts…	3
Saint Vicent…	2
Sierra Leone	3
Tanzania	7
Thailand	1
Togo	4
Ukraine	1
United Kingdom	2
Vanuatu	2
Non indicated	1

The detained vessels with MLC-related deficiencies fell in one of these two groups: 64.1% ships older than 20 years, and 47.8% ships older than 30 years. With regard to vessel types, it can be said that 53% were general cargo/multipurpose ships and 18% bulk carriers.

5 CONCLUSION

Although the data used may be misleading (due to the use of different codes deficiencies with respect to compliance with the MLC, 2006), the results of Paris MoU inspections in the first year of the MLC, 2006 yield a hundred detained vessels in the ports of the European Union (17% of the total number of detentions).

The average profile of these cases is: a 20 or more-year-old open register vessel, whose main deficiencies are related to the lack of documentation or non-existent regulation of wages on board.

Recognized Organizations and Open Registries are required to put into effect more rigorous policies to control the current abuses in the recruitment and manning of crews and with all the main MLC-related aspects described above.

PSC inspections are a necessary tool for an effective stepwise enforcement of new maritime labour policies.

DISCLAIMER

The content of this article does not necessarily reflect the opinion of the European Maritime Safety Agency. Responsibility for the information and views expressed in this article lies entirely with the authors.

REFERENCES

Alderton, T., & Winchester, N., 2002a. "Globalisation and de-regulation in the maritime industry." *Marine Policy* 26 1: 35-43. http://dx.doi.org/10.1016/S0308-597X0100034-3

Alderton, T., & Winchester, N., 2002b. "Flag states and safety: 1997-1999" *Marine Policy and Management* 29 2: 151-162 http://dx.doi.org/10.1080/03088830110090586

Bauer, P.J. 2007. The Maritime Labour Convention: An Adequate Guarantee of Seafarer Rights, or and Impediment to True Reforms *Chi J Int Law* 8:643-655.

Brook, M.R., 1996. "The privatization of ship safety" *Marine Policy and Management* 23 3: 271-288 http://dx.doi.org/10.1080/03088839600000089

Cariou, P., Mejia, J., Maximo, Q. & Wolf, F.C., 2007. "An econometric analysis of deficiences noted in port state control inspections." *Maritime Policy and Management* 34 3: 243-258. http://dx.doi.org/10.1080/03088830701343047

Cariou, P., Mejia, M.Q., & Wolf, F.C., 2008. "On the effectiveness of port state control inspections." *Transportation Research Part E: Logistics and Transportation Review* 44 3: 491-503. http://dx.doi.org/10.1016/j.tre.2006.11.005

Cariou, P., Mejia, M.Q., & Wolf, F.C., 2009. "Evidence on target factors used for port state control inspections." *Marine Policy* 33 5: 847-859. http://dx.doi.org/10.1016/j.marpol.2009.03.004

Chaumette, P. 2010. El Convenio sobre el Trabajo Marítimo: cuarto pilar del Derecho Internacional Marítimo. *Rev Ministerio Trabajo* 82:65-76

EC European Council 1994. Council Directive 94/57/EC of 22 November 1994 on common rules and standards for ship inspection and survey organizations and for the relevant activities of maritime administrations.

EC 1997. Directive 97/58/EC of 26 September 1997 amending Council Directive 94/57/EC on common rules and standards for ship inspection and survey organizations and for the relevant activities of maritime administrations.

EC 2001. Directive 2001/105/EC of 19 December 2001 amending Council Directive 94/57/EC on common rules and standards for ship inspection and survey organisations andfor the relevant activities of maritime administrations.

EC 2007. Council Decision 2007/431/EC of 7 June 2007 authorising Member States to ratify, in the interests of the European Community, the Maritime Labour Convention, 2006, of the International Labour Organisation. OJ L161:63 22-06-2007

EC 2009a. COUNCIL DIRECTIVE 2009/13/EC of 16 February 2009 implementing the Agreement concluded by the European Community Shipowners' Associations ECSA and the European Transport Workers' Federation ETF on the Maritime Labour Convention, 2006, and amending Directive 1999/63/EC.

EC 2009b. Directive 2009/21/EC of the European Parliament and of the Council of 23 April 2009 on compliance with flag State requirements.

EC 2009c. Directive 2009/15/EC of 23 April 2009 on common rules and standards for ship inspection and survey organisations and for the relevant activities of maritime administrations.

EC 2009c. Ruling CE no 391/2009 of the European Parliament and of the Council of 23 April 2009 on common rules and standards for ship inspection and survey organisations.

EC 2012. Proposal for a Directive of the European Parliament and of the Council concerning flag State responsibilities for the enforcement of Council Directive 2009/13/EC implementing the Agreement concluded by the European Community Shipowners' Associations ECSA and the European Transport Workers' Federation ETF on the Maritime Labour Convention, 2006, and amending Directive 1999/63/EC.

EC 2013 Directive 2013/54/EU of the European Parliament and of the Council of 20 November 2013 concerning certain f lag State responsibilities for compliance with and enforcement of the Maritime Labour Convention, 2006. OJ L329:56 10-12-2013

González-Gil, J. 2013. Le contrôle par l'État du port en tant que cadre pour la réuss ite de la CTM 2006. *Revue de droit comparé du travail et de la sécurité sociale.* Centre de droit comparé du travail et de la sécurité sociale.

ICS-ISF International Chamber of Shipping - International Shipping Federation, 2013. *Shipping Industry Flag State Performance Table 2012.* Published annually: http://www.ics-shipping.org/resources.htm

IMO International Maritime Organization, 1958. International Convention on the High Seas. London: IMO Pub.

IMO, 1966. International Convention on Load Lines LL. London: IMO Pub.

IMO, 1969. International Convention on Tonnage Measurement of Ships TONNAGE. London: IMO Pub.

IMO, 1973. International Convention for the Prevention of Pollution from Ships, 1973, as modified by the Protocol of 1978 relating thereto and by the Protocol of 1997 MARPOL. London: IMO Pub.

IMO, 1974. International Convention for the Safety of Life at Sea SOLAS. London: IMO Pub.

IMO, 1982. United Nations Convention on the Law of the Sea UNCLOS. London: IMO Pub.

IMO, 1986. United Nations Convention on Conditions for the Registration of Ships UNCCROS. London: IMO Pub.

IMO, 1993. Guidelines for the authorization of organizations acting on behalf of the Administration. A 18/ Res.739. IMODOCS data base: http://docs.imo.org

IMO, 1995. Specifications on the survey and certification functions of recognized organizations acting on behalf of the Administration. A 19/ Res.789. IMODOCS data base: http://docs.imo.org

IMO, 2002. Responsabilities of Goverments and measures to encourage Flag State compliance. FSI 10/3. IMODOCS Database: http://docs.imo.org

IMO, 2003. Development of provisions on transfer of class. FSI 12/12. IMODOCS Database: http://docs.imo.org

IMO, 2005a. Goal-based standards on agenda at IMO's Maritime Safety Committee. Ship construction rules under the spotlight at MSC 80. Briefing 22/2005, 10 May 2005. http://www.imo.org/Newsroom/mainframe.asp?topic_id=1018 anddoc_id=4878

IMO, 2005b. Survey and certification-related matters: Guidelines for Administrations to ensure the adequacy of transfer of class-related matters between recognized organizations ROs. MSC-MEPC.5/Circ.2. IMODOCS Database: http://docs.imo.org

IMO, 2007. Goal-based new ship construction standards. MSC 83/5. IMODOCS Database: http://docs.imo.org

IMO, 2008. Work programme: Development of a Code for recognized organizations RO Code. MSC 84/22. IMODOCS Database: http://docs.imo.org

IMO, 2009. Development of a Code for Recognized Organizations. FSI 17/14. IMODOCS Database: http://docs.imo.org

IMO, 2010. Development of a Code for Recognized Organizations. FSI 19/14. IMODOCS Database: http://docs.imo.org

IMO, 2011. Flag State Implementation: Development of a Code for recognized organizations and amendments to the Code for the implementation of mandatory IMO instruments. MSC 91/10. IMODOCS Database: http://docs.imo.org

IMO, 2012a. Harmonization of Port State Control activities. FSI 20/6. IMODOCS Database: http://docs.imo.org

IMO, 2012b. Development of a Code for Recognized Organizations. FSI 20/13. IMODOCS Database: http://docs.imo.org

IMO, 2012c. Flag State Implementation: Port State control and recognized organizations. MSC 91/10. IMODOCS Database: http://docs.imo.org

IMO, 2013. Harmonization of Port State Control activities. FSI 21/6. IMODOCS Database: http://docs.imo.org

Knapp, S. & Franses, P.H., 2007. "Econometric analysis on the effect of port state control inspections on the probability of casualty: Can targeting of substandard ships for inspections be improved?" *Marine Policy* 31 4: 550-563. http://dx.doi.org/10.1016/j.marpol.2006.11.004

Li, K.X. & Wonham, J., 2001. "Maritime legislation: new areas for safety of life at sea." *Maritime Policy and Management* 28 3: 225-234. http://dx.doi.org/10.1080/03088830110048880

Li, K.X., & Zheng, H. 2008. Enforcement of law by the Port State Control PSC. *Maritime Policy and Management* 351:61–71

McConnell, M.L. 2011. The Maritime Labour Convention - reflections on challenges for flag State implementation. *WMU Journal of Maritime Affairs* 10:127-151

Paris Mou 2014. Results first year Maritime Labour Convention: 113 ships detained for MLC related deficiencies. *Press release* 17th Nov. 2014.

Perepelkin, M. Knapp, S., Perepelkin, G., and Pooter, M., 2010. "An improved methodology to measure flag performance for the shipping industry." *Marine Policy* 34 395–405 http://dx.doi.org/10.1016/j.marpol.2009.09.002

Piniella, F., Silos, J.M., Bernal, F. 2013 Who will give effect to the ILO's Maritime Labour Convention, 2006? *International Labour Review* 1521:59-83

Ruano-Albertos, S. & Vicente-Palacio, A. 2013. Adapting European Legislation to the Maritime Labour Convention 2006 Regulations in Relation to the State Responsibilities of Both the Flag State and the Control of Ships by Port State Control. *Beijing Law Review* 44:141-146

Schröder-Hinrichs, J-U., Hollnagel, E., Baldauf, M., Hofmann, S. & Kataria, A., 2013. "Maritime human factors and IMO policy." *Maritime Policy and Management* 40 3: 243-260. http://dx.doi.org/10.1080/03088839.2013.782974

THETIS: https://portal.emsa.europa.eu/web/thetis

Vicente-Palacio, A. 2009. El Convenio OIT de Trabajo Marítimo CTM 2006 avanza hacia su entrada en vigor. Últimas actuaciones de España y de la Unión Europea a favor de la protección del trabajo de la gente del mar. *Revista General del Derecho del Trabajo y de la Seguridad Social*, Iustel 20

Yang, Z.L., Wang, J. & Li, K.X., 2013 "Maritime safety analysis in retrospect" *Maritime Policy and Management* 40 4: 261-277. http://dx.doi.org/10.1080/03088839.2013.782952

Maritime Law and Policy
Safety of Marine Transport – Marine Navigation and Safety of Sea Transportation – A. Weintrit & T. Neumann (eds.)

A New International Law to Protect Abandoned Seafarers: Amendments to MLC, 2006

F. Bernal
Department of Labour Law and Social Security, Universidad de Cádiz, Spain

F. Piniella
Department of Maritime Studies, Universidad de Cádiz, Spain

ABSTRACT: The aim of our paper is to study the proposal of a new international law to protect abandoned seafarers, according to the development of the Maritime Labour Convention (MLC,2006), in force from 20th August of 2013. Our paper is addressed to the role of the international social partners appointed by the Special Tripartite Committee in the amendment process and principles adopted by the Joint Group IMO/ILO about abandoned seafarers. Government, employer and worker delegates voted in favour of approving amendments to MLC,2006 in order to better protect abandoned seafarers and to establish binding international law on these issues. Finally we focused our study on two main aspects: a new database about the cases of abandoned seafarers and the financial security instrument for compensation to seafarers.

1 INTRODUCTION

1.1 *Aim and scope*

The magnitude of labour rights violations in the case of abandoned seafarers has had the effect of raising awareness among international maritime institutions. In the present paper, we will focus first on the meetings held by the joint International Maritime Organisation (IMO)/International Labour Organisation (ILO) Ad Hoc Expert Working Group, which was commissioned to examine the issue of liability and compensation regarding claims for death, personal injury and abandonment of seafarers.

Nine meetings concerning abandonment have been held, which will be considered here in chronological order and from which we will draw some conclusions. Second, we will also analyse the international regulations drawn up by the ILO in the form of conventions and resolutions, and by the European Union in the form of directives and decisions.

In Figure 1 we include a list of abandoned vessels from ILO database, in order to indicate the magnitude of this problem. However, these numbers are actually higher, for several reasons: the notice of abandonment is voluntary, so many cases are not reported; most cases are reported by ITF so in those ports where there are not inspectors this abandonment is not reported (Gonzalez Joyanes, 2009a).

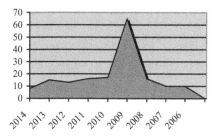

Figure 1. (No. Vessels) Abandonment of Seafarers

1.2 *State of the art*

A literature review reveals that extensive literature exists on maritime labour in general and the Maritime Labour Convention (MLC, 2006), but less has been written about the area of labour law. Traditionally, the legal doctrine on labour has paid scant attention to maritime labour law. However, as indicated by De la Villa Gil (2009): "the previous long cycle of neglect has been broken, and interest is now being directed towards a singularly important legal area". The main studies that have addressed the problem of abandoned seafarers have been conducted by María Arantzazu Vicente Palacio, Olga Fotonipoulou Basurko, Patrick Chaumette, Alexandre Charbonneau, José Ignacio García Ninet, Xome Manuel Carril Vázquez, Denis Nifontov, D.

A. Pentsov and Cristina Sánchez-Rodas Navarro. In the words of Fotinopoulou (2009), *"the abandonment of seafarers can be classed as one of the most common and visible symptoms of the impact on living and working conditions brought about by the deregulation of shipping, the main manifestation of which is the subterfuge of flags of convenience"*.

Most previous studies have addressed the working, living and safety and health conditions of abandoned seafarers. The originator of abandonment is always the ship owner, since this is the person responsible for maintaining satisfactory standards as regards employees' employment conditions, health and welfare. These studies have revealed that seafarers are abandoned every year and for various reasons, which among others include: insolvency of the ship owner, seizure of the ship, bankruptcy of the company, ship damage, wreck or accident, opportunistic management in bad faith on the part of a maritime entrepreneur, immobilisation due to lack of safety measures and, more recently, for breach of the MLC, 2006. Abandoned seafarers are subject to extremely precarious living conditions, with no food, no medical or essential health care, no heating, no financial resources, far from home and isolated. They do not speak the language, nor do they understand the culture of the country or know what steps to take to solve their problem. To make matters worse, seafarers are frequently disadvantaged people who have joined a ship as the only means for them and their dependents to survive.

In addition, as authors such as Pentsov (2001) and more recently Carril Vázquez (2009) have indicated, seafarers work in difficult conditions and are more exposed to specific occupational hazards. Seafarers have longer working hours and are required to work abroad. They are thus highly subject to the shipping company's power of disposition, management and control. These factors increase their risk of being exploited and abandoned when ship owners fail to meet their contractual obligations (De la Villa Gil, 2009). Moreover, the above-mentioned authors have also highlighted the many practical problems that seafarers face in taking legal action in a foreign country: the cost of the security to file a lawsuit, the cost of legal counsel, lack of lawyers on call or the seafarer's own irregular immigrant status as a result of being abandoned.

Those to blame for such abandonment take no responsibility for abandoned seafarers. Neither do the port State authorities provide assistance, sometimes because of a desire to avoid becoming involved in a labour dispute and sometimes because they lack the resources to help them. Abandoned seafarers are thus dependent on the commendable efforts of charitable organisations such as Stella Maris and trade unions such as the ITF. The

phenomenon of abandoned ships and seafarers that has been occurring since 1990 has revealed the limits of international provisions for the protection of unpaid wages and the repatriation of seafarers (Chaumette, 2009, 2014).

Finally, it is important to highlight the work of Gonzalez Joyanes (2009b) and Garcia et al. (2013) on the direct relationship between abandonment of vessels and proliferation of open registers.

2 AN INTERNATIONAL SOLUTION TO THE PROBLEM OF ABANDONED SEAFARERS

2.1 Joint IMO/ILO Expert Working Group

In the absence of adequate international provisions to address this problem, an international instrument was deemed necessary to ensure payment of wages due, alleviation of immediate needs and repatriation of abandoned seafarers. Given the global scale of the maritime industry, such an instrument would need to be international in scope. A human, social and moral issue, as well as a question of human rights, this problem required regulation through an international law.

The Working Group was commissioned to ensure special protection for abandoned seafarers and to improve their working and living conditions. Its remit was to present a series of recommendations to the Governing Body of the ILO and the Legal Committee of the IMO, both deeply concerned about this issue.

The IMO recognises the importance of the human factor for the effective operation of this sector and the safety of navigation. Meanwhile, practically since its inception the ILO has devoted its efforts to protecting maritime workers and improving their conditions of employment. Abandonment is one of the issues addressed in the ILO's Decent Work Agenda, the aim of which is to achieve a reciprocally beneficial situation: if the maritime industry offers decent work, it will attract trained, qualified seafarers; meanwhile, a knowledgeable and experienced crew will ensure the objective of safe navigation (Vicente, 2010a,b).

The origins of the Working Group go back to 1991 and the 26th meeting of the Joint Maritime Commission (JMC). One of the goals of this meeting was to prepare a report which reviewed the working and living conditions of seafarers, bearing in mind the structural transformations that the maritime industry had recently undergone.

The appointment of a joint Working Group was proposed by the IMO to the ILO in April 1998 in the context of the work of the IMO Legal Committee to establish financial guarantees in relation to maritime claims. Agreement was reached on the need for an international instrument that guaranteed the rights of seafarers to compensation for abandonment. The remit proposed for the Working Group would include, but not be limited to: examining the issue of financial security in situations of abandonment, assessing the problem and evaluating the adequacy and effectiveness of existing instruments. Despite the existence of international instruments at that time, cases of abandonment were still occurring and the scant ratification of conventions, insolvency of maritime companies and delays in bureaucratic procedures did nothing to alleviate the situation.

Recognising this to be a serious problem, the ILO authorised the creation of the joint Working Group at its 273rd meeting in November 1998. A total of nine meetings were held between 1999 and 2009, with several intermediate sessions organised between these meetings.

In 1999, the Working Group held its first meeting, at which it was concluded that no international instruments existed which adequately addressed the problem of abandonment. The Working Group considered that in order to resolve this issue it would be necessary for States to implement the agreements and ratify international conventions. It was also proposed that flag States should devise measures to ensure ship owners' compliance with their obligations. A questionnaire was drawn up during this meeting to collect information from States regarding the systems, legislation, practices and plans that they had adopted to address the problem of abandonment. This information would be used to seek possible solutions to the problem of abandonment: national funds, an international fund, compulsory insurance, a system based on bank guarantees or subrogation.

At the second meeting, held in 2000, two stages were proposed. The first, short-term stage would be focused on preparing a draft resolution with measures and guidelines on the provision of financial security, whilst the second, long-term stage would be focused on developing a binding instrument.

At the third meeting, held in 2001, preparation was concluded of the draft resolution and related guidelines on the provision of financial security. Subsequently, these documents were adopted by the IMO Assembly and the Governing Body of the ILO.

2.2 *First resolutions*

Specifically, the IMO Assembly adopted the document as Resolution A.930 (22) on the provision of financial security in the case of abandoned seafarers. Vicente (2009a,b,c & 2010a,b) has stressed the importance of this instrument in relation to protecting seafarers, among other reasons because it proposed the creation of a database on abandoned seafarers. This information would be compiled from cases of abandonment reported by Member States,

and it was envisaged that the information contained in the database would be made public in order to expedite speedy resolution of cases of abandonment.

In January 2001, an important event took place that would have a considerable impact on the work of the Working Group. In parallel with the work carried out by the Working Group, the JMC of the ILO announced a project aimed at revising the regulation of maritime labour, the "Geneva Accord".

In 2002, the Working Group held its fourth meeting, at which the implementation of resolutions was examined to assess the overall situation and determine the need to take action. The creation of a joint database on cases of abandonment proved extremely valuable for evaluating and understanding the evolution of the problem worldwide.

The fifth meeting of the Working Group was held in 2004, at which implementation of the guidelines was evaluated, and the sixth meeting took place in 2005. Despite the years that had passed and the measures that had been taken, the problem of abandonment had not improved. It was decided to direct the work of the Working Group towards the preparation of a mandatory instrument. The debate centred on whether to adopt this instrument as a new regulation or to include it in the existing instrument. The initial conclusion was to consider the text of the MLC, 2006.

The position of the parties was different. It was not considered necessary to develop a binding instrument for ship owners, since it was felt that the MLC would provide the solution to the problem of abandonment, given the low incidence of abandonment reflected in the database. However, seafarers believed that the small number of reports of cases of abandonment was due to the boom currently being experienced by the maritime sector, and that cases of abandonment damaged the industry's image. It was therefore considered necessary to formulate a mandatory instrument that would render the maritime profession attractive to trained workers and which would be effective in the case of future economic crises.

The procedure, operation, components, compilation, maintenance and financing of the database was approved at the sixth meeting.

The seventh meeting of the Working Group was held in 2008, and agreement was reached on a series of principles that would underpin an effective system of financial security.

2.3 *Abandonment under the MLC, 2006*

However, two years earlier, in February 2006, the MLC, 2006 was approved at the 94th Maritime Session of the ILO. At the same session of the ILO, a resolution was adopted stating that the MLC, 2006 did not fully reflect the stipulations relating to the provision of financial security. According to

Charbonneau (2009), it was necessary to raise the more specific question of the impact of the MLC, 2006 on the phenomenon of abandonment. Certain situations were not included in the content of the MLC, 2006, such as the abandonment of crews. Similarly, Vicente (2009a) highlighted the incompleteness of the MLC, 2006, especially as regards the question of abandonment. In effect, some issues were not subject to detailed regulation by the MLC, 2006 due to lack of agreement.

Recognising the work and importance of the Working Group in monitoring adoption of the MLC, 2006, the ILO urged the Working Group to prepare a standard and guidelines for inclusion in the MLC, 2006.

In legal terms, this established that an employer's breach of the employment contract did not extinguish his responsibility towards his employees. Furthermore, in a situation of abandonment, if the employer disappeared or absented himself, the mechanism for financial security provided suitable means to prevent this situation from causing even more harm or injury to the abandoned seafarer.

At this seventh meeting, it was decided that further discussion was necessary to seek a sustainable, long-term and binding solution to the problem of abandonment.

The eighth meeting was held in 2008, in the context of the global economic crisis and its consequent effects on employment conditions in the maritime industry. The Working Group examined the content and administration of the financial security system.

2.4 *Amendments to the MLC, 2006*

The ninth meeting was held in 2009, amidst a worsening of the global economic crisis and the ensuing impact on trading conditions in the maritime industry. International trade declined, leading to a deterioration in the employment and working conditions of seafarers. Fear grew that seafarers would be at greater risk of abandonment and that there would be an increase in cases due to rising insolvency among maritime businesses.

After more than ten years of intense debate in pursuit of a long-term solution to the problem of abandonment, the end result was unsatisfactory. Dissention remained over whether the instrument should be mandatory or not, and the only agreement reached concerned a battery of principles that in the future would facilitate the drafting of a document with binding provisions. However, despite this failure in negotiations, the Working Group recommended that this instrument be adopted in the form of an amendment to the MLC, 2006.

Concerned about the scale of the problem and given the urgent need to finalise enactment of international laws, the Tripartite Preparatory

Committee included this issue in the agenda of the first meeting of the Special Tripartite Committee (MLC Committee). It was agreed that this discussion should be transmitted to the ILO forum, considering the text of the MLC, 2006. The MLC, 2006, came into force on 20 August 2013. By January 2015, it had been ratified by 64 countries which accounted for 80% of the world fleet's gross tonnage. Hence, 64 countries are currently committed to improving the living and working conditions of seafarers in over 80% of the world fleet. A historic milestone, the MLC, 2006 is aimed, among other objectives, at consolidating all current legislation applicable to international maritime work into a single legal instrument, ensuring fairer conditions of competition, improving maritime safety and enhancing the attractiveness of the profession (Sánchez Rodas Navarro, 2009).

The MLC Committee reviewed the proposed amendments and established a simplified amendment procedure for incorporating changes so that the instrument would meet the sector's needs. As designated members of the MLC Committee, Group representatives of seafarers and ship owners jointly presented two sets of proposed amendments to the Director General of the ILO in October 2013.

These were aimed at amending Regulation 2.5 on repatriation (standard and guideline) and its annexes. Amendments to the Code to apply this rule were aimed at better addressing the specific problems entailed in situations of abandonment of seafarers in foreign ports, and to this end established the necessary requirements to ensure the creation of a system for fast and effective financial security. It is interesting to note the solution proposed by Nifontov (2009), which suggested that the ideal and feasible way to meet the requirements envisaged in the MLC, 2006 protection system would be to create financial guarantee insurance. Nifontov argued that given the extensive network of insurance agencies worldwide, the insurance industry represented the least expensive, most flexible and most efficient solution as regards ensuring the speedy repatriation of abandoned seafarers. This is not the only example of the use of private insurance as a means to protect migrant maritime workers. However, and in contrast, Bouza & Carril (2014) have claimed that this regime does not usually solve all the problems of lack of legal protection. Along the same lines, Fotonipoulou (2012) has highlighted the importance of compensation for injury and accidents in maritime work.

The proposed amendments to the Code were supported by the principles agreed upon at the ninth meeting of the Working Group. These principles and the current proposal were based on the IMO/ILO Guidelines on the provision of financial security in the case of abandonment of seafarers, in Resolution A.930 (22).

The Director General of the ILO communicated the proposed amendments to all members, whether or not they had ratified the MLC, 2006, and established a six month period during which they could make comments and suggestions.

The first meeting of the MLC Committee was held from 7 to 11 April 2014, at which the amendments were discussed and adopted. These were then submitted for adoption in the plenary session of the 103rd session of the ILO held in June 2014. On July 18, the approved amendments were submitted to members that had ratified the MLC, 2006, for their consideration. Members were given a period of two years in which they could officially submit their disagreement with the amendments to the Director General.

2.5 *Entry into force of the amendments to the MLC, 2006*

It is expected that amendments to the MLC, 2006 will come into force on 18 January 2017, six months after expiry of the two year period established in accordance with and under the terms laid down in Article XV of the MLC, 2006.

The amendments will not take effect in the event that more than 40% of ratifying members, representing at least 40% of the gross tonnage of the world merchant fleet, formally stated their disagreement with the amendments.

Meanwhile, the MLC Committee also considered the possibility of approving a resolution to provide for transitional measures once the amendments entered into force, either by establishing a period of grace for adjustment to certain obligations, such as documentation, or by establishing deadlines for the entry into force of the amendments. Resolution CTE-1 on transitional measures concerning the entry into force of amendments to the MLC, 2006 distinguishes between those States that have ratified the MLC and those that have not. In principle, the date of entry into force of the amendments will be the same for all ratifying States, except for those that have formally expressed their disagreement with the amendments and those that have accepted them, but with the provision of sending an express statement of acceptance at a later date.

On the basis of these transitional measures, it is expected that non-ratifying members, members who have formally expressed their disagreement with the amendments, and members who have applied for the provision of sending an express statement of acceptance at a later date will be subject to control by the port State authorities of ratifying members who have accepted the amendments.

The measures also establish that ratifying States may postpone the entry into force of the amendments for up to a year, but also that they may it bring forward.

Adoption of these amendments and their subsequent entry into force will constitute a crucial step towards the establishment of a new international law to protect abandoned seafarers. The aim is to provide a binding international law on a fundamental and human rights issue that ensures the welfare of seafarers in situations of abandonment.

As Charbonneau (2014) has recently stated, the amendments strengthen and improve the provisions of the MLC, 2006 because they endow the instrument with a universal scope and should serve as a means to prevent the risk of social dumping to which ratifying States and their fleets might otherwise have been exposed.

3 THE ROLE OF THE EUROPEAN UNION

EU institutions are working intensively to promote ratification of the MLC, 2006 by EU Member States, whilst at the same time working to strengthen legislation governing maritime labour at Community level.

First, most of the provisions of the MLC, 2006 have been subject to regulation by Directive 2009/13/EC of 16 February 2009 governing application of the MLC, 2006.

Second, with respect to the entry into force of the amendments, the European Union considers that these provisions constitute an act of a body set up by an international agreement that will have a legally binding effect. On this basis, in the Decision of the Council of 26 May 2014 on the position to be adopted on behalf of the European Union regarding amendments to the MLC, 2006 at the 103rd Session of the ILC, Article 1 supports the adoption of these amendments and authorises Member States to adopt the position of the Union in order to act together in the interests of the Union.

Previously, the Commission had prepared a draft directive on 23 November 2005. This proposal regulated the provision of specific measures to protect abandoned seafarers, based on and referring to Resolution A.930 (22). The draft directive regulated a specific mechanism of financial security to cover the risk of abandonment, establishing in its Article 6 the obligation of ship owners to subscribe to an insurance and financial guarantee scheme to cover the costs of repatriation of abandoned seafarers. As Fotonipoulou (2009) noted previously, the Community proposal sought to go further than the ILO legislation because the afore-mentioned draft directive incorporated the IMO Guidelines on this matter in their entirety.

However, in December 2008, the Council approved a common position paper on the previous draft directive. The Council noted that the provisions of the proposal contradicted a number of international obligations; specifically, those requiring ratification of an IMO Convention were not admissible on constitutional grounds. Moreover, other provisions would entail unnecessary administrative burdens. In short, the Commission's proposal was ambitious and protective, but apparently also impracticable. Therefore, the proposal was revised, eliminating those provisions deemed inappropriate, which included the regulation and protection applicable to the abandonment of seafarers, and thus the proposal ceased to regulate this question. Indeed, even the title was substantially changed. On 23 April 2009, Directive 2009/20/EC of the European Parliament and of the Council on the insurance of ship owners for maritime claims was adopted. In short, financial security in the case of abandoned seafarers disappeared from the scope of this directive. In the case of Spain, for example, Vicente (2009b) noted that this directive was of interest not so much for what it did cover as for what it might have covered, most specifically with regard to the right to the repatriation of seafarers. In the case of French law, abandonment of seafarers in a foreign port is considered a criminal offence. Repatriation in France is a contractual and legal obligation involving public and private interests (Chaumette, 2009).

4 DISCUSSION AND CONCLUSIONS

In principle, the effective implementation of Directive 2009/13 and MLC2006 Convention (as amended) significantly minimize the scale of the problem. Establishing a system of protection, by way of insurance or an equivalent appropriate measure, to compensate seafarers for monetary loss that they may incur as a result of the failure of a recruitment and placement service or the relevant shipowner under the seafarers' employment agreement to meet its obligations to them is the real challenge we must face.

The amendments to the MLC, 2006 are formulated as an ad hoc instrument prepared by the ILO that is intended to be mandatory for ratifying States. The aim is to strengthen and protect the rights and living and working conditions of abandoned seafarers. Thus, ship owners must comply with the provisions relating to financial assistance and the certificates that prove the availability of such aid.

After more than a decade of negotiations and given the complexity of the issue, consensus had been reached to create a protective mechanism for abandoned seafarers. This consensus was the result of the efforts and compromises made by the parties. The desired aim is to achieve fairer conditions of competition between shipping companies

worldwide. It could thus be a mechanism to combat social dumping.

In the long term, these measures will help to enhance the attractiveness of working in the maritime sector for young, trained people. Workers employed in this industry - which, by definition, is genuinely global - constitute a valuable resource for ensuring the development of the international economy, especially in a context where 90% of world trade is transported by sea.

As a tripartite organisation, the ILO has unquestionably been of service to the shipping industry since its inception, and represents a suitable forum in which to formulate international legislation. Furthermore, assuming that the amended MLC will enter into effect as scheduled, the key issue will be to consider how this instrument will be applied nationally by the ratifying States, i.e. how the States will incorporate these amendments into their national legislation. The question will be to determine the link with the MLC, 2006, considering the consequences for those ratifying States which have failed to implement its provisions.

As regards Community action, the EU has not incorporated the case of abandonment proposed in the draft directive. Therefore, the proposed directive did not achieve its goal, and failed to produce the desired effect of legally guaranteeing the protection of abandoned seafarers. However, it should be noted that the European Union has expressed its position regarding the amendments to the MLC, 2006. This, coupled with the European Union's adoption of a directive encouraging Member States to ratify the MLC, 2006, suggests the creation of a scenario in which the basic labour and social rights of abandoned seafarers are guaranteed. However, it is expected that the European Union will assume a deeper commitment in the coming years.

Therefore, given the questions that remain and whilst waiting for the legislative actions of ratifying States to lead to positive results, it must be concluded that at present, the abandonment of seafarers remains a concern and a human and social problem. The amendments have yet to enter into force; hopefully, once they do, their application will not be limited to the letter of the law but will result in true protection of abandoned seafarers' working and social conditions. This will require social partners and private subjects, as the watchdogs responsible for ensuring and controlling compliance with the legislation, to adopt a proactive approach to the regulation of all aspects that improve the social and working conditions of abandoned seafarers. Lastly, we believe that the terrible situation of abandoned seafarers must not remain unaddressed pending the entry into force of the afore-mentioned amendments, and that States should adopt the commitment to implement Resolution A.930 (22) on the provision of financial security in the case of abandonment.

REFERENCES

Bouza Prego, M.A. & Carril Vázquez, X.M. 2014. Los daños derivados del trabajo a bordo de buques: el accidente de trabajo maritime. In: Fotinopoulou, O. (ed.) *La seguridad marítima y los Derechos laborales de la gente de mar*: 93-126. Bilbao: GomyLex.

Chaumette, P. 2009. L' abandon des marins, une vision generale. Introduction au probleme de Iábandon des marins Dans un port etranger. In: Fotinopoulou, O. (ed.) *Derechos del hombre y trabajo marítimo: los marinos abandonados, el bienestar y la repatriación de los trabajadores en el mar*: 13-24. Vitoria: Eusko Jaurlaritza.

Chaumette, P. 2014. La ratificación y la transposición del CTM 2006. Ley nº 2013-619 de 16 de julio de 2013, por la que se aprueban diversas disposiciones de adaptación al Derecho de la Unión Europea en el marco de un desarrollo sostenible. *Revista General de Derecho del Trabajo y de la Seguridad Social*, 36:143-162.

Carril Vázquez, X.M. 2009. Derechos del hombre y trabajo marítimo: Los marinos abandonados, el bienestar y la repatriación. In: Fotinopoulou, O. (ed.) *Derechos del hombre y trabajo marítimo: los marinos abandonados, el bienestar y la repatriación de los trabajadores en el mar*: 217-229. Vitoria: Eusko Jaurlaritza.

Charbonneau, A. 2009. Bienestar de los marinos: el procedimiento para la interposición de quejas en tierra. *Revista del Ministerio de Trabajo e Inmigración*, 82:357-382.

Charbonneau, A. 2014. El convenio de trabajo marítimo, 2006, de la OIT: de la promoción a la aplicación. *Revista General de Derecho del Trabajo y de la Seguridad Social* 36: 163-179

De la Villa Gil, L.E. 2009. Editorial of the special issue *Revista del Ministerio de Trabajo e Inmigración*, 82:9-13.

EU, 2005. Proposal for a Directive of the European Parliament and the Council on the civil liability and financial guarantees of shipowners (SEC(2005) 1517).

EU, 2009a. Council Directive 2009/13/EC implementing the Agreement concluded by the European Community Shipowners' Associations (ECSA) and the European Transport Workers' Federation (ETF) on the Maritime Labour Convention, 2006, and amending Directive 1999/63/EC.

EU, 2009b. Directive 2009/20/EC of the European Parliament and of the Council of 23 April 2009 on the insurance of shipowners for maritime claims.

EU, 2012. Proposal for a Directive of the European Parliament and of the council concerning flag State responsibilities for the enforcement of Council Directive 2009/13/EC implementing the Agreement concluded by the European Community Shipowners' Associations (ECSA) and the European Transport Workers' Federation (ETF) on the Maritime Labour Convention, 2006, and amending Directive 1999/63/EC.

EU, 2014. 2014/346/EU: Council Decision of 26 May 2014 on the position to be adopted on behalf of the European Union at the 103rd session of the International Labour Conference concerning amendments to the Code of the Maritime Labour Convention.

Gonzalez Joyanes, D. 2009a. La puesta en marcha de una base de datos sobre los casos de abandon de marinos. In: Fotinopoulou, O. (ed.) *Derechos del hombre y trabajo marítimo: los marinos abandonados, el bienestar y la*

repatriación de los trabajadores en el mar: 82-91. Vitoria: Eusko Jaurlaritza.

González Joyanes, D. 2009b. *Abandono de buques y tripulaciones*. Barcelona: Marge.

Garcia, R., Castaños, A. & Irastorza, I. 2013. Abandonment of seafarers in a foreign port. *Archives des Maladies Professionnelles et de l'Environment*, 74(5):579

Pentsov, D.A. 2001. Las normas internacionales del trabajo: Un enfoque global. In: *Gente de mar*. ILO. p.630.

Fotinopoulou, O. 2009. El Convenio refundido sobre trabajo maritimo y el abandono de marinos en puertos extranjeros. *Revista del Ministerio de Trabajo e Inmigración*, 82:219-244.

Fotinopoulou, O. 2012. La indemnización daños y el accidente de trabajo maritime. *Revista General de Derecho del Trabajo y de la Seguridad Social*, 31:117-134.

ILO database:
http://www.ilo.org/dyn/seafarers/seafarersbrowse.home

ILO, 1998. GB. 273/STM.

ILO, 1999. GB. 277/STM/4.

ILO, 2000. GB. 280/STM/5.

ILO, 2001. GB. 282/STM/5.

ILO, 2002. GB. 286/STM3.

ILO, 2004. GB. 289/STM/8/2.

ILO, 2005. GB. 295/STM/5.

ILO, 2006. *Maritime Labour Convention*.

ILO & IMO 2008a. ILO/IMO/CDWG720083.

ILO & IMO 2008b. ILO/IMO/WGPS/8/2008/5.

ILO & IMO 2009. ILO/IMO/WGPS/9/2009/10.

IMO, 2001. Guidelines on provision of financial security in case of abandonment of seafarers. Resolution A.930(22), adopted on 29 November 2001. 17 December 2001.

Nifontov, D. 2014. Seafarer abandonment insurance: a system of financial security for seafarers. In: Lavelle, J. *The Maritime Labour Convention, 2006*: 117-136. New York: Routledge.

Sánchez-Roda Navarro, C. 2009. El Convenio sobre el trabajo maritimo y el Derecho social comunitario. *Revista del Ministerio de Trabajo e Inmigración*, 82:45-64.

Vicente Arántzazu, A. 2009a. El control por el Estado del puerto de las condiciones de vida y de trabajo a bordo de buques que utilizan puertos comunitarios o instalaciones situadas en aguas de la jurisdicción de los estados miembros en la normativa comunitaria. In: Areta, M. & Sempere A.V. *Cuestiones actuales sobre Derecho Social comunitario*: 279-395. Madrid: Laborum.

Vicente Arántzazu, A. 2009b. Obligaciones y responsabilidades en materia de repatriación: grado de adecuación de la normativa española a las previsiones del CTM 2006. *Revista del Ministerio de Trabajo e Inmigración*, 82:291-339.

Vicente Arántzazu, A. 2009c. El CTM 2006 avanza hacia su entrada en vigor. Últimas actuaciones de España y de la Unión Europea a favor de la protección del trabajo de la gente del mar. *Revista General de Derecho del Trabajo y de la Seguridad Social* 20:1-27.

Vicente Arántzazu, A. 2010a. El CTM 2006: respuestas eficaces de Derecho Internacional a un problema global: el trabajo marítimo como la primera actividad profesional globalizada. *Revista Soluciones Laborales*, 3:1-13

Vicente Arántzazu, A. 2010b. El Reglamento 883/2004 y las obligaciones de los armadores en el ámbito de la seguridad social. In: Sánchez-Rodas Navarro, C. (ed.) *La coordinación de los sistemas de Seguridad Social*: 65-90. Murcia: Laborum.

Vicente Arántzazu, A. 2014. El control de las condiciones de trabajo en el ámbito maritimo. Facultades del estado del pabellón y del estado del Puerto. In: Fotinopoulou, O. (ed.) *La seguridad marítima y los Derechos laborales de la gente de mar*: 645-703. Bilbao: GomyLex

Piracy

Effectiveness of Measures Undertaken in the Gulf of Guinea Region to Fight Maritime Piracy

K. Wardin & D. Duda
Naval University, Gdynia, Poland

ABSTRACT: Maritime piracy in the Gulf of Guinea region has not been in the centre of interest for international community from the beginning of the 21st century because of the piracy in the Gulf of Aden. Although the region is not considered as the most important to international sea lanes of transportation, it should be considered as one of the most vital for the region itself and for such countries as the United States of America, France, Great Britain or China. Reach deposits of oil and gas make the region important to international community and enforce proper reaction to any kind of crime and maritime piracy especially. Various local and international organizations take some measures to make maritime transport in the region secure but it is important to assess the effectiveness of those steps in order to set proper future instruments. The purpose of this article is to introduce the measures and to assess their efficiency in the fight of maritime piracy.

1 INTRODUCTION

Regardless of high levels of maritime crime and piracy for a number of years, it was not until 2013 that the Gulf of Guinea drew global attention away from the Horn of Africa. Following this course of interest, many attempted to draw quick comparisons between East and West African pirate groups, which lead to simplification the diversity of piracy in the Gulf of Guinea. A well known model of "hijacking-for-ransom", employed by Somali pirates from the begging of year 2008, is not the only model moved to the Gulf of Guinea. Piracy in this region should not be considered as having one particular "piracy model" but many. This fact, together with the circumstances existing in the area in general, makes the fight of piracy difficult and not exactly the same as in the Horn of Africa. Measures taken to resolve the problem should be chosen and monitored carefully as well as evaluated to make sure that they have a positive impact and limit the problem. The lesson learnt in the Gulf of Aden cannot be forgotten but piracy of West Africa is distinctly different from, and unrelated to, Somali piracy in many significant ways related to historical, legal, and geographical differences between the two regions. The model worked out in East Africa has to be adjusted to western circumstances. The main problem of the article is to present all the activities in this matter first and secondly to evaluate their effectiveness for the past two years.

2 CURRENT SITUATION IN WEST AFRICA

Africa is the continent where most, for as much as 70%, of all States, are classified or known as weak and fragile, as well as those which have serious internal problems. West Africa is one of the poorest and least stable politically and economically part of the continent. Therefore, all sorts of crime in this region of the world are almost a regular part of modern reality. Piracy and marine crime on the West coast of the continent concentrates mainly in the waters of the Gulf of Guinea. It is a vast diverse and highly important region, which constitutes 16 countries[4] that are strung along roughly 6, 000 kilometers of unbroken coastline. From Senegal to Angola provides an economic lifeline to coastal and landlocked West African countries, and is of strategic importance to the rest of the world. Safe

---

[4] There are: Angola, Benin, Cameroon, Central African Republic, Ivory Coast, Democratic Republic of Congo, Equatorial Guinea, Gabon, Gambia, Ghana, Guinea, Guinea-Bissau, Liberia, Nigeria, Republic of Congo, São Tomé and Príncipe, Senegal, Sierra Leone and Togo., C. Ukeje, M. Mvomo Ela, *African approach to maritime security – the gulf of Guinea*, http://library.fes.de/pdf-files/bueros/nigeria/10398.pdf, p. 5, 01.12.2014.

passage to ports in the region and security within its waters are vital for global energy raw materials and the production of their derivatives, as Nigeria and Angola are among the world's 10 biggest crude oil exporters. For West Africa's fishing industry, this provides sustenance and employment for a large swathe of the West African population. It is also essential for the prevention of the trafficking of narcotics, people and weapons into Europe and into fragile regions that are vulnerable to destabilization[5].

While the number of piracy attacks in the Gulf of Aden significantly dropped down to 15 reported attacks according to International Maritime Bureau (IMB) and 23 by an Oceans Beyond Piracy (OBP) report in year 2013, they rose dramatically in West Africa. Again according to different sources there were 46 attacks by IMB and 100 by the OBP report. For the second year in a row, there were more piracy attacks than the Indian Ocean. The data set estimates, as said above 100 attacks off West Africa, of which 56 were successful. In general 1,871 seafarers were attacked by pirates, 279 were taken hostage. 73 were kidnapped and held on average for 22 days, 183 were hostages on average for 4 during oil thefts and 22 were kept hostages while their vessel served as a mother ship for pirates, which lasted on average for 17 day. There were also 58 robbery attempts[6]. The discrepancy between the numbers given by IMB and OBP shows that some of attacks probably have not been reported and the real figures could have been even higher. This may be true particularly for local fishermen as they are often afraid of revenge and do not want to lose their boats, so do not report piracy attacks. On the other hand they are aware of corruption inside the government and different forces responsible for maritime security and little possibilities in changing the situation.

There is some risk in combining the figures for both piracy and armed robbery in West Africa in the one report, but fundamentally, maritime crime and piracy are related to instability on shore. Exceptionally complex geopolitical situation in the area is not favorable in stabilizing the region, and many armed conflicts, civil wars, frequent political upheaval lead to the fact that this part is one of the poorest regions of the world and the most unstable politically. Six of the ten countries (Côte d'Ivoire, Nigeria, Liberia, Cameroon, Equatorial Guinea and Togo) rank with the top 50 on the list of Fragile States Index 2014[7] because of the assessment of the

situation as alarming, meaning that their internal situation, with at least part of the territory being beyond the control of the Central Government, and force departments do not working properly. The internal situation in the other States of the region (Benin, Gabon, Ghana and the Island of St. São Tomé and Príncipe), although rated to as slight better is still worrying or very disturbing and it does not guarantee that will not deteriorate in the near future. The assessment represents the view that in the region of West Africa international society deals with deeply dysfunctional countries, which clearly are not able to cope with the security on their territory, including the waters of the sea. The fact of the local population addressing piracy and maritime crime generally is influenced by widespread, among the local population, poverty and development imbalances and uneven distribution of resources derived from the extraction and export of fossil fuels. In addition, widespread and accepted at every level of government corruption, does not allow managing effectively even available financial resources. The assessment of the potential reaction to piracy is largely dependent on the condition of the individual government departments, especially national forces, responsible for the proper functioning of the State and law enforcement. Unfortunately the evaluation is not satisfactory because according to research assessing the potentials of vessels of navies capable of a real action to combat piracy, it has to be said that there are only 25-units available for interdiction efforts[8]. To sum up, the conclusion is obvious that with such a small potential of the region, the fact of the political decision of the need to combat piracy cannot guarantee any effective counter and antipiracy activities in terms of force multiplying.

3 REASONS FOR PIRACY IN THE GULF OF GUINEA

Generally root causes of maritime insecurity should be searched in several important facts. First of all, the importance of maritime transportation in the world in general and in the Gulf of Guinea in particular is a key factor. The region transits about 50,000 vessels yearly which carry all sorts of cargo but most of all energy raw materials important for the United States, but also European Union countries. For this reason the security of sea lines of transportation is crucial not only to particular countries but to world's economy.

An impact of oil policy of European countries after 9/11 involves the quest to gain access to and

[5] Maritime trade shearing information center – Gulf of Guinea, SAMI, www.security.org, 12.12.2014.
[6] K. Wardin, *Model reagowania na zagrożenia piractwem morskim*, BelStudio, Warszawa 2014, p. 179, 215; *Maritime piracy 2013 report*, Oceans Beyond Piracy, www.oceansbeyondpiracy.org, 03.11.2013.
[7] *Fragile States Index 2014*, http://library.fundforpeace.org/fsi14-overview, 04.01.2015.

[8] J. Bridger, *Countering piracy in the Gulf of Guinea*, http://news.usni.org/2013/07/12/countering-piracy-in-the-gulf-of-guinea, 10.08.2014.

secure control of oil resources all over the world. Last two decades had shown that African continent has become a vital area especially for The United Stated national security interests. Rising demand on oil reflected in the search of new possibilities. This trend has been also noticed in other developing countries such as China. The presence of the investment of Chinese oil and gas companies in some countries on African continent shows growing importance of this region.

Another reason is a profit-based economy organized around the course of oil. Oil and gas-reach countries saw the chance to get the most profits, although it should be remembered about so called "Dutch Disease syndrome"[9] and all consequences of such policy. The research done by P. Collier in his book *The bottom billion*" shows that reliance on income from the exploitation of natural resources greatly reduces chances of overcoming the instability. A doubling of the share of revenues from natural resources in GDP extends the time required to repair the fragile state twice[10].

The negligence of some of the colonial states, led to inability or unwillingness to accomplish even the most basic sovereign duties and responsibilities. States organized by colonial powers in the past virtually were not able to meet expectations of their citizens in term of economy, security and social needs.

This situation reflected in poverty and has been the main reason to engage in piracy and maritime crime activities by local people.

The reaction of international society to that problem, especially in the Horn of Africa, but also in the Gulf of Guinea resulted in the presence of Private Military Companies (PMC), which offer protection services. Some of the local companies in the Gulf of Guinea tend to overuse arms which are not legal according to the Montreux Doccument[11], and may bring even more threats to maritime regions than security. It should be mentioned that local PMC are very often corrupted and have interest in keeping their eyes closed for local gangs even though they are paid for having vessels secured.

Another problem lies in poor delimitation of maritime boundaries. Along with the discovery of hydrocarbon deposits it is responsible for permanent presence of risk of conflict and violent border confrontations between neighboring countries, and it does not help to organize effective piracy and maritime crime fight. Luck of everyday cooperation is the most important problem. The states understand the importance of maritime transportation for the individual states and the region in general but are not able to achieve a certain level of trust and cooperation in this matter. Although there is a slight improvement but there are still many issues to tackle.

The rights and social, environmental justice issues tend to feed into broader questions about identity, inclusion and exclusion which have become a significant reason for instability in the region both on land and at sea. Foreign Oil Multinationals exploit the region, take lands and expose it to oil pollution along with environmental degradation, while the indigenes and owners of the land do not benefit from a billion dollars' business created from their region, nor do they get an adequate compensation for the destruction of livelihoods or the "loss" of lands. These are the most common allegations of local piracy and organized crime gangs but also ordinary people. They cannot explain any illegal activities but cannot be either totally neglected by international community in planning and building possible solutions to this problem.

Weak legal framework for effective maritime security management seems to be a logical outcome of the complex situation. This problem is again connected with inability or unwillingness of individual states to domesticate treaties freely signed. Even when the need to cooperate within existing legal framework in response to new maritime security imperatives is self evident different levels of governmental agencies are not able to take up a proper cooperation to tackle the problem appropriately and successfully.

Listed difficulties seriously hamper regional collaboration to fight piracy and maritime crime. This can be seen first of all in the fact that although there have been some kind of activities since 2010, the efforts taken seem to be ineffective because the number of attacks have been growing for the past two years and the model of piracy has been evolving as well.

4 SHIFT IN PIRACY MODEL IN WEST AFRICA

Piracy in the Gulf of Guinea is not simply about cargo theft from tankers, but in the last two years more broadly relates to theft, robbery, kidnapping and hijacking, involving a variety of vessel types in a variety of locations. While certain groups do target

---

[9] Dutch disease is the negative impact on an economy of anything that gives rise to a sharp inflow of foreign currency, such as the discovery of large oil reserves. The currency inflows lead to currency appreciation, making the country's other products less price competitive on the export market. It also leads to higher levels of cheap imports and can lead to deindustrialization as industries apart from resource exploitation are moved to cheaper locations. The origin of the phrase is the Dutch economic crisis of the 1960s following the discovery of North Sea natural gas, http://lexicon.ft.com/Term?term=dutch-disease, 12.01.2015.

[10] P. Collier, *The Bottom Billion*, Oxford University Press 2008, p. 38-44.

[11] The Montreux Document is an intergovernmental document intended to promote respect for international humanitarian law and human rights law whenever private military and security companies are present in armed conflicts, https://www.icrc.org, 10.10.2014.

product and chemical tankers in order to steal refined cargo ("hijacking-for-cargo") the biggest increase in activity in 2013 actually related to offshore kidnapping-for-ransom, focused around the Niger delta region. Most significantly, attacks off West Africa occur in both territorial waters (those extending 12 nm, from the coast) and international waters (those waters beyond the 12 nm territorial zone). Given that the majority of piracy-related incidents in West Africa during 2012 and 2013 occurred within 12 nm of a coast, those incidents fell within a specific country's jurisdiction and protection. West African maritime crime is also different because the region has many ports actively involved in both international and regional maritime trade. For this reason, vessels not only transit through the region, but also enter into ports to load or unload cargo. West African pirates may attack ships passing through the region or ships berthed or anchored and waiting to berth. This changes the character of the challenges faced by ships dealing with piracy off West Africa in that they must have security systems in place to address threats both while under way and while stopped[12]. Certainly the security conditions in these circumstances are far more complicated than in East Africa.

Generally when we talk about the Gulf of Guinea region, especially the territorial waters and Exclusive Economic Zones (EEZs) of Nigeria, Benin and Togo we should have in mind that these areas are considered at greatest risk of the following types of criminality:
– Piracy and armed robbery at sea;
– Theft of oil and other cargo;
– Illegal, unreported and unregulated fishing;
– Trafficking of counterfeit items, people, narcotics and arms.

There was a 30% increase in overall activity and taking into consideration only first two activities it is important to underline that pirates were mostly involved in kidnapping-for-ransom and hijacking-for-cargo in years 2012 and 2013. Both IMB and OBP organization noted a 355% increase in maritime kidnapping-for-ransom and a 12.5% decline in hijacking-for-cargo incidents.

The trend most associated with piracy in the Gulf of Guinea is based on a well known modus operandi. Usually product and chemical tankers carrying refined fuel products are targeted while they are anchored or drifting off major ports. They are easy targets for well armed and informed piracy groups. Once hijacked the tankers are moved to an offshore location, typically south-west of the Niger delta. Communications and tracking is disabled while crew is forced to move the vessel to a transfer position. The next step is to move some or all of cargo to another tanker via ship-to-ship transfer. Before the

vessel and the crew are set free all cash, equipment and crew valuables are stolen and then pirates leave. In most cases pirates are not interested in taking hostages, for the fact that the stolen cargo can be sold successfully. Occasionally crew members were forced to help pirates to prepare successful transfer and if it does not happen, crew can be in some kind of danger and pirates might want to keep them and ask for ransom. So far selling cargo has been more lucrative business. Unfortunately thought this type of attacks become less often practiced the range is expending suggesting that vessels are better prepared for this kind of activity and it is not easy, as it used to be in 2011, to conduct such attacks.

Kidnapping-for-ransom was particularly dangerous and most common activity in 2013, in one of the

highest risk areas, the Niger delta region in the south-east of the country, where there is an environment made up of interconnecting waterways and isolated communities.

It follows a very different methodology to that of tanker hijackings elsewhere across the region. Vessels are attacked and boarded largely accidentally. After stealing cash, equipment and crews belongings, pirates usually select between two and five crew members that are taken back to the coast. Frequently they are senior officers and are held in camps, hidden in the depths of the Niger delta away from the patrolling Nigerian authorities. Worrying should be the fact that as Somali piracy from 2008 kept extending the range, Nigerian piracy gangs extended noticeably the range of their attacks from 30-40 nm from Nigerian coast in 2012 up to 150 nm from the cost in 2013[13]. Kidnapping groups have used mother ships to move further from the coast, and, local fishing trawlers have been used as platforms to mount offshore attacks in the Gulf of Guinea. This model was exercised by Somali pirates in the Horn of Africa and very quickly became a serious problem for international maritime trade. For that reason international society cannot ignore the threat and should start to cooperate widely with interested countries, regional organizations and international community as a whole to defend our interests.

5 MEASURES TAKEN IN THE GULF OF GUINEA REGION TO FIGHT MARITIME PIRACY

Revenues from the export of natural resources such as oil and natural gas are major source of budget revenue in the Central and Western African

---

[12] Vide: *Maritime piracy 2013 report...* op. cit., p. 51.

[13] *Insecurity in the Gulf of Guinea: assessing the threats, preparing the response,* International Peace Institute, January 2014, http://www.isn.ethz.ch, p. 4-6, 20.11.2014.

countries. In 2013, the State, which was significantly dependent on oil supplies from the Gulf of Guinea were the United States, which companies, in 2011, imported 29% of its demand for oil from Nigeria, and according to estimates of the National Intelligence Council of the United States by the end of 2015 should grow to about 35%[14]. European Union countries, especially Great Britain or France are also interested in natural resources from this region as well as countries from Asia. Security of international maritime transportation in the region is one the highest priorities for domestic or national level in the Gulf of Guinea, for the region itself and for the national community. This would be also the best way to analyze and access activities and cooperation of different parties on different levels. It should be mentioned thought that the action undertaken in the past was very often initiated by decisions on regional level but was implemented by forces of particular or combined States, so clear distinction might be quite difficult.

In response to the growing threat of piracy in the Gulf of Guinea, the States have begun to propose a number of initiatives which could be developed by individual countries to address the challenges posed by maritime insecurity. Such initiatives include efforts to increase national attention to the maritime domain, as well as to improve equipment and training of the relevant security personnel. Moreover, awareness-raising and the ratification and domestication of international instruments on maritime security, including the 1988 *Convention for the Suppression of Unlawful Acts against the Safety of Maritime Navigation*[15], could help reduce maritime insecurity. Nigeria, for example in 2004, started bringing together the army, navy, air force, and the mobile police in the Niger Delta to restore order and help to reduce oil theft. In January 2012 the Nigerian government transformed its Joint Task Force *Operation Restore Hope*, which was initially established to combat militancy in the Niger Delta into an expanded maritime security framework, known as *Operation Pulo Shield*.

Benin, in 2011, implemented a number of measures, involving both government structures and civil society, which aimed at developing an integrated response to the comprehensive security challenges. These activities take into account the threats posed by security both at sea and on land, and local monitoring mechanisms have been established in various communities.

The threat posed by Somali piracy moved the region to closer cooperation in order not to allow piracy in the region to grow stronger. Earlier, in 2009, there was made an attempt of cooperation of

four countries, Cameroon, Equatorial Guinea, Gabon and the Islands of São Tomé and Príncipe. The minister of defense of Cameroon stated that they wanted to organize a combined naval force to patrol the coasts, while Angola proposed the establishment of joint mechanisms to combat threats. In October 2009 Maritime Transport Ministers of the African Union (AU) States attended the Conference in Durban in South Africa, which has resulted in the resolution of Durban. The representatives of the Member States declared the cooperation in terms of problem-solving in the field of maritime safety, maritime transport security and threats to the environment. The Executive Council of the AU was to take any necessary action to deploy UN recommendations on maritime security, transport and environmental protection. In September 2011, Nigeria and Benin opted for joint sea patrols after exceptionally high, number of attacks in the region reported to local authorities. In August 2011 in London, was held a conference on maritime security, in particular the prevention of piracy and combating pirates' activities on land. The Conference was attended by a number of government officials of the West African States, and participants discussed plans for regional cooperation on this burning issue. In October of the same year, there was a real action in the form of the operation under the code name PROSPERITY, which consisted of the joint patrols by the navies of Benin and Nigeria.

Unfortunately continuous insecurity forced the United Nations Security Council in October 2011 to express its deep concern about the security of the sea in view of the persistent threat in the waters of the Gulf of Guinea from piracy in the Resolution No 2018[16] and to call for rapid cooperation countries in the region. Understanding the seriousness of the threat, UN Secretary General Ban Ki-moon sent a team of experts, which was intended to assess the situation in the area. The final report stated that the pirates resorted to sophisticated ways of attacks with the use of heavy weapons. In these circumstances, under pressure from growing concerns about maritime security of the region, the UN Security Council adopted Resolution No 2039 in February 2012, which called for organizing a regional summit under the aegis of UN, in order to facilitate regional and comprehensive approach to find the right response to the growing problem of maritime piracy. These efforts culminated in the March 2012 ministerial conference on maritime security in the Gulf of Guinea held in Cotonou-Benin. It was organized in partnership with the Economic Community of Central African States (ECCAS), the Economic Community of West African States (ECOWAS), and the Gulf of Guinea Commission

[14] *Rekordowy import ropy z Afryki Zachodniej*, Fundacja Europa Bezpieczeństwo Energia, http://ebe.org.pl/, 3.03.2012.

[15] There are countries, which has not ratified that convention: Angola, Democratic Republic of Congo, Gabon, Cameroon and Congo.

[16] *Resolution 2018*, UN official documents, http://www.un.org/, 20.01.2014.

(GGC) in response to Resolution 2039. At the meeting of the Member States the participants proposed the implementation of the Gulf of Guinea Code of Conduct (GoGCoC)in order to eliminate marine piracy and proposed signing of the Joint Declaration relating to maritime security in the Gulf of Guinea. The Cotonou conference paved the way for the June 2013 summit in Yaoundé, Cameroon, which brought together twenty-five countries from the Gulf of Guinea to formalize the adoption of an integrated response to a comprehensive security challenge in the region. The New York roundtable meeting to prepare the summit in Cameroon emphasized the need for individual Gulf of Guinea countries to take full responsibility in tackling maritime insecurity, before regional organizations are able to come together. Putting in work the outcomes of the Yaoundé summit depends on the political will of member States. Regular interaction and the development of an overall plan, as well as national action plans with clear, realistic, and measurable objectives are essential[17].

The African Union, at the continental level, on the meeting of top maritime security experts in Addis Ababa in December 2011 proposed a draft document entitled *Integrated Strategy of Maritime Security in Africa by 2050*, which proposed a framework for cooperation by all States of the region on this issue.

Commercial and economic interests of such countries as, the USA, France, the United Kingdom (UK) and China, do not allow neglecting the issues related to the security of maritime transport. In this situation the problem was often discussed at G-8 summits, starting from 2011, when the number of attacks in the region grew significantly. These activities are can certainly be considered as an international or even global level of reaction to maritime piracy in the Gulf of Guinea.

In 2012 the Navy of United States Africa Command (AFRICOM) organized the joint exercises under the code name OBANGAME EXPRESS, which was carried out in order to improve cooperation between Cameroon, Ghana, Gabon, São Tomé and Principe and Spain. In addition, the United States issued from 2009 for training of naval forces about 35 million US $[18].

The United Kingdom Government, supports the region in much the same way as other Western countries, also puts great emphasis on launching Maritime Trade Information Sharing Centre in Ghana, whose task would be the effective exchange of information and coordination of measures to combat piracy. The Centre would operate mostly under the auspices of the private sector oil industries involved in the extraction of fossil fuels in the Gulf

of Guinea, and especially interested in elimination of piracy in the region. Such countries as France, Norway, the Netherlands and Austria have officially agreed to finance this initiative and other Western countries also have expressed a desire to support the Center[19].

International exercises are organized in the Gulf of Guinea more often and are associated with the presence of the ships of the navies of such countries as China, India or Brazil, which proves the importance of this region especially in the context of the ever-increasing demand of these countries in energy fuels derived from Western and Central Africa. The involvement of the Chinese Government in investments in this region of the world can suggest that they will also become more interested in the protection of their interests by strengthening bilateral and multilateral cooperation in the field of maritime transport security.

One of the good examples could be the Gulf of Guinea Code of Conduct. It was formally adopted by the heads of State at the meeting in the capital of Cameroon (Yaoundé) and March 2013 and signed by the Ministers of the 22 States. The code contains a number of elements from the Djibouti Code of Conduct (DCoC), which is a regional agreement in the fight against pirates of East Africa. Drawing from the experience of DCoC, GoGCoC does have a much wider scope, because it also applies to cooperation in the field of any illegal activities at sea, including illegal fishing, drug trafficking, as well as piracy. GoGCoC uses the existing findings of an integrated system for coastguards in West and Central Africa, which was created in 2008 on the initiative of IMO and Maritime Organization for West and Central Africa (MOWCA). It was developed by ECOWAS and ECCAS and GGC in cooperation with IMO and in accordance with the UN Security Council Resolutions 2018 and 2039.

GoGCoC defines how the State will carry out activities to prevent and combat acts of piracy and armed robbery, and other illegal activities at sea. The Trust Fund has been established, which resources will be used for the implementation of IMO projects in the field of maritime safety. These include various types of exercises, aimed at improving cooperation at government level and with the agencies. They use a wide range of scenarios related to the evolving situation in order to determine the obligations of the participants, the scope of cooperation and procedures for both routine of relevant services and reported pirates' attacks. The implementation of this agreement should lead to long-term improvement of maritime safety, which will promote the development and the future economic growth of the region. It should be noted that while the importance of maritime security issues

[17] *Insecurity in the Gulf of Guinea... op. cit.*, p. 7.
[18] K. Wardin, *Model reagowania ....op. cit.* p. 195.

[19] Ibidem, p. 196.

at the regional level is noticed and widely promoted, it is not a priority at the level of individual States. It is extremely important to local communities to ensure their participation in the efforts to tackle the problem of piracy, and thus give a sign that they understand the need for strong action in this direction and that these actions have a direct effect on local communities, corporations and international mining companies. Such awareness can be achieved only by sending clear information about the local bad effects of the lack of security at sea. Developing a common strategy for the activities of the Member States has committed themselves:

- to change in national law and regulations, in order to introduce criminal penalties, piracy and armed robbery at sea;
- to develop a regional framework for the fight against piracy and armed robbery at sea, including the establishment of mechanisms to inform Parties on the current situation and operational coordination in the region;
- to progress and strengthen national law and regulations, in accordance with international law;
- to strengthen international cooperation at all levels in the fight against maritime piracy and to ensure safety of oil mining infrastructure[20].

With regard to the piracy the States have decided to implement in addition some procedures:

- to arrest, investigate and prosecute individuals who have committed piracy or are suspected of committing the act;
- to seizure of pirated vessel or aircraft and property on board such units;
- to rescue ships, persons and property, which have been attacked by pirates[21].

Although these statements remain so far (as of January 2015), only and exclusively in the form of a declaration, it should be said that the States through the planned actions try to take the lead in fighting piracy and do not want to admit to this problem and allowed Western countries to solve it.

6 EFFECTIVENESS ASSESMENT OF MEASURES TAKEN IN DIFFERENT AREAS IN THE REGION

The effectiveness of all measures taken to stop piracy in the Gulf of Guinea is not satisfactory in relation to expenditures and the effort spent so far to eliminate it. None of the interested parties can be satisfied with the effects as there were still over hundred reported piracy attacks in 2014. It is also difficult to point at one particular reason why this cooperation does not lead to full success. The main

factor which complicates activity in the waters of the Gulf of Guinea is the fact that most of the illegal activities, including piracy, take place within the territorial sea and formally remain under the jurisdiction of the coast guard, police or possibly navies. Unfortunately, in most countries of West Africa, these services are not well equipped, and so their ability to anti-piracy operations is limited.

Another obstacle in the effective fight against piracy is the lack of a clear legislative scheme related to the transfer of suspects of maritime crime and piracy to relevant authorities on the mainland. Poorly functioning judicial systems indicate that the pirates and hackers are practically set free immediately after the hearing. This problem requires a systemic solution, similar to what was used in the respect of the Somali pirates, however, in 2014, was not in the agreement between the Member States of the Gulf of Guinea

Apart from that geopolitical situation is responsible for poverty and anger of local communities. They consider piracy as a natural way of fighting with governments, as the root causes the problem stems from the decades of bad governance. Stability must start with human security. However, many communities in the Gulf of Guinea region suffer from chronic insecurity in their daily lives.

The youth populations of many of the countries in the region are left with few opportunities for profitable employment providing a fertile recruiting ground for pirates and criminal networks who offer work, financial incentives, status and in some cases, basic services and protection to communities in need, effectively replacing the State. Economic discrimination is worsened by social disproportion as ethnic and tribal divisions exclude some groups and women from decision making, both formal and informal representation, and many employment sectors. Aggressive natural resource exploitation has degraded the environment and polluted ecosystems. The high levels of water and air pollution are significant contributing factors to instability in the region, not only in terms of environmental and economic impact, but also by their contribution to a climate of desperation among the population[22]. To sum up, the obvious conclusion is that with such a small potential of the region, the fact that the political decisions to combat piracy are taken, cannot guarantee any effectiveness of anti-piracy activities.

The involvement in fighting piracy should be measured in four areas: political, military, economic and legal, this will show a real situation and would allow evaluation of security actions in the region. In terms of the political area the countries in the region are aware of the need to take multilateral cooperation from 2011, and this fact is expressed by

---

[20] P. J. Heyl, *West and Central African Leaders Unite Against Piracy in the Gulf of Guinea*, 3rd UAE Counter Piracy Conference Brochure, http://www.counterpiracy.ae/, p. 6-8, 21.01.2014.
[21] Ibidem, p. 9.

[22] *Insecurity in the Gulf of Guinea... op. cit.*, p. 4.

political decisions of the States. Regional organizations initiate appropriate action at the ministerial levels. In addition, the UN adopted Resolution No 2018 from 2011 and UN Resolution No. 2039 from 2012, which generally draw the plan how to deal with piracy. Unfortunately individual States and the level of organizations in the region as well as international efforts are definitely not sufficient to eradicate piracy. Policy actions taken by regional organizations, although show some commitment to cooperation, are rather stimulated by the external pressures of international community that perceive piracy as a threat to the global maritime transport and its own economic interests.

The military area and efforts of the States lack the ability to launch extensive patrols in the Gulf of Guinea. While international partners can donate patrol vessels and provide training and advice, local navies cannot be forced to put to sea and to conduct expensive naval or aerial patrols. Organizing expensive operations such as the one in the Horn of Africa is out of the question as different countries are not prepared to extra expenses on another costly military, naval operation, which will not eradicate the problem but rather appease the effects. Due to the specific geopolitical situation (zone of influence the US and other countries), and limited confidence in regional partners, this is the area in which cooperation in the field of analyzed problem, still leaves a lot to be done to bring positive results.

While considering the economic area and all activities, the primary problem is an uneven development of individual regions, poverty and economic collapse of many of regional States. The sense of social injustice, felt by people, is even more intensified by the desire to revenge on foreign companies, exploiting natural resources. This causes that piracy has become one of the forms commonly accepted by local communities. The lack of adequate investments that produce alternative forms of living and partially incurred losses of the local communities, is the main cause of the situation. This situation cannot be improved without help of international community, for economic conditions of the countries in the Gulf of Guinea have no financial potentials or the right instruments to change the situation. If the international community does not want to spend money on expensive military operations it has to invest some money on creating a number of economic programs dedicated to the poorest societies in the region of Western Africa.

Analyzing the legal area, the most significant problem seems to be a very poor assessment of the States of the region in respect to the judiciary system. Despite the fact that they are all members of the UN, IMO and MOWCA, not all of them are signatories to the SUA Convention[23] (not signed up by the DRC, Gabon, Cameroon and Congo). The lack of treaties relating to extradition between the States of the region is another negative aspect which should be taken under evaluation. It ought to be underlined that the legal aspect of the cooperation in the fight against piracy is a big challenge and it remains an open card. Pre-existing disputes over borders and control over EEZs at sea greatly impede from the legal point of view and hamper proper cooperation.

Taking everything into consideration it must be said that an overall situation in the Gulf of Guinea is not satisfactory and all the measures taken, although noticeable, cannot be evaluated as positive enough and fully effective.

7 CONCLUSION

The process of globalization in the 21st century is so advanced, that no State can afford insulation or breaks in the supply of energy materials. For this reason the problem of piracy in the Gulf of Guinea seems significant not only for countries exploiting natural resources there, but it is also directly associated with any developed county. Fluctuations in the prices of raw materials on the world's stock markets are often the result of different activities occurring in very remote places of the globe, such as piracy.

The dispute over the abilities of neighboring countries in the Gulf of Guinea to coordinate naval or capacity-building efforts to fight piracy is still in progress. All criminal activities inland and off the coast of West Africa, including armed robbery, piracy, oil theft and illegal refining, and illegal fishing are connected. Initiatives to stem piracy must therefore be intimately intertwined with those that address other forms of maritime and land-based crime. Although progress has been made in the discussion of regional initiatives, little has been done, even at the State level, to coordinate and implement them[24].

The piracy and other maritime crime in the Gulf of Guinea must be in the center of attention of the whole community because for the past year attacks off West Africa not only continued to spread toward international waters, but they also spread geographically, leading the Joint War Committee's expansion of the War Risk Area in 2013 to include waters off Togo[25]. This is the best example that all

---

[23] *The Convention for the Suppression of Unlawful Acts against the Safety of Maritime Navigation* (SUA) or Sua Act is a multilateral treaty by which states agree to prohibit and punish behavior which may threaten the safety of maritime navigation. IMO official documents, http://www.imo.org, 12.10.2013.
[24] Vide: *Maritime piracy 2013 report...* op. cit., p. 75.
[25] Ibidem, p.73.

we do as an international community to stop piracy in the region is not enough.

In January 2015 the community must admit that a helpful hand, we offered to the countries in the Gulf of Guinea has not been enough to solve the problem. The international community, as the most important participant of international relations, should put more emphasis on eliminating the direct causes of piracy, which are rooted in presence of weak, dysfunctional and fragile states. The involvement in the reconstruction of the internal structures of such States, financial and legal assistance, help in building proper conditions to invest for foreign entrepreneurs as well as native ones, will be motivation to rebuild their economies. Such actions have a chance to start a significant improvement of the living conditions of local communities from which pirates are recruited. In addition, the restoration of weak or fragile states and strengthening regional capacity (in every aspect), will develop an appropriate model in the future how to respond to any potential threat type such as terrorism, which in the XXI century is one of the most significant problems in the world. The effectiveness of all actions is not as successful as it could be so there are a lot of things that need to be done if we want to help the people living in the region and make secure our interests. Hopefully the lessons learnt in the Horn of Africa will bring immediate, positive changes in the Gulf of Guinea States.

REFERENCES

Bridger J., *Countering piracy in the Gulf of Guinea*, http://news.usni.org/2013/07/12/countering-piracy-in-the-gulf-of-guinea, 10.08.2014.

Collier P., *The Bottom Billion*, Oxford University Press 2008.

Dutch disease, http://lexicon.ft.com/Term?term=dutch-disease, 12.01.2015.

Fragile States Index 2014, http://library.fundforpeace.org/fsi14-overview, 04.01.2015.

Heyl P. J., *West and Central African Leaders Unite Against Piracy in the Gulf of Guinea*, 3rd UAE Counter Piracy Conference Brochure, http://www.counterpiracy.ae/, 21.01.2014.

Insecurity in the Gulf of Guinea: assessing the threats, preparing the response, International Peace Institute, January 2014, http://www.isn.ethz.ch, 20.11.2014.

Maritime piracy 2013 report, Oceans Beyond Piracy, www.oceansbeyondpiracy.org, 03.11.2013.

Maritime trade shearing information center – Gulf of Guinea, SAMI, www.security.org, 12.12.2014.

Rekordowy import ropy z Afryki Zachodniej, Fundacja Europa Bezpieczeństwo Energia, http://ebe.org.pl/, 3.03.2012.

Resolution 2018, UN official documents, http://www.un.org/, 20.01.2014.

The Convention for the Suppression of Unlawful Acts against the Safety of Maritime Navigation IMO official documents, http://www.imo.org, 12.10.2013.

The Montreux Document https://www.icrc.org, 10.10.2014.

Ukeje C., Mvomo Ela M., *African approach to maritime security – the gulf of Guinea*, http://library.fes.de/pdf-files/bueros/nigeria/10398.pdf, 01.12.2014.

Wardin K., *Model reagowania na zagrożenia piractwem morskim*, BelStudio, Warszawa 2014.

Counter Piracy Training Competencies Model

G. Mantzouris & N. Nikitakos
University of the Aegean, Department of Shipping Trade and Transport, Chios, Greece

D. Huw
Flir Systems, United Kingdom

ABSTRACT: Competencies is a study that tries to identify and set the relevant associated factors for maritime professionals to perform above a specific status / level during the protection of merchant vessels in high risk / counter piracy operations. Maritime professionals need to have specific competencies in order to accomplish their mission efficiently and effectively, which in other words is eventually the success of the protection of their merchant vessel transiting through a high risk area.

Without a minimum required training status the mission is precarious and the level of risk is being elevated to an absolute maximum. Further more training lessen the cost associated with counter piracy operations taking into account that trained people minimize the use of costly counter measures.

This work evaluates the measure of effectiveness that training has and give effectiveness indexes in order to materialize and quantify daily routine.

A multidisciplinary analysis has been performed in order to measure the available effectiveness levels and produce quantifiable indexes. Indexes create a minimum and/or maximum level of competence that a maritime professional should perform before his/her participation into a counter piracy operation / passage with the ship from a counter piracy related region. Finally a training competence model has been proposed. It is important also to note that the study covers all related personnel with a counter piracy mission, which is officers and crew on board a merchant vessel and back office associated personnel.

1 INTRODUCTION

Competencies is a study that tries to identify first and then set the relevant associated factors in order for the maritime professionals to perform above a specific status / level during the protection of merchant vessels in counter piracy operations. Following the below mentioned analysis and essentially the steps and following the order of training for countering piracy maritime professionals will need to have specific competencies in order to be able to accomplish their mission which is eventually the protection of their merchant vessel through transiting of a high risk area.

Without a minimum required training status the mission is precarious and the level of risk is being elevated to an absolute maximum. Further more training lessen the cost associated with counter piracy operations taking into account that trained people minimize the use of costly counter measures.

This work will evaluate the measure of effectiveness that training has to the evaluation of available options and give an effectiveness index in order to materialize and quantify the reality.

The objective of this work is to define the training competence / status of the people that are going to use the counter measures in order to effectively counter maritime piracy when necessary. If a merchant mariner / seafarer is not trained effectively and efficiently on how to react in a counter piracy operation e.g. use the available and suitable counter measures, have the available psychological preparedness in order to react positively and also generally to follow the procedures then the use of counter measures will not be effective and this person instead of helping the defense of a merchant vessel may deteriorate the situation of his fellow seafarers instead of supporting them.

By this study we should evaluate / give the available tools to evaluate the measure of effectiveness that training has to the use of options during the counter piracy operation.

It is not under the scope of this feasibility study to spend critical time in order to investigate and finally create methods that would not be used in the future operational environment. Innovative results would be produced taking into account stakeholder's workshop and operational experiences collected from counter piracy professionals. Therefore, the scope is limited to the use of available means that will simulate counter piracy training reality and support professionals in the selection process of their personnel. Of course the results we are going to produce / provide at the end of this research will be a prototype and consequently will create hopefully a new roadmap in the training selection / evaluation process for counter piracy operations or at least support them as much as possible in this direction.

2 METHODOLOGY

In this work we firstly performed a literature survey in order to evaluate all (to the extent of our resources and knowledge) the available training materials / roadmaps / guides and competencies methods that robust / rigorous organizations like IMO or government courses, like Australia, Greece or UK (traditional navy nations) are using for training their fellow mariners.

Then a use of a questionnaire during the NATO Maritime Interdiction Operational Training Center (from now on referred as NMIOTC) Workshop in Crete, to Souda Bay, was disseminated to stakeholders in order for them to evaluate the most critical factors during the training process taking into account their operational and/or training experiences on the field.

As a last step a multidisciplinary analysis was performed in order for us to measure the available effectiveness levels and produce quantifiable indexes that will help create safe results. Under the above context we should also mention that tables in the form of check of list have been created in order for a seafarer to be evaluated by his/her superiors before participate as a crew member of a merchant vessel to a counter piracy operation.

Having searched thoroughly the available literature (as per selective references literature at the end of this study) we first identified that all international maritime organizations are using similar techniques and methods where superiors are quantifying the level of competence either using indexes of training effectiveness either through other more sophisticated methods. On the other side maritime oriented universities or generally universities that perform research on the above field are creating models through complex factor systems and try to evaluate training status and approximate reality as more as possible. All the below referenced documents and more created the mindset / mind lock so as to create a starting point for this study which

has a more operational than traditional academic focus.

A workshop was also held to benefit from stakeholders and partners knowledge, professional experience and expertise.

After literature and stakeholders surveys, partners contributed with their final remarks and comments that were included / embedded and modified positively the whole operational research effort.

3 COMPETENCIES ANALYSIS

3.1 Competencies Quantitative Analysis Background

Protection of Merchant Vessels at Sea (from now on referred as ProMerc)is a study that tries to analyze all Piracy Counter Measures during different realistic scenarios and provide to the European Community valid results for the best available use of them during a real operation, so as to eventually support merchant mariners during transiting a high risk area. Therefore, in this training competence study is of utmost importance to have in the back of our minds that the competence that we are planning to acquire should be that required for a person to perform effectively in the below mentioned scenarios. Additionally the individual evaluation should fulfill all the different requirements that he/she needs to acquire in order to efficiently undergo a situation similar to the below mentioned realistic piracy scenarios. We must mention here that these scenarios were developed by ProMerc participants through rigorous analysis / workshops and research.

1 Primary Scenario Candidates (International waters)
   – Large vessel with high freeboard at high speed during day in international waters.
   – Large vessel with low freeboard at slow speed at dawn or dusk in int'l waters.
   – Small vessel with low freeboard at low speed during day in international waters.
   – Small vessel with low freeboard drifting at night in international waters
2 Secondary Scenario Variants (Territorial Waters)
   – Large vessel with low freeboard at slow speed at dawn or dusk in territorial waters)
   – Small vessel with low freeboard steaming at low speed during day in territorial waters.
   – Small vessel with low freeboard drifting at night in territorial waters
   – Small vessel with low freeboard at anchor at night in harbor area

Coming to the competencies baseline requirement is essential to understand that in appointing a maritime professional in a Counter Piracy measures environment, consideration should be given to their overall competency to undertake the mission role (in accordance with the associated work position - for example if he/she is an officer or belongs to the lower crew members) and their physical capability to perform these duties.

The Baseline Scenario we are going to use in order to design the minimum acceptance competence for a maritime professional during a counter piracy operation / mission (with the use of counter measures available on board a vessel) is inevitably the worst case scenario since if someone is successful on completing the task during the worst operational situation he / she should be also effective in any other operational situation that may arise. It is very important to mention here that this material refers only to the operational level professionals and not to maintenance, support or management levels. For those, other criteria should be applied and it is not under the scope of this study. The competencies referred here are only applicable to seafarers that operate counter piracy measures during counter piracy missions / passages from high risk areas.

In our case the baseline scenario (as mentioned above and proposed from the University of the Aegean to the Consortium) and having interviewed maritime security professionals from various operational environments (Gulf of Aden, West Africa, Malacca Strait etc), is to have a Large vessel (low maneuverability capability) with low freeboard (full of cargo – the value of cargo is an alluring to aspiring pirates) moving in international or territorial waters at dawn or dusk (in this specific timeframes of the day the illumination is very low and therefore pirates can act more freely without being noticed (especially when transiting with high speeds).

Under this scope, we will try to evaluate the measure of effectiveness (MOE) that training competence has to the evaluation of the available options.

In order to fulfill the above critical task a detailed design and outline of minimum training and qualifications requirements must be performed. This is referenced below as follows:

The Person:

1 must hold at least a Certificate in Security from a class Organization that is in force (it is not sufficient to hold a certificate from their own shipping company or any other private non certified organization).
2 must have undergone theoretical training of at least 2 days with exams
3 must have undergone simulated training scenarios of at least 2 days with exams
4 must have undergone practical training of at least 2 days of using available CMs that will be provided

5 must have undergone a total training period of at least 5 days (including theory, simulation and practice)
6 must have a working experience on board vessels having transited a high risk area of at least 3 consecutive days during one single passage.

All these above training requirements have been associated with Knowledge (K), Understanding (U) and Processing (P) factors that an individual must acquire.

In order to perform the quantitative analysis the mathematical approach selected was the Multi-Criteria Decision Analysis tool / weightings, as the most appropriate taking into account all accumulated operational experience collected. The most important reason to that was the fact that in this analysis, stakeholder's opinions were included based on their different criteria. Each Subject Matter Expert (from now on referred as SME) ranked its preferences in accordance to training levels and competence and supported the process in a sense that their operational experience was accumulated in the mathematical function, so as to create a more realistic quantifiable training competence index. By SMEs we selected knowledgeable professionals either from operational fields (e.g. shipping companies) or from training entities in NATO and generally from entities that are dealing with counter piracy operations for years. At the same time NMIOTC the sole sound organization for training in NATO provided their invaluable feedback and with other selected SME's opinions made the survey state of the art and unique.

3.2 Competencies Quantitative Analysis Methodology

$$C = K \times U \times P \tag{1}$$
$$= (TT + ST + P) \times (ST + PT) \times (WE + RWE) \tag{2}$$

Competence is being defined here as the multiplication function of Knowledge (K), Understanding (U) and Processing (P) (function (1)). Whereas K, U and P respectively are further being analyzed / subdivided in Theoretical Training (TT), Simulated Training (ST) and Practical Training (PT) for Knowledge, Simulated and Practical Training for Understanding and Working Experience (WE) and Real World Evaluation (RWE) for the Processing part of the function.

Further to this analysis and taking now into consideration the SME questionnaires we are more dividing the above mentioning factors from function (2) into important subcategories where weight factoring is going to take place taking into account all available thematic areas that a trainee should perform in order to finally acquire the minimum desirable training competence. These weight factors are going to be different of course for officers and the crew members of a merchant vessel as well as back office personnel since their duties and responsibilities on board a vessel is way different on these levels of hierarchy. NMIOTC workshop SMEs replied to the questionnaire and by collecting all of their ideas / operational knowledge and experiences regarding the training level that a maritime professional should have, we then moved to the analysis part and defined Competence Level (C) index. The weight factor used was a scale from 0 to 10 on 20 different questions that SMEs were asked to answer referring to competence for officers, crew and back office personnel. We should also indicate here that the thematic areas / fields that were being covered in the questionnaires as a part of the qualitative factor process has been generated from consolidated experiences stemming from SMEs and the authors.

Weightings using Multi Criteria Decision Analysis have put to the following function (3) and by using an excel spreadsheet Competence Level results have defined the future roadmap regarding competence training levels.

Therefore, the final competence level will be calculated from the below function as follows:

$$C = \prod_{i=1}^{n} w_i A_i \qquad (3)$$

where,
C = Competence Level
w_i= criteria weight
A = Criterion Average across SMEs
n = number of total criteria

In this study each criterion (Theoretical, Practical, Simulated training etc) is considered to be of the same importance, except if SMEs comment on that with a different approach. If this happens then we will fill this gap in our analysis and evaluate the total competence with the respect of the following additional function (4) before we use

$$w_i = I_c \times w_c \qquad (4)$$

where,
w_c= specific sub-criteria weight
I_c= Relative Importance of the training level in accordance to the other training levels.

As an aftermath anyone who wants to testify his/her own maritime professional should use the above mentioned Competence index as a rule of thumb in order to select individuals and have the safety that this individual will be able to perform on a worst case scenario (counter piracy situation) as mentioned in detail above function (3).

$$C = K \times U \times P$$
$$= (TT + ST + P) \times (ST + PT) \times (WE + RWE) \qquad (5)$$

1 for 50% then weight is 5 out of 10 so C=(5+5+5)*(5+5)*(5+5)=1500 units of competence (approximately for crew and back office personnel)

2 for 70% then weight is 5 out of 10 so C=(7+7+7)*(7+7)*(7+7)=4116 units of competence (for officers)

3 for 100% - maximum possible value will be 10 out of 10, so C=(10+10+10)*(10+10)*(10+10) =12000 units of competence which means that we have a professional / excellent and full competent expert in counter piracy operations. In this case as we see the ability is becoming 4 times stronger than in the second case and 11 times bigger than in the first case. So competency increases drastically from first to second case (approximately 3 times) and 6 times approximately for the maximum possible case. Thus by training a person on how to handle a counter piracy operation we increase drastically the possibilities to perform the mission successfully.

3.3 Competencies Questionnaire Analysis

The desirable objective aim of the training process of a maritime security professional should be in general terms the following:

1 Acquire the capability to assess global piracy risk / threat.

2 Use all available tools to design an operational passage from a high risk area by implementing risk mitigation techniques in order to complete the mission / voyage successfully.

3 Increase ability of measuring piracy operational risk so as to act proactively.

4 Enhance knowledge in risk management procedures to react effectively to counter piracy situations when on board a merchant vessel.

For performing the above mentioning general tasks Knowledge, Understanding and Processing we have created different questionnaires that have been answered promptly from workshop participants. Also rating of training levels have been considered to be theoretical, simulated, practical, working experience, real world evaluation and other area.

All collected Questionnaires from workshop participants analyzed accordingly. There were 14 SMEs, stakeholders - partners that participated in this research that were coming from different cultural environments. 8 of them were coming from the Operational environment whereas 6 were academic oriented. Therefore the analysis was balanced among experts from operational and academic world giving the desired reality to the below extracted results. Questionnaires are available upon request.

3.4 Analysis - Results

During the analysis all answers received from questionnaires plus all received comments (oral or written) from the NMIOTC stakeholders workshop have been accumulated to the below results. The analysis has been divided into different steps as follows:

Firstly all SMEs inputs were tabulated into excel spreadsheets as the one below. As we can see there are three different tables associated with the three main categories that we analyze in our research. These are the officers on board the merchant vessels, the crews and finally (as was being commented from SMEs during the workshop) the back office people that need to be educated well in order to support real counter piracy operations. On the horizontal axis we have all the inputs from 14 SMEs whereas on the vertical axis we have the levels of training / competence that a person must acquire. There are also three columns at the end, the Total Sum points, the average (points with respect to 10) and final the % percent for each associated training level. The numbers that are listed in EXCEL cells under each SME is the weight factor associated per SME stemming from the respective questionnaire.

Taking into account the used tabulated results we proceeded to analysis and graphs that were being created in association with each of the above mentioned tables. For example figure 1 shows officer training level percentages - Ic factor (similar figures have been created for back office and crew personnel).

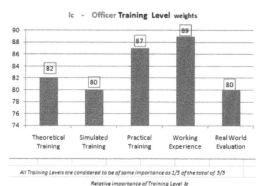

Figure 1. Officer Training Level percentages - I_c

On the above figure 1 one can distinguish the different levels for officers training during the different training levels. The index depicted here is the Ic factor (function 4 above). For example SMEs decided that working experience training and practical training are the most important aspects for officers training with a collected average of 89% and 87% respectively. Additionally it is important to mention from the above graph that officers training

is important enough to collect more than 80% overall in any different level. This is not the case in the crew and back office personnel training as we can see in the next graphs.

For the crew the case is not the same. Here SMEs decided that training is not so important as in officers and back office personnel level with averages to almost reach 80%. But it is apparent that practical training and working experience are also the most important aspects of the training environment with theoretical training here to reach only 58% which practically means that crew can receive 30% less theoretical training that the officers and the operation could still be successful.

Finally back office training / competence is almost at the same levels as for the officers. This is alone a very important result since we understand that it is more significant for a company / organization that deals with counter piracy operations to train its officers on board the merchant vessels and its back office personnel in order to effectively support the operation. The crew can still perform adequately even if receives 30-50% training less than the others. This practically means decrease of the training costs for the company which is a driving factor in nowadays world.

The generated graphs can be combined into a single representation and by that way a cumulative result can be shown on a more effective way as follows.

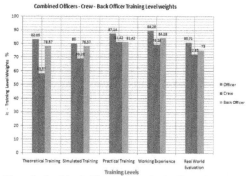

Figure 2. Combined officers - crew - back office training Level weights

Training Competence Level %

Figure 3. Overall Competence Training Level (%) officers - crew - back office

In the above figures 2 and 3 we depict in % how much training officers, back office personnel and crew could receive. For example If we have to train a number of personnel participating in a real counter piracy operation we should train 39.5% of our officers, 33,81% from our back office personnel and only 26,68 % of our crew. This result depict that if we select not to train our crew at all the mission will still be accomplished with well trained personnel from our officers on board the vessel and of our support officers at the back office ashore.

3.5 *Competence Training Index*

The desired result from this analysis is of course, as we stated above, to find out an index where we will have an indication on how competent is an individual before participating on a high risk counter piracy mission (ship's voyage etc). By using SMEs inputs we tried to split a counter piracy mission / operation into different thematic areas and to ask subject matter experts what is their opinion. Is it needed for a specific individual belonging to one of the three distinct above categories (officers, crew and back office personnel) to have knowledge on the available information and how much of it is needed in knowledge depth and time? The results were indeed very interesting taking into account that all interviewed SMEs are leaders in the counter piracy world environment either from an academic perspective either from an operational / commercial view.

On figure 4 below we have quantified the need for an individual belonging on the crew to have knowledge on specific thematic areas relevant to counter piracy. As one can see areas like Lessons Learned, BMP implementation procedures, QA sessions relevant to real case study scenarios and at least one real voyage passage from the area of operations is acquiring the most attention of the SMEs. In other words the most important thing is for a crew member of a vessel to have a onetime experience from the area and to have studied previous lessons learned and BMP implementation methods. All other theoretical areas are considered to be under an average level of knowledge. In this

graph if one can add all the thematic areas the total percent is 100%. So for a person to be capable as crew member, one need only to have acquired knowledge from the red column thematic areas plus some other selective areas in order to reach the 26.68% that the research computed on the above paragraphs graph. Having these metrics we can now infer on how competent is a crew member of a vessel for performing and contributing to a counter piracy mission effectively.

To be more specific and give a real example let's consider that we have an individual and we want to measure its effectiveness prior to board a vessel and participate in a counter piracy mission. In the below graph we see all the thematic areas that a person should know theoretically in order to have a training competence 100%. The red columns, as we mentioned also above, are those that the SMEs pinpointed as the most important thematic fields and gave the fact that without the knowledge of these thematic areas one should not participate in a counter piracy mission. These areas are:
- Lessons Learned - Safety precautions and BMP implementation
- Exercise in a real scenario
- One passage from the counter piracy area
- Five days of training during a passage of a high risk area

The above fields have in the total competence of a trainee 7,83 - 9,06 - 13,85 - 7,72 percent respectively with the most important to be the "At least one passage from the high risk area". All the above thematic areas have a total sub-percentage of 38,46% of the total training competence that one must have. One member of the crew though to be considered as counter piracy competent must collect a total of 45,56% and that directly means that except the red column training thematic areas must at least take two other thematic areas (from the blue columns with the associated percents) in order to meet the least requirement of 45,56% and be considered competent, in accordance with SMEs opinion.

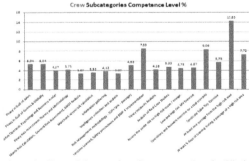

Figure 4: Crew Subcategories Competence Level (%) in accordance with training thematic field

228

Figure 5 is the digest - epitome of all the above performed research analysis since it directly shows to us competence level percentage comparison for officers, crew and back office personnel. From this graph SMEs defined that officers should acquire no less than percentage of 69,28 % when being trained in counter piracy thematic areas whereas this percentage is 55,35 for back office personnel and 45,92 for the crew of the vessel.

Therefore it is more important and necessary to train officers and back office personnel and be safe that your mission will be successful when dealing with counter piracy. On figure 6 all these different subcategories of training thematic areas we analyzed above and that give the overall competence have been depicted on a comparison graph for officers, back office personnel and crews and one can infer the importance in any different sub - training thematic field that his/her personnel must acquire.

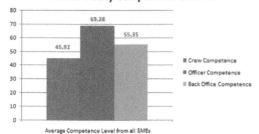

Figure 5: Overall Piracy Training Competence Level (%)

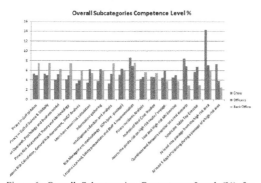

Figure 6: Overall Subcategories Competence Level (%) for officers, back office personnel and crews.

Having completed the research analysis from the received SMEs opinions we now return to our theoretical calculation of the training competence level of an individual (quantitative competencies analysis). We tried to find a way of resembling the real training environment - competence level with the use of a simple mathematical model since it is the first time that this approach is performed for a counter piracy operation. Based on the model that

we used Competence (C) is a function given by the multiplication of Knowledge, Understanding and Processing. Therefore and in order for an individual to complete a specific percentage in the above mentioned thematic areas (as we have shown in the above analysis with graphs) we have assigned, based on the weights that the SMEs provided, metric units and calculated through functions 1 to 5 the competence level units in order to give a specific quantifiable metric for an individual competence level. These metrics (in thousand competence level units) have been calculated and produced figure 7 which is the quantification epitome of our research. On the graph we have depicted an example with competence level units for three different individuals (crew, back officer and officer) where they must specific units in order to be able to be considered competent for a counter piracy mission. The crew member has accumulated 1500 units which is a little more from the 45,92% that the research gave us as a limitation percentage. The back officer respectively has 4116 units a lot more than the 55,35 % that he should acquire in order to be considered competent and finally the Officer has acquired 12000 units which is way more than 69,28% that he should have in order to be considered confident.

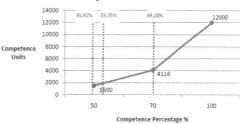

Figure 7: Competence Level Units (Quantification approach) for a crew, back office and officer (example case)

To move one step further we should note that 1500 units for a crew member is respective to have participated in 5 of the most important thematic areas and get a 50% total training competence. 4116 units is for a back officer to have acquired 7 of the most important courses - thematic areas and 12000 is for an officer to have acquired 10 of the most important thematic areas. With this simple quantification process we now have a metric on how a person is competent enough to participate in a counter piracy mission / operation / trip / voyage or not.

4 DISCUSSION AND CONCLUSIONS

Having performed the above analysis we deduce the following important conclusions that are useful for the design and execution of a counter piracy mission / operation when a merchant vessel is to go through / transit high risk area:

1 Officers are the individuals that must be trained the most when we have to participate in a piracy operation.
2 Back office personnel is the next team that a shipping company focus on its training efforts since those people need to support effectively and in most cases to understand thoroughly the situation in the ocean since they are the decision makers ashore.
3 Crew needs to be competent on a relevant percentage of around 50% of the thematic areas.
4 SMEs believe that training is important not only in officers and crew members of the vessel but also for back office personnel where in most cases they are not individuals coming from the shipping world but from science oriented environments.
5 Quantification of training competence has been performed and we now have a metric system on measuring the level of competence of a person dealing with a piracy - high risk area situations.
6 Multi Criteria Decision Analysis and weightings from SMEs interview give us similar results and a safe method of measuring the competence of an individual.
7 The above method gives results based on worst case scenario which is the passage of the vessel with low freeboard, with low visibility and low speed from a high risk area.

Finally and to sum up this research and analysis study it is worth to mention that a quantification method in place gave logical and important results for those that deal with counter piracy missions daily. It is also important that the above methodology can be performed in any high risk mission taking into account that training thematic areas will be defined from the user first in order for the methodology to quantify a person's competence level to participate in a high risk - maritime security mission such as counter weapons of mass destruction, counter narcotics, counter illicit trafficking, maritime theft etc.

ACKNOWLEDGEMENT

This work was carried out in the framework of the project ProMerc: Protection Measures for Merchant Ships, which has received funding from the European Union's Seventh Framework Program for research, technological development and demonstration under grant agreement no 607685.

REFERENCES

1. STCW 2010/03 AUG 2010, MANILA AMENDMENT
2. Non Lethal Weapons Effectiveness Assessment Development and Verification Study, Final Report of Task Group SAS-060, CMRE Technical Report, October 2009
3. Technology Achievements in Maritime Educational Procedures: Behavioral Assessment Framework, Vasilakis Panagiotis, Prof. Nikitas Nikitakos, University of the Aegean, Dept. Shipping Trade and Transport.
4. Malone 1990. "Theories of learning: A historical approach" . Bel-mont, CA: Wadsworth.
5. Jeffrey D. Doyle, and Eric M. Webber,and Ravi S. Sidhu,2007 "A universal global rating scale for the evaluation of technical skills in the operating room." The American Journal of Surgery : 551–555.
6. IMO, 2000 "Sub-Committee on Standards of Training and Watchkeeping. Validation of model training courses." Vol. 32nd session. London.
7. Marzano and Robert J. 1989"A Theory-Based Meta-Analysis of Instructions, USA Colorado: Mid-continent Regional Educational Laboratory.
8. IMO STCW/ISCG 2/5/3, Training of Personnel operating in ice - covered waters, Norway, 27 July 2009
9. STW 43/3/3, IMO Subcommittee on standards of training and watchkeeping, 43rd session, Agenda item 3, Validation of model training courses for Ship Security Officers, 4 August 2011
10. STW 43/3/2, IMO Subcommittee on standards of training and watchkeeping, Model courses – Security awareness training for seafarers with designated security duties and Security awareness training for all seafarers, 25 July 2011
11. STW 37/10/1, IMO Subcommittee on standards of training and watchkeeping, Development of Competencies for Ratings, 18 October 2005
12. IMO Model Course for Maritime English Professionals
13. Competencies for Maritime Security Guards (Baseline Requirements), Australian Government
14. Maritime Transport Education and Competence, Development in a Maritime EU, FP7, Press Transport Consortium, Cybion Srl et.al, Maritime Development Center Europe, www.press4transport.eu

Ship's Operations

Experimental Study for the Development of a Ship Hull Cleaning Robot

K. Watanabe
Tokai University, Shizuoka, Japan

K. Ishii
Kyusyu Institute of Technology, Kita Kyusyu, Japan

K. Takashima
Tokai University, Shizuoka, Japan

ABSTRACT: In this paper, we present the system components of the robot and recent experimental results which is executed at Shimizu Port in Japan using Tokai university research and training ship Bosei-Maru. We tested the cleaning robot while Bosei-Maru is berthed at a pier. As the operation cost of the robot should be reduced as much as possible, so we designed the robot very small and light weighted (30kgf in air) that it can be handled by only two persons. The cleaning speed of the robot is also important for cost reduction, we designed the robot as fast as possible. The average speed is around 0.2m/s during the experiments. The cleaning result was examined to compare the difference between the cleaned surface and the non-cleaned surface in the dockyard when Bosei-Maru was under the annual inspection.

1 INTRODUCTION

Recent trend towards increased cost of fuel will continue further, so the reduction of fuel consumption will be required more severely as well as the severe requirement of reduction of CO_2 emissions. To achieve these requirements, the reduction of resistance is essential for efficient ship navigation. As there has been made many technological challenges to improve fuel efficiency, we believe the prevention of marine biofouling is one of the promising options. Anti-fouling paint is effective to prevent biofouling to the ship hulls like barnacles, however, even the painted hull acquires slime-like biofouling caused by marine alga on its surface easily. The cleaning of the ship hull is generally carried out during inspection in dockyard once a year. Frequent cleaning while the ship is berthing is desirable to keep good fuel efficiency. If frequent ship hull cleaning is possible while ships dock at berth with ease, ships can keep good fuel efficiency that makes the transportation costs and CO_2 production less. Cleaning by divers costs much and is a high risk task. One possible solution for this issue is to introduce underwater robots for ship hull cleaning.

Underwater environment is one of the extreme environments because of high pressure, low visibility by turbid water, low communication density from radio wave attenuation (Ura 1989). So it's not been easy to develop an underwater robot.

However, as technology in mechatronics advances the cost of computers, actuators and sensors became very low. As a result, recently, underwater robots are applied to various missions such as observations of underwater structures oil-well drilling rig on Deep Ocean, bridge piers, discovery of unknown creatures, ecology investigations, maintenance of underwater cables, investigation of hydro-thermal vents and underwater volcanoes (Ura 2005, 2006, 2007, Miyake 2005, Suzuki 2004, Ito 2004).

By the same token, the ship hull cleaning robot can be developed and realized relatively low cost, which we suppose is competitive with services by divers. In this paper, we introduce our robot which we are developing and several results on cleaning experiments executed using Bosei-Maru as well as discussion from our experience through this experiments.

2 THE SHIP HULL CLEANING ROBOT

2.1 *Development concept of the robot*

We developed a prototype of ship hull cleaning robot as shown in Figure 1. In Figure 1, the yellow line is an armored umbilical cable which contains 200V maximum 2A power supply cable and an Ethernet cable for PC communication as well as underwater video camera monitoring. The umbilical cable is armored enough strong to bear the weight of the robot in air so that we can suspend the robot from the deck or a pier. The body of the robot consist of two pressure hulls which contain a video camera, actuator driving system, sensor system and computer system to control the total system. The robot is controlled by three thrusters, one is for vertical motion, others are for horizontal motion, that is, xy-coordinate and heading. The pressure hulls and actuators are attached to the frame shown in Figure 1. Once the robot reached a wall surface of a ship, it runs on the wall using four wheels attached to the frame. These wheels also enable the robot to keep the distance between its center of gravity and the wall surface constant, which results in keeping the brush pressure force against the wall constant. The frame works as a bumper and protects the ship hull and the robot body from contacting each other not to damage one of them.

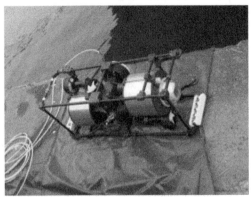

Figure 1. External view of ship hull cleaning robot

The design concepts of this robot are as follows,
1 The users can control the robot by remote operation with observing the ship surface real-time.
2 The robot size and weights are suitable for two persons handling to launch and recovery the robot between air and underwater.
3 The brush device has function not only cleaning the ship hull but also it yields suction force to attach the robot on the ship hull surface.
4 Center of gravity (CG) and center of buoyancy (CB) are as close as possible to implement the control system keep arbitrary attitude of the robot.
5 Robot has the front and rear cameras to record and monitor underwater images while and before/after cleaning.
6 The robot is equipped with even number of brushes. The brushes rotate in reverse direction each other to cancel the reaction moment, e.g., one brush rotates in CW to have suction force and the other in CCW. Therefore, the propeller pitch is designed to be opposite in a pair of thrusters.

2.2 Components of subsystems

The computer system of underwater robot consists of the low-level control board inside the pressure hull cylinders and a host PC outside of the robot to monitor inside status of the robot. The role of lowlevel control system is to measure sensor data, control thrusters, control brush units and communicate to the host PC including video camera monitoring. The host PC is for high-level control to decide the robot behavior and the video camera control. Users operate a joystick and send commands such as go forward, go backward, up, down, rotate, and brush start. The measured sensor data are displayed on the monitor(s). And the network cameras record the video image of underwater ship hull. The camera system has pan/tilt/zoom functions, which are also controlled from the host PC.

2.2.1 Low-level control system

As the low-level control system, two 6-axismotordriver-control boards has been developed and mounted in a pressure hull cylinder. One board (master board) handles the control of 6 thrusters, sensor data acquisition, and the other (slave board) does the control of 2 brush units. The communication between two boards is I2C communication protocol and that between master board and the host PC is based on RS232 signals, which is converted into fiber optic transmission. The board has a Microchip dsPIC (dsPIC30F6014) which has 16ch12bit ADC of which 6 ch are used for current sensors of motors and 1 ch for depth sensor, 4-ch timer, 8-ch PWM of which 6 ch are for motor speed control, 2 uart serial communication ports of which 1 uart for PC communication and 1 for the attitude sensor (PNI TCM-XB), and I2C commutation.

2.2.2 Host PC

The behavior of robot is decided based on calculation of the host PC according to algorithm implemented. The sensor data is transmitted by the low level control system and if the robot is operated as automatic mode, desired control forces will be yielded by the low level control system according to calculated values from the host PC. We can control the robot manually using a joystick. The obtained sensor data (roll, pitch, heading, depth, electric currents, target values) are shown on the GUI on PC. The images of 2 network cameras are stored on a PC on the network, and camera parameters such as zoom, pan and tilt can be also controlled from a remote PC.

2.2.3 Pressure hull cylinder and frame

The main pressure cylinder is composed of 2 cylinders and a housing part in the middle of cylinders. The diameter, length and thickness are decided based on mounted device sizes, the positions of CG, CB and neutral buoyancy balance. The thickness of cylinders was calculated based on the paper (Ito 2004), under the condition of maximum depth: 50 m, material: aluminum, safety factor: 2, therefore design pressure is 1 MPa. The minimum thickness in calculation was 1 mm, however, 4mm thickness aluminum cylinder was common and cheaper as commercially available one, so we fabricated using this thickness.

2.2.4 *Sensory system*

As a sensory system, an attitude sensor: PNI TCMXB to measure roll, pitch and heading, a depth sensor: YOKOGAWA FP101A and 2 network cameras: CANON VB-40. TCM-XB sends the binary data to the master board by RS232 serial communication every 0.2 ms. TCM-XB is an electric compass which detects the magnetic field direction and calculates heading direction.

2.2.5 *Power system*

AC 200 V is provided to the underwater robot through the tethered cable, and AC 200 V is converted to DC 24 V, 12 V and 5 V. DC 24 V is given to thrusters and actuators, 12 and 5 V are for sensory system and isolated from 24 V to prevent EM noises from actuators.

3 CONTROL ALGORITHM OF THE ROBOT

As mentioned earlier, the host PC and the low level controller which is implemented to the On-Board Computer in the cylinder communicate each other to control the robot. The low level controller algorithm is shown in Figure 2. The control status is updated in 10Hz. The controller waits a set of command for the procedure of Decision Making & Control in Figure 2 from the host PC. This command set includes voltage signals for thrusters and brushes to yield desired control forces which are calculated by the host PC. In this robot, we are controlling roll, pitch, yaw angles while cleaning, and the control algorithm is based on simply LQR negative feedback whose state variables are angles and speed of angles.

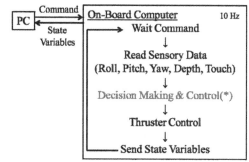

Figure 2. Low level control schematic

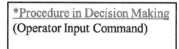

Figure 3. Decision making command set from the host PC

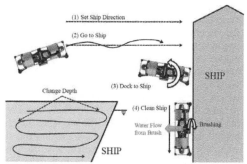

Figure 4. Robot motion according to command set

Procedure	Condition	Surge	Depth [m]
(2) Go to Ship	|Yaw| < 30 deg	GO	Docking Depth
	else	STOP	
(3-1) Docking	-70 >Yaw > -110	STOP	Docking Depth
	else		
(3-2)After Docking	Num of Touch Sensor "ON" >=3	STOP	Docking Depth
(4)Clean Ship	|Depth Error| < 0.2	GO	Target Depth
	Else	STOP	Full Thrust

Procedure	Pitch [deg]	Yaw [deg]	Brush
(2) Go to Ship	0	0	OFF
(3-1)Docking	0	-90 deg	ON(Full) OFF
(3-2)After Docking	0	No Control	ON(Full)
(4)Clean Ship	0	No Control	ON(Full)
	No Control		ON(Full)

Figure 5. Control parameters with command set

The command set also includes what we call Decision Making parameters which we designed as shown in Figure 3. An operator should input numbers from (1) to (4) from the console of the host PC to change the procedure mode of the robot. This helps the operator to control the robot intuitively. Figure 4 shows how a robot moves corresponding to the command set. At first, when the robot has just put into water, we check the robot's depth, roll, pitch and yaw angles which are displayed on the console. Then select (1) to change heading angle of the robot from the initial position toward the ship. The rotation angle of heading must be input by the operator and this value is sent from the host PC as a control value with this command set to the on-board computer. Then robot starts rotating to change its head toward the target ship. The rotation angle is determined by the operator by looking the head orientation of the robot floating near the surface to change its orientation toward the ship. Then the operator puts the command (2) and the robot starts to move forward to reach the ship. This command also needs control parameters as an example is shown in Figure 5. In an example shown in Figure 5, the robot moves toward the ship if the yaw (heading) angle is less than 30 degrees, otherwise it stops. If the robot approaches the ship hull successfully, then the operator puts command (3) with control parameters like Figure 5. This procedure has two phases as shown (3-1) and (3-2) in Figure 5. If the touch sensor counts more than three, the robot considers it successfully docked the surface. To fit on the surface of the hull, the brush must be ON because the suction force which attaches the robot to the hull surface is yielded by the propeller attached behind the brush. As the robot cannot find the ship automatically, we need to implement this procedure. Then we put the command (2) to move the robot near the ship hull where we want to start cleaning.

Because we have not developed a position sensor which can detect the robot's 3D coordinate in the water, this command and control value/status set is very useful to drive the robot in the water. The coordinate determination near the surface using ultrasound ranging system is not easy because the multi-path echoing of the sound cannot be ignored due to sound reflection from berth wall, water surface and the ship hull. Plus, the ultrasound ranging system like LBL is very expensive. The detection of underwater coordinate relative to the ship using ultrasound ranging sensors is our next work in the near future.

4 CLEANING EXPERIMENTS USING BOSEIMARU

Bosei-Maru is the research and training ship of Tokai University. Its international gross tonnage is 2174t, length is 87.98m, breadth is 12.80m, depth is 8.10m and the maximum draft is 4.80m. It is the only oceanographic research vessel which is owned by a private university in Japan. It is used for oceanographic research and education including training course of navigation officers in the school of marine science and technology. The mother port is Shimizu port in Shizuoka Prefecture. It sails almost every weeks from March to November not only in Japanese coastal sea area but also in international sea like Polynesian islands.

We executed the cleaning experiments on November 17th and 18th while Bosei-Maru was at its berth in Shimizu port just before it was going to dock for annual maintenance and inspection at Miho Shipbuilding Company whose yard is at Shimizu.

Figure 6. The robot docked on the hull wall

Figure 7. The robot handling in the narrow space

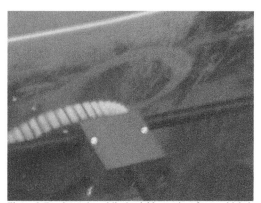

Figure 8 Brush marks while it yields suction force which is used to attach the robot on the wall

Figure 6 shows the robot docked to the hull surface.

In this case, though we put the robot directly to near the ship hull because space between BoseiMaru and the berth was very narrow as shown in Figure 7, the control procedure is all the same with the procedure when the robot starts from open space. We can see a circle mark in Figure 8. This mark is drawn by the brush while the robot was trying to attach the surface wall at the initial position to start cleaning.

During experiments, we found the electric compass TCMXB didn't work well because of magnetic distortion due to not only ship hull iron but also something inside the berth concrete like reinforcing steel structures. It has a kind of magnetic distortion compensation function, but it didn't work well. When the robot was in open space where was far from both the ship and the berth, the heading control worked correctly. However, when the robot was near the berth or the ship, sometimes the attitude sensor values became not true. This didn't matter when we control the robot manually using joystick with video camera monitoring. This would be serious when the robot is used in autonomous mode. We sometimes observed thrust forces became excessively high in some cases and the robot yaw motion was not controlled correctly in autonomous mode because false heading sensor value was used in the feedback control force calculation in the host PC.

.Development of autonomous cleaning system is our final goal. Through this experiments we felt it is essential for us to develop a series of sensor system which we can detect the robot's rigid body rotational and translational coordinates even though under the condition that there exists iron which degrades compass sensor as well as multi-path environment which degrades ultrasound ranging sensors. We will consider this issue further as well as better autonomous control algorithm.

We set up our host PC system on the starboard side deck of Bosei-Maru. The power was supplied from Bosei-Maru's 200V electricity. It was easy to carry the robot on land by two students. Although lowering down an underwater robot to the water from the berth is very difficult task in general, we had no difficulty in this task because the robot is designed as light weighted as possible. We slipped the robot along a stepladder as shown in Figure 7. Once the robot is in water, the handling of the robot becomes very easy because it's neutrally buoyant.

It turned out that the suspended umbilical cable from the deck affects the motion of the robot because the robot is so small that the tension force of the catenary becomes relatively higher compared to the forces which the robot can yield to move in the water. We handled the cable manually from the deck not to affect the motion of the robot. We have to solve this issue in the near future especially for development of autonomous control mode.

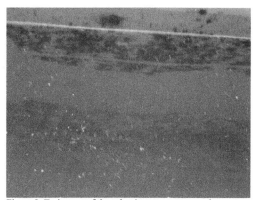

Figure 9. Trajectory of the robot in autonomous mode

In the experiments, we tried to control the robot to move from the side wall to the bottom wall but it was not successful. One reason is we didn't get the attitude sensor data correctly as mentioned earlier. As the curve of the ship hull from the side wall to the bottom is under the depth of 4.5m so that we cannot see the robot from the surface. The robot needs to move away from the side wall then it comes around behind the bottom wall. In this case, only camera image was not enough. We had a renewed sense of the importance of implementation of autonomous mode.

Although the attitude sensor's reliability degraded sometimes, as we implemented filters in the program to compensate the motion of the robot, it successfully cleaned along the side wall as shown in Figure 9 in autonomous mode. The forward velocity is around 0.2m/s. A wider range cleaning trajectory of the robot is shown in the next section.

5 EXPERIMENTAL RESULTS

We examined the cleaning results on November 25th at Miho Shipbuilding Company when Bosei-Maru is dried in the dock. Figure 10 shows a part of the cleaned hull. Whiter part of the hull shows the trajectory which the robot moved straight while it was cleaning the wall. Darker part in the picture like right hand part around the bow thruster is a kind of slime which we can see in Figure 9. Figure 11 shows an enlarged picture of Figure 10 focused on the bow thruster part. We can see contrast between the cleaned part and the bio-fouling part. The slime was perfectly cleaned by the brush which we deployed to attach the robot in our experiments.

Figure 12 shows the bow part of Bosei-Maru before it is cleaned using jet of water for cleaning service in the shipyard. On the other hand, Figure 13 shows after it is cleaned by water jet cleaner for annual cleaning procedure.

Figure 12. Bow part before water jet cleaning in the dock

Figure 13. Bow part after water jet cleaning in the dock

Figure 10. Contrast between the cleaned and not cleaned part

Figure 11. Enlarged picture around bow thruster

We can see whiter mark which the robot made during its cleaning even after the hull wall was cleaned. This means the antifouling self-polishing coating was ground off by the brush. We have several types of bristles of brush and it turned out that the bristle we used in this experiment was too hard. This indicates we need to be careful when we choose the bristle. Once the ship is in service, it is difficult for the ship to be docked in a shipyard just because its coating is removed. So the bristle must be chosen not to damage the coating. We need to find the best bristle which doesn't affect antifouling self-polishing coating. We will continue experiments further from this view point.

6 CONCLUSION

In this paper, we introduced a prototype of ship hull cleaning robot and cleaning experiments using Bosei-Maru. The cleaning experiments were executed successfully applying our control algorithm and we could confirm the designed system functioned very well. We examined cleaning results a week after our experiments while Bosei-Maru was dry docked in the yard of Miho shipbuilding co. for maintenance.

From the experiments, we found several issued to be solved. Firstly, the electric compass like TCMXB is not a good sensor because magnetic distortion cannot be ignored. Secondly, coordinate determination of the robot is important to make the robot control autonomous, thus effective ultrasound ranging system should be developed. The underwater position detection is also essential for the robot to be able to make its trajectory determined in the control algorithm. We'd like the robot to move from the side wall to the bottom and move over some devices like a bow thruster not to damage its structure. In addition, umbilical cable suspended from the deck affects the motion of the robot in service. The robot needs a power line and a communication line to be used like camera image transmission. So we cannot remove an umbilical cable and we need to develop efficient cable handling system. Finally, the bristle we used turned out that it damages antifouling coating of a hull wall. We need to implement a bristle which never affects the coating but is able to clean the wall peeling biofouling creatures.

REFERENCES

Ura T., et. al 2007, Dives of AUV "r2D4" to Rift Valley Central Indian Mid-Ocean Ridge System, CD-ROM Proc. of Oceans'07, 6 pages

Ura T., et. al. 2006, Experimental Result of AUV-based Acoustic Tracking System of Sperm Whales, CDROM Proc. of OCEANS'06 Asia, 6 pages

Ura T., et. al. 2006, Dive into Myojin-sho Underwater Caldera, CD-ROM Proc. of OCEANS'06 Asia, 6 pages

Ura T 2005, Two Series of Diving For Observation by AUVs-r2D4 To Rota Underwater Volcano and TriDog 1 to Caissons at Kamaishi Bay-, Proc. International Workshop on Underwater Robotics 2005, Genoa, Italy, pp.31-39

Miyake, et.al, 2005, Cephalopods observed from submersibles and ROV-II. A gigantic squid in Sagami Bay, Chiribotan, Vol.36, No.2, pp.38-41

Suguki M. 2004, Development of Remote Operated Vehicles for Underwater Inspection in Nuclear Power Plants, Journal of Robotics Society of Japan 22(6), 697-701

Kameyama M. 2004, Emission of air-pollution substance by air and ship transportation of trade goods, National Maritime Research Institute report, PS-37

Ito K. 2004, Development of Small Under Water Vehicle, Journal of Robotics Society of Japan 22(6), pp.702-705

Obara T. et. al. 1997, Autonomous Underwater Robot "R-One Robot", Journal of Japan Society of Mechanical Engineer, Vol. 100, No.943, pp.665

Ura T. 1989, Free Swimming Vehicle PTEROA for Deep Sea Survey, Proc. of ROV'89, pp. 263-268

Assessment of Variations of Ship's Deck Elevation Due to Containers Loading in Various Locations on Board

P. Krata

Gdynia Maritime University, Poland

ABSTRACT: The paper deals with ship motions in sea ports aroused by the weight of containers loaded by a gantry crane. The result of ship motion is a variation of her deck elevation. The realistic range of such variations are assessed for a variety of cargo location on-board. The common formulation of ship motion equations is applied and rolling, pitching and heaving are estimated to enable carrying out further computations aiming at an assessment of specified deck spot elevations.

The described results are a part of wider study focused on improvement of gantry control algorithms with regard to faster operations thanks to more accurate estimation of the moment of cargo release from a gantry hook or spreader. The study may be the contribution to the development of a planned new gantry control systems in sea ports.

1 INTRODUCTION

The main commonly discussed features of maritime transport are usually its safety and effectiveness. The effectiveness of a transportation process can be considered in both an economical and a technical aspect. Considering cargo vessels related issues one may point out ship stability as one of her crucial characteristic. The stability of a vessel belongs to operational characteristics enabling cost effective and safe operation [2]. Moreover, stability as an ability of ship's body to withstand and counteract external and internal heeling moments, shall be considered in terms of both: sailing and cargo operations undertaken in ports too. The second aspect is taken into account in this study.

There is a variety of causes for stability problems on board and they can be divided into several typical groups like: cargo and ballast operation, cargo shifting, icing, forces of the sea etc. [5]. The two first mentioned ones may effect a ship during cargo operations which are crucial component of carriage of goods overseas.

In case of dry cargo transportation an essential part of the cargo handling is its loading and discharging operation by means of cranes and gantries. Since the global containerization trend has reached a significant share of the market and it is still in progress, there is a point to focus especially on gantry operations. Such kind of equipment is in common use in sea ports worldwide due to its high loading and discharging rate and a precise cargo handling.

As a gantry is firmly established on the ground, usually on a dedicated rail system, it seems to allow smooth cargo shipment down to ship's holds and tween decks. However, the surface of decks and tank tops persists in permanent movement with the whole body of ship's hull, creating control challenges resulting from this relative motion of the gantry and the cargo destination position. This motion is an integral part of cargo operations on-board and thus has to be dealt with as precisely as it is reasonably possible. Any increase in the accuracy of gantry control in terms of relative motion compensation, improves the overall performance of the cargo handling process. The essential problem of gantry control is considered by many authors, nevertheless they omit the problem of a moving base of cargo destination. Even when the dynamic modeling and adaptive control of a gantry is researched and applied the efforts are aimed at the tracking errors reduction with no consideration dealing with unstable position of ship's deck or cargo hold [8].

The contemporary gantry control systems found in sea ports might be relatively advanced and sophisticated but they do not capture any external data describing ship rolling, heaving and pitching. Even is such extreme conditions like cargo transfer carried out on gas and oil offshore fields the moving ships transmit no information enabling her motion estimation. The lack of ship motion estimation during cargo operations in ports is evident too. Both remarks result from authors' sea service experience onboard ships and series of reviews with ship masters and chief mates responsible for cargo loading and stowage.

Generally, the interaction between a ship and cargo being loaded on-board can be consider from one of two essential points of view. The first one is an influence of cargo weight on the transverse and longitudinal ship stability. Any extra weight discharged or loaded on-board changes vessel's weight distribution and as its result, it modifies her stability performance. Moreover, cargo being stored within the whole space of a hold or a tween deck has to be loaded in all available room in a cargo space. Therefore it generates an extra moment heeling and trimming the ship. The angle of heel is significantly more important due to much lower ship resistance against transverse heeling than longitudinal trimming and for that reason the vessel's transverse stability assessment has to be carried out prior cargo operations in ports, principally in case of heavy lift operations.

The second aspect of interactions taking place between loaded or discharged cargo and a vessel reflects possible hazards to ship and cargo resulting from too impetuous placement of a piece of cargo, for instance a container, on deck or tank top [6]. This may cause some damages to ship construction or loaded cargo and always generates an economical loss. The explanation of such a phenomenon is based on a simple remark, that cargo is smoothly lowered by a gantry to be released when in contact with deck, while the vessel rolls and pitches due to some external excitation or other cargo influence. However, the problem of moving base which impedes and slows down cargo operations in sea port, can be solved by means of gantry control improvement and an application of proper compensation [6].

2 CONTEMPORARY APPROACH TOWARDS SHIP STABILITY ISSUES IN HARBOR CONDITIONS

From the very beginning of sea navigation some stability problems occurred and they take place nowadays as well. Taking into account cargo operations in ports one could mention a couple of typical scenarios leading to the stability accident. One of the most important and dangerous reason for stability accidents in ports is a stability loss due to cargo operation carried out by own cranes. The vertical center of gravity of a vessel may rise so excessively that the righting arm curve is reduced causing ship's capsizing.

The example of such an accident may be capsizing of the Dutch heavy lift M/V Stellamare at Port Albany, NY, dock on Hudson River in December 2003. The was equipped with 2 x 180 tons deck cranes so the total lifting capacity was relatively high as per such ship being 88.20 meters long overall and 15.50 meters wide overall [9]. The heavy heave - causing vessel to capsize – was the reason for major injury to some crew members and massive financial loss. Eight crew members were thrown into partly frozen water, seven were rescued from ship, some by helicopter. According to a witness report the ship "began to roll slowly, almost silently and turn away from the dock, sinking on its side in the ice-choked waters of the Hudson River; once she went to rollin' never stopped". Due to the insufficient care on pending stability characteristics the extremely costly cargo of electrical generators was broken due to this stability and operation failure [9].

Although the stability loss accident can be easy explained in the case of heavy lift operations, a similar scenario was noticed during container loading with the use of a shore-based gantry crane. One of the example can be stability incident of 101 meters long Germany based company owned container ship M/V Deneb [10]. She partially capsized at Juan Carlos I dock of the Port of Algeciras in June 2011. The ship's load was to be boarded from forward to aft, separating the containers with destination to one or the other port in bays. In order to correct the excessive trim to the ship's bow, the port and starboard number 1 double lined tanks were to be deballasted [4]. during the loading the ship had experienced heeling as much as 10° to each side and, therefore, the loading of containers on the sides was alternated. As the operation was close to being completed, the Deneb began to heel towards the pier and instead of stopping at 10°, she continued heeling without stopping until she impacted and ended up resting on the pier, at a permanent 45° heel angle. In barely 30 seconds, the vessel went from floating upright to lying on the pier, with an approximate heel angle of 45° causing significant disturbances in port operation and massive costs of repair [4].

The mentioned stability incidents suffered by ships in ports are some examples of a general issue related to ship stability management. The phenomena of ships heeling and trimming is known for a long time and officers responsible for loading and discharging cargo on-board are generally aware of the importance of this issue. Thus, the ship stability assessment is a routine part of chief mate's duties. Nevertheless, the contemporary methods applied in the course of standard stability assessment carried out on-board are rather simple and statics-based.

Nowadays the transverse stability assessment during cargo operations takes into account the GZ curve (righting arm curve) being a graph presenting the lever of moment restoring the ship to her initial upright position. The angle of heel due to an external moment, for instance the heeling moment due to cargo load, can be easy obtained. Having calculated the values of righting arm (GZ curve) in the full range of ship's angles of heel, the heeling arm is plotted on the graph to find out the expected angle of heel to due cargo operation. The moments balance method is thus applied. Also in the course of stability management the positive value of the metacentric height needs to be maintained.

The stability assessment carried out by cargo officer takes into account the angle of heel due to cargo operation and only some simple characteristics, like:

− the metacentric height obtained as a first derivative of GZ curve at an angle of heel equal zero ($GM_{(\varphi=0)} = \dfrac{dGZ}{d\varphi}$);

− the maximum value of righting arm GZ;

− the range of positive values of GZ.

It is worthy to be noticed that there are no standards for ship stability performance in harbor condition, although the set of standards for sea conditions is established. Thus, the sole criteria applied for stability assessment in such a case is officer's experience and rarely ship owner recommendations.

The basis of contemporary approach to ship's transverse stability calculation and evaluation is static one. The method is focused on ensuring ship safety against capsizing and two mentioned examples of stability incidents in ports reveals the main scope of stability assessment performed on-board and some serious results of failures in such calculations. The motion of a ship excited by loaded cargo is not taken into account in any contemporary stability-related issues.

A lack of time-dependence aspect of stability characteristics makes this traditional method useless in terms of real-time approximation of vessel's heeling and pitching which is desirable to monitor an elevation of ship deck at any spot of cargo loading. Thus another method needs to be applied to provide time-domain calculations.

3 MODELING OF SHIP MOTION DUE TO CARGO LOADING IN PORT

Ship motion under excitation forces due to cargo loading in a port needs to be modeled to provide the time-dependent information about momentary ship's deck elevation at any spot. Once a container is loaded on-board its weight acting at a specified location arouses rolling, pitching and heaving of the ship. As a result of these motions any spot of subsequently loaded container alters its elevation in the course of natural oscillation damped by water friction. However in case of relatively high rate of cargo loading and especially with two or more gantry crane working at the same time, it is likely to transfer a container from a quayside to the ship hold in pretty short time one by one. If so, the formerly loaded box excites variations of ship's deck and the successive box needs to be put in its destination elevated higher or lower then it would be found with no motion of the ship. Moreover, the elevation depends not only on the location of previous container but also on time delay between these two gantry operations. The described idea is shown in figure 1.

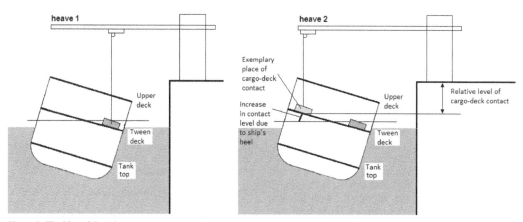

Figure 1. The idea of time dependent variations of ship's deck elevation due to contained loading

243

The dynamic behavior of a free floating vessel in a port or at the sea is affected by a set of forces and moments, both external and internal ones. The most accurate analysis of ship's motion should be based on the differential equations of motion which are given by the complex of six differential equations [1]:

$$m\frac{dV_x}{dt} + m(\omega_y V_z - \omega_z V_y) = F_x$$

$$m\frac{dV_y}{dt} + m(\omega_z V_x - \omega_x V_z) = F_y$$

$$m\frac{dV_z}{dt} + m(\omega_x V_y - \omega_y V_x) = F_z$$

$$I_{xx}\frac{d\omega_x}{dt} - I_{zx}\frac{d\omega_z}{dt} + \omega_y\omega_z(I_{zz} - I_{yy}) - \omega_x\omega_y I_{zx} = M_x \quad (1)$$

$$I_{yy}\frac{d\omega_y}{dt} + \omega_z\omega_x(I_{zz} - I_{xx}) - (\omega_z^2 - \omega_x^2)I_{zx} = M_y$$

$$I_{zz}\frac{d\omega_z}{dt} - I_{zx}\frac{d\omega_x}{dt} + \omega_x\omega_y(I_{yy} - I_{xx}) + \omega_y\omega_z I_{zx} = M_z$$

where:
F_i – resultant external force along appropriate axes ($i = x, y, z$);
M_i – resultant moment about appropriate axes referred to the center of gravity G;
I_{ii} – moment of mass inertia about appropriate axes ($i = x, y, z$);
I_{ij} – moment of mass deviation about appropriate axes ($i \neq j$);
m – mass of the ship and added masses;
V_i – velocity of the center of gravity about appropriate axes ($i = x, y, z$);
ω_i – ship's angular velocity about appropriate axes ($i = x, y, z$);
t – time.

The solution of such general formulated ship motion equations is problematic and for practical applications further simplifications and assumptions are required [7]. By neglected coupling, for the sake of simplicity, the ship's rolling is usually analyzed by the single degree-of-freedom system or three degree-of-freedom system [7]. In this paper three uncoupled equations of ship motion are implemented, e.g. roll, pitch and heave motion is taken into account.

The question on justification of such simplification may be considered, however in the studied phenomena the applied uncoupling of equations is rational. The first reason is relatively small amplitudes of expected motions at which the linear theory may be fair applied. In addition in case of ship rolling, being the most significant motion in the analyzed matter, the strongest coupling could occur with yaw and sway motions. They both may be neglected due to ship mooring forces so any effects of such coupling are very limited or next to

zero. The remaining motions, e.g. pitching and heaving are expected to be one order of magnitude smaller so the potential effect of coupling between them may bo omitted as well.

The governing differential equation of rolling, as the result of equilibrium of moments in direction usually signed "4" (about ship's x axis) is following:

$$I_4\ddot{\phi} + D_4(\dot{\phi}) + R_4(\phi) = M_4(t) \quad (2)$$

where:
I_4 – transverse moment of inertia of a ship and added masses;
D_4 – roll damping moment;
R_4 – restoring moment;
M_4 – heeling moment;
ϕ - angle of heel;
$\dot{\phi}$ - angular velocity of rolling;
$\ddot{\phi}$ - angular acceleration of rolling;
t – time.

The equation of ship pitching (motion in direction "5") e.g. about y axis is described by the formula:

$$I_5\ddot{\psi} + D_5(\dot{\psi}) + C_5 \cdot g \cdot M_j \cdot L \cdot \Delta\psi = M_5(t) \quad (3)$$

where:
I_5 – longitudinal moment of inertia of a ship and added masses;
D_5 – pitch damping moment;
C_5 – unit calculation factor;
g – gravity acceleration;
M_j – moment to change trim;
L – ship's length between perpendiculars;
M_5 – trimming moment;
$\Delta\psi$ – change of an angle of trim;
$\dot{\psi}$ – angular velocity of pitching;
$\ddot{\psi}$ – angular acceleration of pitching.

Consequently, ship heaving (linear motion taking place in direction "3") is governed by the formula:

$$m \cdot g = D \cdot \ddot{T} - D_3(\dot{T}) - C_3 \cdot TPC \cdot g \cdot \Delta T \quad (4)$$

where:
m – weight of loaded cargo exciting heave motion;
D – displacement of a ship;
D_3 – heave damping coefficient;
C_3 – unit calculation factor;
TPC – weight to change draft by 1 centimeter;
ΔT – increase in draft (a difference between momentary dynamic draft and a static one resulting from ship displacement);
$\dot{T}$ – linear velocity of heave motion;
$\ddot{T}$ – linear acceleration of heave motion.

244

It is assumed for the purpose of the research that ship motion excited by a loaded container can be described fair enough by the formulas (2) to (4). Surge, sway and yaw motions are neglected due to the fixed position of a moored ship. The resultant motion of a ship may be obtained by superposing of roll, pitch and heave motions governed by the given formulas (2) to (4).

4 COMPUTATION OF ROLL, PITCH AND HEAVE MOTIONS OF A SMALL CONTAINER VESSEL

Ship motion due to loading of a container in a port can be modeled according to the formulas (2), (3) and (4) described in previous section of the paper. However, it is essential to estimate the possible range of ship deck elevation variations. In other words the considered issue could be significant in case of relatively large decreases in the elevation or it may be neglected in an opposite case.

The weight of a typical container ranges from a few tons up to about thirty five tons regardless the size of a vessel carrying the containers. Therefore the influence of one loaded box on container carrier motions has to be significantly different for huge Malacca-max ship and small coastal feeder. For the sake of estimation extreme ship motions due to container loading one rather small size ship is taken into consideration. The ship chosen as an example is Polish semi-container vessel project B-354. One typical case of loading condition is considered (cond. No 11 according to the B-354 stability booklet). It reflects distinctive arrangement of containers on board. The particulars of the vessel are following:

- length between perpendiculars $L=140\ m$;
- breadth $B=22\ m$;
- hulls' height $H=12\ m$;
- displacement $D=14124\ t$;
- mean draft $d=6,55\ m$;
- longitudinal center of floatation $LCF=-0,67\ m$;
- moment to change trim $Mj=18044$ [tm/m];
- weight to immerse by 1 cm $TPC=24,15$ [t/cm];
- longitudinal center of gravity $LCG=-0,42$ m;
- vertical centre of gravity $VCG=8,88$ m;
- free surface moment $\Delta mh=2439$ tm.

The general view of the B-354 ship is shown in is shown in figure 2.

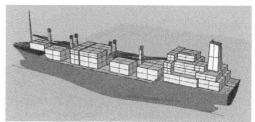

Figure 2. 3D numerical model of the considered vessel (project B-354)

The formulas (2) to (4) were implemented in Matlab script and the exemplary calculation was carried out. The assumed location and weight of a container to be loaded are following:
- container weight $m_k=35$ [t];
- longitudinal co-ordinate of cargo location $x_k=0,4L$ (40% of ship's length from a midship towards a bow);
- transverse co-ordinate of cargo location $y_k=0,4B$ (40% of ship's breadth to a starboard side).

The result of a computation is a history of ship motion. As the formulas (2), (3) and (4) describe three uncoupled motions, the solution is also given in the form of three time-domain curves tracings which is shown in figure 3 for ship's rolling, in figure 4 for her pitching and in figure 5 for heave motion respectively.

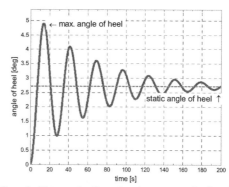

Figure 3. History of roll motion due to cargo loading (still water, linear damping) – sample for $m_k=35$ tons, $x_k=0,4L$, $y_k=0,4B$

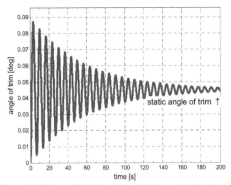

Figure 4. History of pitch motion due to cargo loading (still water, linear damping) - sample for m_k=35 tons, x_k=0,4L, y_k=0,4B

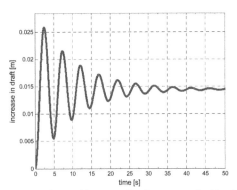

Figure 5. History of heave motion due to cargo loading (still water, linear damping) - sample for m_k=35 tons, x_k=0,4L, y_k=0,4B

Computation of ship motions history is only a pass-piece of the research. The main aim is to estimate variations of a deck elevation. However it could be emphasized that the results presented in figures 3 to 5 truly justify the earlier assumption regarding uncoupling of particular ship motions, enabling separate computations for the equations (2), (3) and (4). According to the linear theory of ship motion which is applicable for such small movements as presented in the discussed graphs, the influence of couplings may be neglected [1].

5 VARIATION OF SHIP'S DECK ELEVATION DUE TO CARGO LOADING

The sample result of the computation of ship motions due to cargo loading is described in section 4 of the paper and shown in figures 3, 4 and 5. Performing a sequence of runs of the prepared Matlab script one can obtain a set of such results for different cargo locations covering the whole area of a deck of the ship. An analysis of variations of deck elevations requires some further processing of the motion histories. The elevation of any specified cargo location on ship's deck can be derived on the basis of basic trigonometric functions. The final change of the elevation may be found by superposing of elementary components due to rolling, pitching and heaving. The results of sample calculations carried out for two different cargo loading spots are shown in figure 6.

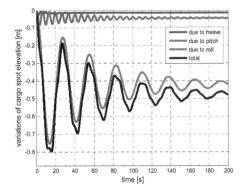

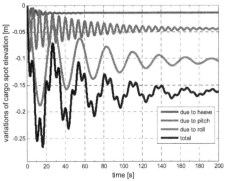

Figure 6. History of variations of ship's deck elevation at a cargo spot; calculation for container weight mk=35 tons and coordinates yk=0.4B, xk=0.4L (upper plot) and yk=0.2B, xk=0.4L (lower plot)

The sample graphs presented in figure 6 reveals a strong dependence of the deck elevation on location of loaded container. Thus a parametric approach was performed. In one case the longitudinal coordinate of container's location was fixed (x_k=0) and the transverse coordinate y_k was variable and ranging from -0,4B to 0,4B. The result of the computation in a form of 3D graph is shown in figure 7. The graph presents the history of variations of deck elevation (variable Δz at the vertical axis) versus time.

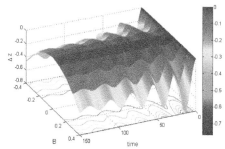

Figure 7. History of variations of ship's deck elevation at a cargo spot – fixed longitudinal coordinate of container's location $x_k=0$ and variable transverse coordinate y_k ranging from -0,4B to 0,4B

In the second considered case the reverse assumption was applied regarding fixing of coordinates. The transverse coordinate of the box loading spot was fixed ($y_k=0$) and the longitudinal coordinate x_k was variable and ranging from -0,4L to 0,5L. The time-domain result of these computation is shown in figure 8.

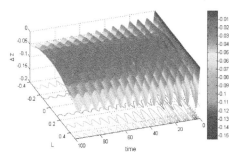

Figure 8. History of variations of ship's deck elevation at a cargo spot – fixed transverse coordinate of container's location $y_k=0$ and variable longitudinal coordinate x_k ranging from -0,4L to 0,5L

The comparison of the graphs 7 and 8 shows that ship's rolling is much more intense in terms of deck elevation alternation then her pitch motion. The momentary values of considered elevation decreases due to rolling are about four times higher. Moreover, the influence of cargo location is significantly higher for roll motions and corresponding to it transverse shift of container destination position. Both motions are damped so they decay due to the water friction and this reflects the decay of deck elevation variations amplitudes.

Another vital characteristic seen in the figures 7 and 8 is the extreme value of alteration of deck elevation. It takes place during the first cycle of motion and it is distinctive for the location of loaded container (considering a specified loading condition of a ship). Therefore further calculations were carried out for the location of the container covering the whole available area of ship's deck. This area

extend reflects the range of container coordinates x_k from -0,4L to 0,5L and y_k from -0,4B to 0,4B. In every single case of computation the extreme value of deck elevation alternation was recorded. The distribution of such extreme values is shown in figure 9.

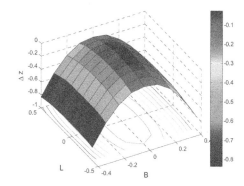

Figure 9. Extreme values of ship's deck elevation variations (Δz at vertical axis) at a cargo loading spot covering whole available space of the deck – transverse coordinate of container's location y_k ranging from -0,4B to 0,4B and longitudinal coordinate x_k ranging from -0,4L to 0,5L

The graph in figure 9 clearly shows that the considered variation of deck elevation due to container loading is mainly governed by the transverse coordinate of the loaded box. The extreme values are significant and reaches about 0,8 m which should not be neglected in terms of gantry crane control.

6 CONCLUSION

The study presented in the paper is focused on the estimation of ship's deck elevation variation due to container loading. This issue is strictly related to the ship stability characteristics. The dual approach towards ship stability in sea ports mentioned in the first section of the paper consists in searching for a compromise solution between safety of operation and effectiveness of cargo handling, especially from the economical point of view. The faster movement of a gantry crane the higher loading rate can be achieved and the ship stay in port is shorter and in consequence cheaper. On the other hand the fast gantry operation is more risky in terms of cargo damage due to excessively impetuous contact of the container with ship's deck.

The calculations based on modeling of ship motion in a port reveals that the considered matter may be important during loading of relatively small ship. The extreme values of alternation of some cargo spot elevation reaching 0,8 m shall be taken into account by the gantry control system. Otherwise the loaded container would come into contact with ship's deck at quite high velocity causing massive gravity load which can be destructive to the cargo inside the box.

The presented result of computations were obtained with the use of prepared Matlab script run on a standard PC-class desktop. The time of computation of on single case was imperceptibly short which is an important remark in terms of potentials to commercial realization. The real-time calculations carried out by the gantry control computer are feasible. Probably the code optimization could allow to run such computation even on quite slow machines which is essential in terms of cost effectiveness of the implementation.

The hereby proposed solution is to increase in cargo handling rate achieved by gantry cranes in a port combined with an extra control module allowing application of appropriate corrections to the cargo trajectory. Such upgraded gantry control system shall meet the requirements of fast operation simultaneously ensuring satisfactory level of cargo security against damage.

The presented results shall be found as a certain stage of the research so they do not fully comprehend all issues related to modeling of ship's deck elevation variations and their effect on cargo handling. The future work will be focused on some effects of external forcing such as the wave forces by other moving ships, winds or currents. Moreover the influences of ship model and adopted numerical schemes will be under future consideration.

ACKNOWLEDGMENT

The research project was funded by the Polish National Science Centre.

REFERENCES

[1] Dudziak J., Teoria okrętu, Wydawnictwo Morskie, Gdańsk 1988.
[2] Final Report and Recommendations to the 22nd ITTC, The Specialist Committee on Stability, Trondheim, Osaka, Heraklion, St. John's, Launceston 1996 - 1999.
[3] International Code on Intact Stability 2008, edition 2009, IMO, London 2009.
[4] Investigation of the capsizing of merchant vessel DENEB at the Port of Algeciras on 11 June 2011, Marine Casualties and Incidents, EMSA 2011.
[5] Kobyliński L., Goal-based Stability Standards, 10th International Ship Stability Workshop, Hamburg, Germany 2007.
[6] Krata P, Szpytko J., Weintrit A., Modelling of ship's heeling and rolling for the purpose of gantry control improvement in the course of cargo handling operations in sea ports, Solid State Phenomena, Vol. Mechatronic Systems and Metarials IV, Trans Tech Publications, pp. 539-546, 2013.
[7] Surendran S., Venkata Ramana Reddy J., Numerical simulation of ship stability for dynamic environment, Ocean Engineering 30 (2003) pp. 1305–1317, www.elsevier.com/locate/oceaneng 2003.
[8] Teo C.S., Kan K.K., Lim S.Y., Huang S., Tay E.B., Dynamic modeling and adaptive control of a H-type gantry stage, Mechatronics 17 (2007), pp. 361–367, www.sciencedirect.com 2007.
[9] http://www.cargolaw.com/2000nightmare_singlesonly5.html
[10] http://www.seanews.com.tr/article/ACCIDENTS/64468/Deneb

Tworty Box to Reduce Empty Container Positionings

U. Malchow

Hochschule Bremen – University of Applied Sciences, Bremen, Germany

ABSTRACT: A significant share of all empty container positioning is resulting from imbalances with regard to container sizes (20ft/40ft). In order to reduce the shipments of 'containerised air' a new type of container has been developed by the author: *Tworty Boxes* can either be used as a standard 20ft or in coupled condition as a 40ft container. The outside appearance resembles any standard 20ft container. However the *Tworty Box* is unique in that it has an additional door at the front side that opens to the inside. This door can be fixed to the ceiling and with the use of bonding elements another *Tworty Box* can be joined up, thereby creating the full 40ft inside space. Operated as a single 20ft box the additional door remains locked, access is only through the existing standard door. The *Tworty Box* does not require any additional components and fulfils all ISO and CSC requirements.

1 TASK

The commodity most often shipped in containers across the seven seas is pure air: Approximately 20% of all worldwide shipped containers are empty! According to estimates related handling costs alone are more than US$ 15 Billion p.a.! Furthermore carrier's box fleets have to be much bigger than actually needed to satisfy shipper's demand. This results in containers standing empty or idle in average approx. 60% of the time which consequently causes additional costs in ports and at depots. Moreover empty boxes void valuable slots on board the vessels. Hence cost effective container management has become the key issue for the profitability of container lines! It is estimated that each empty positioning is valued at approx. 450 US$/box mostly in terms of handling costs in ports.

The high portion of unproductive and costly empty positionings is caused by:

Figure 1. A pair of *Tworty Boxes* is loaded in Hamburg in coupled condition on board the "OOCL Montreal"

Figure 2. Front side with additional door (to be opened to the inside) and bonding elements adjacent to the corner castings.

Figure 3. Inside view of closed additional front side door (cables to open the door are visible).

1 structural imbalances of the general cargo flow (general trade imbalance),
2 seasonal impacts of dominating commodities in specific trades,
3 imbalances of the 20ft:40ft ratio between both trade directions.

A significant share of the empty positioning is resulting from carrier's internal imbalances in container logistics with regard to box sizes (20ft/40ft). Carriers note strong ups and downs in supply and demand of different container sizes in certain areas/ports especially if several services of different trades are calling the same area. Local dispatchers often report: "Too many 20s, not enough 40s", or reverse. Not always the situation can be balanced in time. Not seldom the grotesque situation occures that a carrier has to leave laden (low paying) boxes behind in order to reduce the empty stock of a certain size and position them empty to another port of the world where they are urgently needed for high paying cargo.

2 THE TWORTY SOLUTION

In order to significantly reduce the shipments of 'containerised air' the *Tworty Box* has been developed. Its outside appearance resembles any standard 20ft container. However the *Tworty Box* is unique in that it has doors at each end, the second door opens to the inside and can only be locked from the inside. This door can be fixed to the container ceiling and with the use of special bonding and sealing elements another *Tworty Box* can be joined up, thereby creating a watertight 40ft unit of full value (with standard doors at both ends).

Thus *Tworty Boxes* can either be used as a standard ISO 20ft or coupled as a 40ft container:

Twenty + Forty = ***Tworty***

If two *Tworty Boxes* are coupled to form a 40ft box the additional doors will be opened supported by cables and fixed to the container ceiling to receive the full 40ft inside space. Operated as a 20ft box this door is locked, access is only through the existing standard door. Two coupled *Tworty Boxes* make a 40ft container with doors at both ends. The system does not require any additional components and the coupled boxes remain watertight. The *Tworty Box* complies with all ISO requirements for containers and has successfully passed the full CSC testing procedure with DNV-GL.

The coupling is carried out by bonding elements which guarantees that two *Tworty Boxes* can be handled like a single 40ft container. The connection of two *Tworty Boxes* can only be released from the inside. Four coupling elements are located adjacent to the corner castings (Figure 2). Each Tworty Box has two male and two female bonding elements. They also keep the distance of 76 mm between the boxes in order to comply with the ISO regulation for the length of a 40ft container.

The *Tworty Box* concept is protected by international patents. Following main design targets have been followed:

– minimum changes compared to a standard 20ft container
– robustness
– easy handling
– (almost) no loose parts

Compared to single standard 20ft/40ft boxes the losses of the *Tworty Box* with regard to payload and capacity (if any) are marginal (Table 1). The only loose parts are the flat surrounding sealing ledges which are screwed after the coupling process from the inside into the gap between both boxes providing the necessary watertightness. As each *Tworty Box* carries a set of seals under the ceiling but only one is needed for the coupling of two *Tworty Boxes* there is enough redundancy if one sealing element was missing or damaged leaving enough time for a replacement. The seal fully complies with international customs regulations.

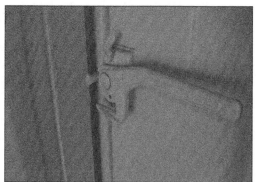

Figure 4. Bolting mechanism with handle for hinged door (from the inside)

Figure 5. Box for bottom male coupling element.

By operating *Tworty Boxes* empty positionings caused by the need to balance different supply and demand of 20ft and 40ft container sizes can be significantly reduced. Even if empty *Tworty Boxes* have to be positioned (e.g. due to inevitable imbalances of the general cargo flow) they can be coupled and immediately 50% of the lift on/lift off charges are being saved.

Table 1. Comparison of payload and capacity.

	20 ft (8'6" high)		40 ft (8'6" high)	
	Payload [t]	Capacity [m³]	Payload [t]	Capacity [m³]
standard container	21.8...28.2	32.8...33.2	25.8...26.7	65.3...67.7
Tworty Box	27.8	33.1	25.2	63.5

On account of global forwarding company DHL a pair of prototype boxes which had been stuffed with commercial cargo has already made a trial trip in 2013 on board of OOCL and Hapag-Lloyd vessels

from Hamburg to Montreal and v.v. to the full satisfaction of the forwarder.

3 ECONOMICS

Approx. 205 Mill TEU (Twenty Foot Equivalent Unit) have been shipped across the seven seas in 2011. Thereof 21% were empty (42 Mill TEU).

In 2011 the world container fleet consisted out of 30 Mill TEU (of which 27 Mill TEU were standard 20ft/40ft dry cargo boxes), i.e. in average each standard TEU has been shipped only 6.8 times throughout the year – thereof 1.4 times empty (empty share of 21%). In average each shipment (full or empty) has caused 2.9 port handlings (the average value of more than '2' is caused by transhipments).

Already in 2001 each empty positioning has been valued at approx. 400 US$/box mostly in terms of port handling costs. It is assumed that this amount has now increased to at least 450 US$/box. Hence in 2011 with a global 20ft/40ft split of 1.53 TEU/box within the standard dry cargo box fleet each TEU had caused at least approx. 410 US$ just for its empty positioning (1.4 x 450 US$/box : 1.53 TEU/box ≈ 410 US$/TEU).

As the average life time of a container is 8 to 9 years it is obvious that each container causes empty positioning costs during its entire life time which are exceeding its current newbuilding price (approx. 2,000 US$ for a standard 20ft container) by far. Hence focusing on reducing empty positioning is much more important than achieving the lowest possible purchase price of standard container equipment!

3.1 *Focus on single boxes*

Maximum savings can be achieved when 2 x *Tworties* are substituting 2 x 20ft and 1 x 40ft standard boxes which are normally due to be empty positioned in opposite trade directions. Table 2 illustrates the economics of operating *Tworty Boxes* compared to standard 20ft/40ft containers.

For the comparison three relevant cost items have been considered whereby the costs of crane moves and the costs for coupling/de-coupling have been varied within a realistic bandwidth. Considering the slightly higher investment for the *Tworty Box*, its daily capital costs have been set more than two times (!) the value of a standard 20ft box, which is by all means much more than the additional door will realistically cause (this leaves some reserve for an eventually slightly higher daily M+R allowance):

Table 2. 2 x *Tworty Boxes* are replacing 2 x 20ft and 1 x 40ft standard container.

Duration RV [days]	Unitcosts [US$] Move	Coupling	Trans-shipment
100	100,-	100,-	0

	Capital Costs [US$]	Moves [US$]	Coupling [US$]	Voyage Total [US$]
2 x 20' (full)	85	400	0	485
2 x 20' (empty)	85	400	0	485
1 x 40' (empty)	68	200	0	268
1 x 40' (full)	68	200	0	268
Total				1.506
2 x Tworty (full)	200	400	100	700
2 x Tworty (full)	200	200	100	500
Total				1.200

Difference	306
Percentage of conv. costs	20%

- daily capital costs:
  - standard 20ft container: .... 0.85 US$/box/day
  - standard 40ft container:[26] .... 1.36 US$/box/day
  - **Tworty Box (worst case): ....2.00 US$/box/day**
- costs per crane move: .... 100 - 200 US$
- costs per (de-)/coupling:[27] .... 20 - 100 US$

In addition the duration of the container voyage and the number of transhipments have also been varied in order to analyse all relevant impacts on the profitability of *Tworty Box* operation:
- duration of container round trip: ......50 - 100 days
- no. of transhipments (during 1 voyage): .... 0 - 2

3.2 *Results of single box view*

Table 2 illustrates the most unfavourable case for the operation of *Tworty Boxes* within the given range of parameters (arrow in Figure 6 & 7), i.e.:
- minimum lift on/lift off rate
- maximum coupling/de-coupling rate
- longest duration of container voyage
- no transhipment

Nevertheless compared to conventional box operations savings of 306 US$ have been revealed for a round trip of the boxes, i.e. two shipments.

This amount represents savings of 20% compared with the operation of standard containers. This magnitude exceeds by far the industry's average profit margin per container shipment (especially at present times). Saved costs for slots on board which do not need to be used for empty positioning have not even been considered.

Savings would logically increase if costs for coupling/de-coupling were decreased. However considering the wide range of this parameter the impact on the *Tworty Box* profitability is not dramatic. It can be revealed from Figure 6 that the impact of the duration of the single container voyage is rather negligible. It is much more the applicable average lift on/lift off rate which is of significant influence on the savings. For general guidance the following rough amounts can be applied (according to the specific trade a respective average out of both ends has to be considered):
- Europe:............................approx. 100 US$/move
- N.America: ....approx. 200 US$/move
- Asia:................................approx. 300 US$/move

---

[26] According to the industry standard the capital costs of a standard 40ft container are defined to be generally 1.6 times higher compared to a standard 20ft container.
[27] These costs may also cover the efforts to track the *Tworty Boxes* to ensure that two boxes are always available to be coupled if needed.

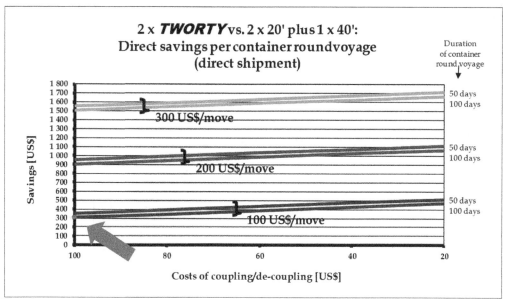

Figure 6.

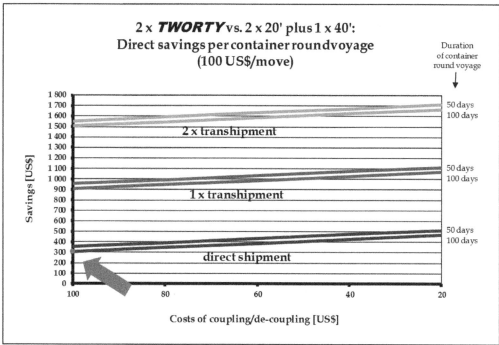

Figure 7.

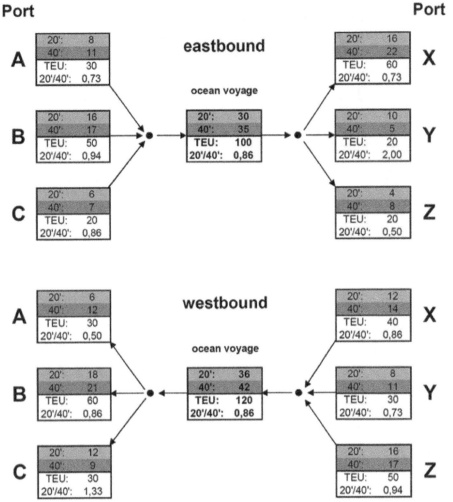

Figure 8. General trade imbalance (low scale example).

As more and more container lines are following the "hub-and-spoke" strategy the influence of transhipment has to be considered as well. In the meantime the share of transhipped boxes in port's global container throughput has risen from 10% in 1980 to more than 30% in 2011! It can be clearly revealed from Figure 7 that the savings the *Tworty Box* can provide become higher the more often the containers are transhipped! Furthermore savings are even much higher if average lift on/lift off charges in excess of 100 US$/move meets with the necessity to tranship empty containers at least once.

It can be concluded that the introduction of *Tworty Boxes* can provide dramatic savings if they are operated and kept in certain imbalanced trades where their advantages can be fully utilised. Thus

contrary to standard containers they have to be individually tracked and treated as special equipment like e.g. flats, reefers etc.

A simple low scale example as per Figure 8 demonstrates that even trades with an almost balanced cargo flow and an identical general 20ft:40ft ratio both ways would very much benefit from the *Tworty Box*. It is assumed that in a hypothetical trade 100 TEU have to be shipped eastbound whereas 120 TEU are due to be carried westbound. Contrary to reality and not beneficial for *Tworty Box* operation both volumes shall exactly have an identical 20ft:40ft ratio (= 0.86) on their ocean leg. In reality this ratio is however varying more or less around an average figure among the various loading and discharging ports involved.

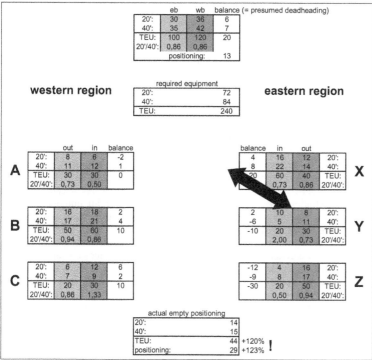

Figure 9. Empty positioning caused by general trade and local imbalance (example).

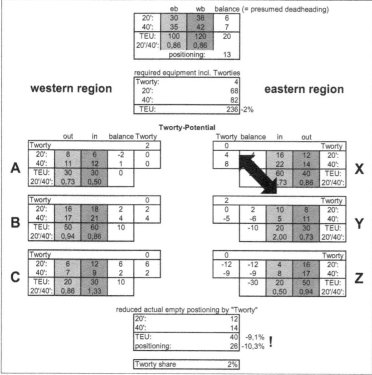

Figure 10. Empty positioning caused by general trade and local imbalance reduced by Tworty Boxes (example).

Hence deadheading is not only required to compensate the general imbalance in required equipment flow between both regions but also to balance the various requirements for different container sizes among the ports within a region. The required box fleet is determined by the dominant trade direction. In this case for both sizes the westbound leg is stronger. Hence at least 240 TEU of equipment would be required to ship both volumes simultaneously.

Just due to the apparent general trade imbalance additional 20 TEU (6 x 20ft + 7 x 40ft = 13 boxes) seems only to be necessary to be empty positioned eastbound (Figure 9). However considering also the various local imbalances at each port 44 TEU (14 x 20ft + 15 x 40ft = 29 boxes) have actually to be shipped empty (also within the regions) in order to compensate the imbalanced supply and demand of container sizes. This is 120% more (in terms of TEU) than one would expect from the pure general trade imbalance, resulting in 123% more empty box movements! Also the box fleet has to be slightly larger than originally anticipated as boxes which are due for an additional intra-regional deadheading cannot be immediately stuffed after having been stripped.

3.3 Result of carrier's entire fleet view

At all ports which suffer from a sudden or permanent lack of one size and a surplus of the other size (e.g. port "A" and "Y") the operation of Tworties would be very advantageous (Figure 10). If in the example only 4 x Tworties were introduced (replacing 2 x 40ft standard boxes) and these boxes were kept plying only between port "A" and "Y" just only 40 TEU (including the Tworties) would have to be empty positioned (instead of 44 TEU). Hence a Tworty share of only 2% (in terms of TEU) within the box fleet could theoretically lead to a reduction of deadheading costs by 10%! Furthermore the entire fleet could be reduced by 4 TEU (-2%)!

Thus a homogeneous container fleet existing completely out of Tworties is not necessary. The huge majority of the fleet can still consist out of standard 20ft and 40ft boxes. As it can be derived from the example even a small number of Tworty Boxes which are kept plying between ports where a chronic surplus of one size meets with the lack of the other size can significantly contribute to improved economics of a carrier's container fleet.

Although the Tworty Box cannot supersede all repositioning necessities, there are many trades where the 40ft:20ft ratio of equipment is varying among the ports and where the Tworty Box concept can help substantially.

Because 2 x Tworty Boxes are destined to replace approx. 2 x 20ft standard boxes and 1 x 40ft standard box the capital costs of the entire box fleet do not increase as the additional expenses for one Tworty Box would not exceed half the costs of a 40ft standard box.

Table 3. Expected effect of Tworty Box operation on deadheadings and container fleet size

		just by general trade imbalance (theory)	reality (due to locally imbalanced sizes)	including 2% Tworty share	savings in reality
necessary	boxes	13	29	26	-10 %
deadheadings	TEU	20	44	40	-9 %
required fleet	TEU	240	240 plus	236 plus	-2 %

4 CONCLUSIONS

The Tworty Box is most advantageous for container trades which suffer from a clear imbalance with regard to the container seizes, i.e. where the 20ft/40ft split of both trade directions differs significantly.

However the calculations have revealed that significant savings can even be realised in case of imbalances with regard to the pure trade volume, i.e. when coupled Tworty Boxes could replace 2 x 20ft standard boxes which otherwise would have to be empty positioned individually. By using Tworty Boxes the empty movements can be realised as one unit, i.e. the respective handling costs can be cut by 50% which exceeds the additional expenses for coupling/de-coupling by far.

Hence the Tworty Box can avoid empty positionings caused by having not the right container sizes available and even if empty positionings are unavoidable it can cut the costs for empty movements of 20ft containers almost by half.

Who is benefiting? It is the container lines which would directly take advantage from operating Tworty Boxes. Presently 53% of the world container fleet is operated by container lines, thereof 90% are of standard 20ft/40ft dry cargo type. However it is not necessary that a container line replaces its entire container fleet by Tworty Boxes to gain maximum savings. Only the portion equivalent to the lines' individual (average) imbalance needs to be replaced.

It is not expected that leasing companies which presently control approx. 44% of the world's container fleet would be immediately interested to operate Tworty Boxes. They are only reacting to the demand of the container lines and therefore are expected to be interested only at a later stage. However big forwarders with shipper's owned containers might be interested as well as it has been already proven by global forwarder DHL which have successfully tested two prototype boxes on occasion of a commercial trial trip.

According to Boedeker, Global Head Ocean Freight, DHL Global Forwarding, (2013) the *Tworty Box* is a very attractive solution which ensures flexible container management and cost efficiency by eliminating empty positioning due to structural imbalances in the general cargo flow or seasonal fluctuations in the dominant commodities in specific sectors. It was quoted to be a smart alternative for customers that note strong ups and downs in supply and demand of different container sizes in certain areas, especially if several services of different trades are calling the same country or region.

Taking the fact that 21% of all containers shipped are empty for reasons of imbalance of whatever kind it is assumed that the potential market volume for the *Tworty Box* might be 20% of the existing global standard 20ft/40ft dry cargo container fleet which is presently operated by container lines, i.e. presently 27 Mill TEU x 0.53 x 0.2 = 2.9 Mill TEU. Hence with an average life time of 8.5 years 34.000 TEU of *Tworty Boxes* would be needed to be introduced annually.

REFERENCES

DHL, Press Release, June 25, 2013.
Drewry Maritime Research, Container Market Review and Forecast, Annual Report 2012/13.
Drewry Maritime Research 2013. Container Forecaster.
Drewry Shipping Consultants 2012. Container Census – 2012.
Hapag-Lloyd AG 2010. Container Specification, Hamburg
Konings, R.& Thijs, R. 2001, Foldable Containers: A New Perspective on Reducing Container Repositioning Costs, TU Delft.
Malchow, U. 2013. Aus zwei mach eins. In: Deutsche Verkehrs-Zeitung. Dec. 10, 2013.
Port of Hamburg Magazine, 3/2013: 34-39
TWORTY BOX GmbH & Co. KG, Hamburg, website: www.tworty.com, last accessed in Feb. 2014.

Ship's Operations
Safety of Marine Transport – Marine Navigation and Safety of Sea Transportation – A. Weintrit & T. Neumann (eds.)

Consideration on Dynamic Modelling of Ship Squat

J. Artyszuk
Maritime University of Szczecin, Poland

ABSTRACT: The paper deals with surveying and evaluating the state-of-the-art in the development of a dynamic model of squat. Special attention is paid to the limited availability of squat data and the related hydrodynamic vertical force and trim moment, especially in varying water-depth. First, hydrostatics related to squat is reviewed and presented in dimensionless form to facilitate the mathematical model engineering and gathering of input data. Next, having the excitations determined for shallow-water stepped bottom conditions, as published in literature, a simple dynamic model is run to evaluate the performance of the squat transient.

1 INTRODUCTION

Keeping the under-keel-clearance (UKC) positive yet with some proper allowance is very crucial for safe navigation. One of contributions to the reduction in UKC is the squat phenomenon when a ship is advancing in shallow water or canal with high speed. The squat is quantified by sinkage (amidships) and trim.

However, the greatest concern among civil engineers and mariners (operators) is just with the constant-depth and steady-state phase of squat, which is commonly referred to as the quasi-steady kinematical approach. This quasi-steady method of treatment is surprisingly widely adopted in most applications, though it introduces of course some errors. The constant-depth squat relationships might be expected inadequate also in the sense of where within a ship 'to sense' the appropriate depth for input.

The situation is complicated through considerable lack of data in the available literature on the transient kinematic data for a rapid, arbitrary change of the fairway bathymetry in the longitudinal direction. In contrast to huge amount of scale model research on constant-depth squat, here practically exists just a few references – Haatainen et al. (1978), Edstrand & Norrbin (1978), Ferguson et al. (1982).

The most recent and abnormal Renilson & Hatch (1998) is completely against the former ones, in which the maximum sinkage of a model ship (related to her bow in the tested case) during the transient over a stepped-bottom is almost three times lower than that achieved from the constant-depth data corresponding to the lowest encountered depth. Also, the explanations served by the authors are not convincing and further verification/diagnosing of these extraordinary results are needed.

In addition, one can only find numerical results of forces inducing the squat in simple yet representative cases of depth variation and those are essentially limited to Drobyshevskiy (2000) and Gourlay (2003), though both are very similar to each other.

Model measurements of forces are not undertaken at all.

Based on the computed forces, the calculation of squat, as more or less correctly inspired by or attributed to Tuck (1966), is often conducted hydrostatically for its steady-state values. In such calculation the following point is likely omitted. If squat is statically computed for a given depth-to-draft ratio h/T, especially under an extremely low value of the latter, then the UKC reduction due to such squat provides for a new h/T. Hence, a recalculation must be done of both the input forces and the resulting squat, since they are not valid anymore. This process seems to be recursive until a full convergence is achieved.

The experimental results of Haatainen et al. (1978) and Ferguson et al. (1982) are not fully consistent. The former, partly confirmed by Edstrand & Norrbin (1978), show the corresponding variation of bow and stern sinkage with distance of either of these ship's extreme points versus the shallow edges. On the other hand, in Ferguson et al.

(1982), as well as in the quoted theoretical works of Drobyshevskiy (2000) and Gourlay (2003), one can notice that both the bow and stern sinkage experience jumps at the same moment.

Some authors, including Haatainen et al. (1978), prove that maximum sinkage for nautical safety evaluation (at bow or stern as dependent on ship's hull fullness) can be calculated from the even bottom conditions for the lowest h/T.

The aim of this paper is to prepare a background for implementation of squat dynamics into ship simulators. As a rule, inhomogeneous with the commonly adopted 6-DOF body dynamics architecture, the squat motion is the only one statically considered in the simulator models. This squat 'motion' is directly incorporated through well-known pure geometric data on sinkage and trim. The corresponding heave and pitch velocities are not present, partly owing to the lack of experimental data. As mentioned before, the steady-state and constant-depth assumptions are also common here.

Although in many applications the existing approach is useful, it requires a separate algorithm and different efforts to mathematically model and calibrate the data. Besides, some essential dynamic/transient effects are lost.

The paper is divided into several topics. First, the statics involved in squat is systematised and the inverse problem of converting the known squat data into the exciting forces is put forward. This would initially provide a major database of force data required by dynamic model. In the light of the aforementioned problems with numerical calculation of forces and squat for very low h/T, the 'identified' forces are more adequate as being equivalent (or average) to give the prescribed original squat.

Second, a simple dynamic model is simulated for the purpose of testing some transient effects involved in squat. The lessons learned therein extend our knowledge on the dynamic modelling of the squat phenomenon.

2 SQUAT STATICS

Let us introduce the right-hand cartesian reference systems as shown in Figure 1. Since the motions relating to squat consist of heave and pitch (a linear movement along Oz axis and an angular movement around Oy axis), a ship can be treated to undergo planar motions in the vertical fore-and-aft plane Oxz, constituting the ship's centreplane. This is depicted in Figure 2, where y-axis is perpendicular to the Figure's plane and marked with an encircled dot, representing the arrow pointing towards the reader's eyes. The other coupling motion (likewise in classical seakeeping theory) to these two motions, namely surge, is omitted in the present considerations. It is of less nautical interest.

Hence, the positive vertical displacement of a ship's origin O, lying at the intersection of the midship and the initial (for a ship at rest) waterplane lines, means downward movement and is referred to as the mentioned midship sinkage (z). The pitch (trim) angle, denoted by θ, and the resulting trim t_T (in meters) of the vessel, are positive for stern-down/bow-up attitude The same convention applies to corresponding forces and moments.

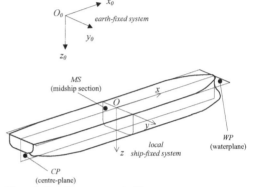

Figure 1. Coordinate systems in 3D.

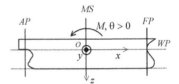

Figure 2. Local coordinate system in side view (2D).

For a ship in steady-state condition of squat we have the following equilibrium of generalized forces (forces and moments):

$$\begin{cases} F_{zSQ} + F_{zHS} = 0 \\ M_{ySQ} + M_{yHS} = 0 \end{cases} \tag{1}$$

where F_{zSQ} and M_{ySQ} represent the exciting hydrodynamic force and moment that cause squat. Those are balanced by hydrostatic terms (HS).

The latter can be decomposed into direct (dir) and coupling (cpl) contributions:

$$F_{zHS} = F_{zHSdir} + F_{zHScpl} \tag{2}$$

$$M_{yHS} = M_{yHSdir} + M_{yHScpl} \tag{3}$$

$$F_{zHSdir} = -\rho g A_{WP} \cdot z, \quad M_{yHSdir} = -\rho g I_{yWP} \cdot \theta \tag{4}$$

$$F_{zHScpl} = \rho g A_{WP} LCF \cdot \theta, \quad M_{yHScpl} = \rho g A_{WP} LCF \cdot z \tag{5}$$

where ρ = water density; g = gravity acceleration A_{WP} = waterplane area; I_{yWP} = inertia moment of waterplane area; LCF = location of longitudinal centre of floatation (centre of waterplane area, counted from the midship, positive if positioned forward).

Since the ship's gravity (G) and nominal buoyancy (B), resulting from the hydrostatic pressure distribution on the underwater hull with no flow, compensate each other, these are omitted in Equation 1, as usual. The shown hydrostatic contributions through Equations 4 and 5 are thus the additional buoyancy/restoring forces and moments as associated with change in z displacement of a ship from her nominal position (i.e. if at rest and in still water). They can take either positive or negative values. With normal positive sinkage z (downward), the hydrostatic force F_{zHSdir} is pointing up. The direction of hydrostatic moment M_{yHSdir} depends on the sign of pitch angle θ. If the latter is negative (e.g. in the case of full form ships that experience bow-down squatting), we have $M_{yHSdir} > 0$.

The moment M_{yHSdir} is roughly related to the inertia moment of the waterplane area. Though less strict, this approach is more convenient (as requiring less data) and sufficiently accurate. The use of expression developed on the basis of the alternative, fully covering quantity - longitudinal metacentric height (GM_L) - can also be encouraged. If unknown, the inertia moment I_{yWP} in Equation 4 can be deduced from the available longitudinal metacentric radius (BM_L - boyancy-to-metacentre distance) or even approximated from the given metacentric height (due to small trim angle):

$$I_{yWP} = BM_L \cdot \nabla \approx GM_L \cdot \nabla \qquad (6)$$

where ∇ = volume of displacement (ship's mass per water density).

The sign of the other, hydrostatic coupling terms (Eq. 5) is additionally dependent on the positive or negative value of LCF.

Combining Equations 1-5 leads to:

$$\begin{cases} F_{zSQ} - \rho g A_{WP} z + \rho g A_{WP} LCF \cdot \theta = 0 \\ M_{ySQ} + \rho g A_{WP} LCF \cdot z - \rho g I_{yWP} \cdot \theta = 0 \end{cases} \qquad (7)$$

that dimensionally resolves as follows:

$$z = \frac{1}{\rho g} \frac{I_{yWP} F_{zSQ} + LCF \cdot A_{WP} M_{ySQ}}{I_{yWP} A_{WP} - LCF^2 A_{WP}^2} \qquad (8)$$

$$\theta = \frac{t_T}{L} = \frac{A_{WP}}{\rho g} \cdot \frac{M_{ySQ} + LCF \cdot F_{zSQ}}{I_{yWP} A_{WP} - LCF^2 A_{WP}^2} \qquad (9)$$

The above gives for a ship's rectilinear underwater hull:

$$z_{FP} = z - 0.5 t_T, \quad z_{AP} = z + 0.5 t_T \qquad (10)$$

where z_{AP}, z_{FP} = local sinkages at fore and aft perpendiculars, respectively (see Fig. 2).

Introducing nondimensionality to the waterplane area-related geometric data in terms of:

$$A_{WP} = c_{WP} L B = \frac{L^2 c_{WP}}{\dfrac{L}{B}} \qquad (11)$$

$$I_{yWP} = A_{WP} r_{yWP}^2 = A_{WP} L^2 r_{yWP}'^2 \qquad (12)$$

$$LCF = LCF' \cdot L \qquad (13)$$

where L, B = ship's length (between perpendiculars) and beam; c_{WP} = waterplane area fullness coefficient; L/B = length-to-beam ratio; r_{yWP} and r'_{yWP} = dimensional and dimensionless inertia radius of waterplane area; LCF' = dimensionless location of longitudinal centre of floatation, we finally get

$$z = \frac{1}{\rho g A_{WP}} \frac{r_{yWP}'^2 F_{zSQ} + LCF' \dfrac{M_{ySQ}}{L}}{r_{yWP}'^2 - LCF'^2} \qquad (14)$$

$$\theta = \frac{t_T}{L} = \frac{1}{\rho g A_{WP} L} \cdot \frac{\dfrac{M_{ySQ}}{L} + LCF' \cdot F_{zSQ}}{r_{yWP}'^2 - LCF'^2} \qquad (15)$$

Dimensional analysis (or so-called non-dimensionality for a physical quantity representation) always improves performance in mathematical modelling, especially in the case of hydrodynamic phenomena. Namely, it enables an easy data transfer between ships similar in shape (more or less) but various in size. Furthermore, it allows establishing an optimal structure of the governing physical laws. Such laws may comprise either kinematical relationships, between input factors and resulting motions, or dynamic relationships. The latter consist of expressions for the hydrodynamic forces and moments involved. The best model can be sought, for example, with regard to the least number of input parameters and/or the widest range of application.

The impact of LCF in Equations 14 and 15, because of its rather low dimensionless value (just a few % of L) and irrespective of being positive or negative, seems not to be large. So one can attempt, according the his/her application objectives, to simplify the above expressions by neglecting all the terms containing LCF. In this way, sinkage z and trim angle θ are mostly governed by the direct hydrostatic contributions F_{zHSdir} and M_{yHSdir} (Eq. 4). I_{yWP} in Equation 12 is not written fully nondimensionally, i.e. up to the L factor. The explicitly maintained waterplane area A_{WP} in this and the subsequent expressions of the paper keeps them little brief. The reader can easily expand them further.

Up to sign of certain terms (due to different coordinate system used), Equations 14 and 15 are consistent with Tuck's (1966) formulas, since both the approaches follow from the well-known and very old hydrostatic theory of integro-differential form.

However, making also dimensionless the exciting force and moment, in the way of Tuck (1966) or other method, is not attempted at this stage. They are still represented by general parameters F_{zSQ} and M_{ySQ}.

Transforming to local fore or aft sinkages, as often used in research on squat, we get:

$$z_{FP} = \frac{F_{zSQ}\left[r_{yWP}^{'2} - 0.5LCF'\right] + \frac{M_{ySQ}}{L}\left[LCF' - 0.5\right]}{\rho g A_{WP}\left[r_{yWP}^{'2} - LCF'^2\right]} \tag{16}$$

$$z_{AP} = \frac{F_{zSQ}\left[r_{yWP}^{'2} + 0.5LCF'\right] + \frac{M_{ySQ}}{L}\left[LCF' + 0.5\right]}{\rho g A_{WP}\left[r_{yWP}^{'2} - LCF'^2\right]} \tag{17}$$

Conversion of squat kinematical data into forces

The inverse problem how to convert the known geometric data on squat into averaged (over sinkage and/or trim) steady-state hydrodynamic force and moment is directly deduced from Equation 7:

– based on the input sinkage z and trim t_T:

$$F_{zSQ} = \rho g A_{WP}\left(z - LCF' \cdot t_T\right) \tag{18}$$

$$M_{ySQ} = \rho g A_{WP} L\left(r_{yWP}^{'2} \cdot t_T - LCF' \cdot z\right) \tag{19}$$

– based on the input z_{FP} and aft z_{AP} sinkages: via the above Equations 18 and 19, but with the following substitutions:

$$z = 0.5\left(z_{FP} + z_{AP}\right) \tag{20}$$

$$t_T = z_{AP} - z_{FP} \tag{21}$$

The above, of course simple formulas can facilitate the future data aquistion process for the dynamic modelling of squat in ship manoeuvring simulation.

3 SIMPLE SQUAT DYNAMICS

In these preliminary investigations the following simple dynamic model of squat is exploited:

$$\begin{cases} \dfrac{dv_z}{dt}\left(m + m_{33}\right) = F_{zD} + F_{zHS} + F_{zSQ} \\ \dfrac{d\omega_y}{dt}\left(J_y + m_{55}\right) = M_{yD} + M_{yHS} + M_{ySQ} \end{cases} \tag{22}$$

supplemented by kinematical relationships:

$$\frac{dz}{dt} = v_z, \quad \frac{d\theta}{dt} = \omega_y \tag{23}$$

where t = time; v_z = heave velocity; ω_y = pitch (angular) velocity; m = ship's mass; J_y = ship's moment of inertia around y-axis; m_{33}, m_{55} = added masses in heave and pitch directions, accordingly; F_{zD} = damping heave force; M_{zD} = damping pitch moment.

The problem in Equation 22 exists with regard to the added mass m_{33} and added moment of inertia

m_{55}, as well as with damping terms. All of them are functions of the performed ship motions, which essentially can be arbitrary and just constitute the variables being sought. The linear theory of seakeeping, still recognized as the basic source for non-manoeuvring added masses and damping/drag forces, considers harmonic external (wave) excitations and the resulting harmonic motions of small amplitudes. Both the added masses and damping terms are thus functions of frequency.

For slow motions, like in ship manoeuvring (of horizontal plane) and in ship squatting (of vertical plane), the frequency tends to zero. So, we should use the values of added masses and damping terms for null frequency, and hold them constant throughout the whole simulation. This is the case widely accepted only in ship manoeuvring. The nature of squat-related excitations is significantly different, so we should revert to the fully equivalent so-called time-domain simulation. In the latter, based on the control theory, the arbitrary motions are modelled by means of impulse response functions using convolution integrals and the Fourier transform, see e.g. Pawlowski (1999). Moreover, the values of m_{33} and m_{55} approach infinity with frequency tending to zero, while at the same time the corresponding damping terms assume null values.

We will leave the stated modelling difficulties and possible application of more advanced methods for future concern. In the present analysis, the damping components in Equation 22 will be ignored, so the dynamics is mostly governed by the inertia terms, either of the rigid body or of the hydrodynamic added masses. Consequently, the usual coupling of motions due to other (mix-type) added masses and damping terms has been also neglected in Equation 22,

Below we adopt the constant added masses, treated as somehow average. Since there is essential uncertainty with regard to their values, also because of shallow water effects, a certain amount of sensitivity analysis is going to be performed. For deep water conditions, the lateral added mass and the yaw added moment of inertia are in the order of a ship's mass and her yaw moment of inertia, respectively. For ship manoeuvring in shallow water, the 'manoeuvring' added masses rapidly increase with decrease in water depth. For depth-to-draft ratio $h/T = 1.1$-1.2 (where T = ship's draft) we observe the added masses equal to several ship's masses or moments of inertia, accordingly. It is believed that similar multiples are also valid for squat-related added masses m_{33} and m_{55}.

The examined numerical scenario of simulation consists of moving a ship: a) into shallower water ('approaching'), and b) into deeper water ('leaving'), both with a step depth change, see Figure 3.

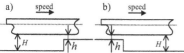

Figure 3. Simulation scenarios (a - approaching, b - leaving a bottom step)

The higher depth-to-draft ratio $H/T = 2.5$ and the lower depth-to-draft ratio $h/T = 1.2$ are assumed. They correspond to the most extreme case - the most perturbed bottom - considered by Drobyshevskiy (2000), for which Figure 4 reproduces the originally computed dimensionless heave force and pitch moment data. Please note the opposite sign adopted versus the original in the lower sub-chart of this Figure as to account for the current coordinate system. These coefficients are defined as follows:

$$c_{fzSQ}(s') = \frac{F_{zSQ}(s')}{\rho v^2 L^2}, \quad c_{mySQ}(s') = \frac{M_{ySQ}(s')}{\rho v^2 L^3} \tag{24}$$

where s' = dimensionless distance, in ship's length units, of a ship's midship point versus the bottom step (the negative value denotes a situation before reaching the step); v = ship's forward speed.

An evaluation of the above non-dimensioning scheme for forces does not lie within the scope of the present paper.

Together with disregarding the mentioned damping force and moment in our model (as to provide a kind of provisional simulation), we can introduce some dimensionless quantities to the left side of Equation 22 and obtain as follows:

$$\begin{cases} \dfrac{dv_z}{dt} m(1 + k_{33}) = F_{zHS} + F_{zSQ} \\[2mm] \dfrac{d\omega_y}{dt} mL^2 r_y'^2 (1 + k_{55}) = M_{yHS} + M_{ySQ} \end{cases} \tag{25}$$

where r_y' = dimensionless ship's gyration radius around y-axis; k_{33}, k_{55} - dimensionless coefficients of appropriate added masses.

Ship data used in the simulation will be those of Drobyshevskiy (2000) and are gathered in Table 1, both dimensionally and nondimensionally.

Table 1. Ship's exemplary data

ship's type - crude oil tanker		
L	284.75 m	
F_{nL}	0.1167 [-]	($v = 6.167$ m/s)
ρ	1025 kg/m³	
g	9.806 m/s²	
L/B	6.328	(B = 45 m)
B/T	2.647	(T = 17 m)
c_B	0.7944	(block coefficient, m = 177372 t)
c_{WP}	0.8585	(A_{WP} = 11000 m²)
r_{yWP}'	0.2452	(GM_L = 310 m, see Eq. (6))
LCF'	0.0026	(LCF = +0.74 m)
r_y'	0.24	(estimated, I_y = 8.284·108 tm²)

The numerical solution of Equation 25 is realized by the first-choice yet powerful Euler method, in view of velocities v_z and ω_y, and the trapezoid method, very efficient in computing displacement-type data, i.e. z and θ. The both latter rely on the corresponding velocities acquired in the current and preceding time instants. The integration time of 0.1 s has been carefully selected to avoid numerical instability in the computed squat data. This aspect is also discussed below.

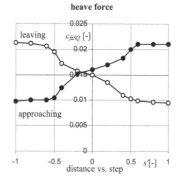

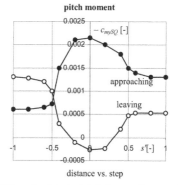

Figure 4. Squat-related heave force and pitch moment dimensionless data for approaching and leaving the bottom step by a tanker (Drobyshevskiy 2000)

The reference value for dimensionless added masses with account of shallow water, k_{33} and k_{55}, is 3 for both. They also go in parallel when we attempt to vary them (on the global, time- and space-invariant basis) for the purpose of a certain sensitivity analysis. Thus additional, supportive simulations will be run for the values 1 and 0 of these ratios.

The dots in Figure 4 denote discrete points that have been digitized from the original plots. They are stored in a lookup-table and the intermediate points are linearly interpolated and input to Equation 25.

The ship is assumed to be initially in equilibrium, i.e. running with certain static sinkage and trim resulting from the forces relevant to the even bottom condition ($s' = -1$).

The results of dynamic squat simulation under the stated simplifying assumptions are presented in

Figures 5 to 7. They depict sinkage at the bow and at the stern (Eq. 10), but in the dimensionless form versus ship's draft, to provide easier account for a decrease in the under-keel-clearance. The solid, oscillating lines represent the dynamically varying squat data according to the performed numerical simulation. The dot-marked lines depict the abstract steady-state squat data (Eqs. 16 and 17). However, these steady-state data are still useful, since they provide for somehow average run of both sinkages. This is rather surprising and interesting issue.

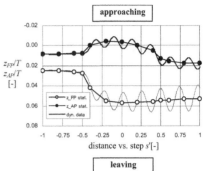

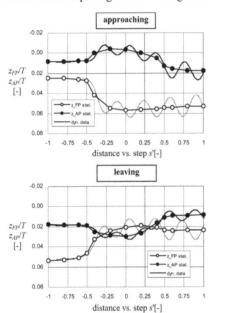

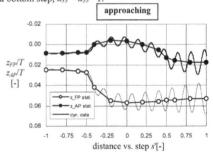

Figure 5. Bow and stern sinkage of a tanker while running over a bottom step, $k_{33} = k_{55} = 3$ (reference)

For investigating the observed oscillations in the squat data of Figures 5 to 7, it is purposeful to call the basic relationships from classical linear model of mechanical vibrations. The natural periods of undumped oscillations for a ship, T_{z0} (heave) and $T_{\theta0}$ (pitch), can be determined from the following expressions:

$$T_{z0} = 2\pi \sqrt{\frac{c_B L(1+k_{33})}{g c_{WP} \frac{L}{B} \cdot \frac{B}{T}}}, \quad T_{\theta0} = 2\pi \sqrt{\frac{c_B L r_y'^2 (1+k_{55})}{g r_{WP}'^2 c_{WP}}} \qquad (26)$$

Associated with restoring forces, these are proportional to $L^{1/2}$ and dependent on the input added masses. Table 2 provides for an impression on their magnitudes, including dimensionless natural periods (in units of time needed for traveling one ship's length, i.e. being equal L/v). The knowledge of dimensionless natural periods thus appears essential in studying ship's performance in longitudinally varying depth, because they well explain the actual period of oscillation.

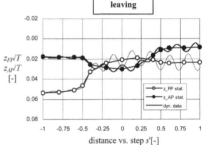

Figure 6. Bow and stern sinkage of a tanker while running over a bottom step, $k_{33} = k_{55} = 1$.

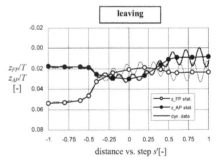

Figure 7. Bow and stern sinkage of a tanker while running over a bottom step, $k_{33} = k_{55} = 0$.

The bottom-step proximity induced forces per Figure 4, as the ship's relative position-dependent, upset the ship's equilibrium, that in turn instantly involves the balancing action of restoring forces. Dominating here appears to be the heave natural period T_{z0} that well fits the actual oscillations seen in Figures 5 to 7. Such result is not trivial. Similar oscillations seem to be also revealed in model tests of Edstrand & Norrbin (1978), Haatainen et al. (1978), and Ferguson et al. (1982). More detailed and quantitative comparison, also covering some remarkable difference between the quoted model tests themselves, however, is not challenged in the present investigation.

The achieved maximum amplitude of oscillations in our simulation experiment is in the order of $0.01\,T$ (absolutely 0.17 m) for the higher bow sinkage. This decreases UKC. For the lower stern sinkage, this is nearly proportionally less.

Table 2. Natural periods in heave and pitch

$k_{33}=k_{55}=$	3	1	0
T_{z0}	15.9 s / 0.34 L	11.3 s / 0.24 L	8.0 s / 0.17 L
$T_{\theta0}$	63.7 s / 1.38 L	45.1 s / 0.98 L	31.9 s / 0.69 L

Trials of simulation with the integration time step of 0.5s, under our three conditions of added masses, lead to numerical instability that produces unaccepted local sinkage oscillations increasing with time. The latter should be obviously constant due to removed damping. The simulation performance for the reference added mass ratios (all equal to 3) with the time increment 0.2s, not shown in the paper, is much better and acceptable, as only slightly different from that of Figure 5.

However, a feeling of such unreliable oscillations one may already have in Figure 7, especially in the upper chart connected with the step-approaching phase. The adopted 0.1 s integration step is also not valid in this abstract (null added masses) case of simulation, since it seems too large with respect to the reduced actual natural period in heave.

It shall be remembered that for smaller ships the numerical time step must be much less.

4 CONCLUSIONS AND FINAL REMARKS

Based on the performed literature survey and the present simulations, it can be indirectly concluded that 'an impulse' in the hydrodynamic forces arising from passing over a bottom-step is too small as to significantly disturb the movement of an advancing ship. For this purpose, we need a better injection of kinetic energy, equivalent to much higher heave and/or pitch velocities, which are rather. This way, the usually published, purely hydrostatically derived squat data, as supplementing and interpreting numerical computations of the excited forces in rapidly and longitudinally varying depth, still remain helpful. Otherwise, we should see a significant bias in the plots of squat data between the static conversion and the dynamic model – this has not been however revealed in Figures 5-7.

According to the existing numerical computations, the expected rather moderate jump in excitations while passing over a bottom-disturbance only occurs with reference to the pitch moment. The heave force is varying here stepwise. At this point, we shall also remember that the computation of these forces are undertaken in steady flow conditions beneath the ship's hull. For development of this steady flow (stabilisation) a certain amount of time is required that is not necessarily the case if a ship is moving fast.

In contrast to widely available experimental results on squat kinematics in transient conditions, the direct scale model measurements of hydrodynamic forces in varying-depth movement of a ship practically do not exist. The dynamic model of squat, even in its more sophisticated version, is relatively simple. Thus, it is being encouraged to identify in the future the seabed-generated forces based on published kinematical records.

The presented simplified dynamic model of squat can provisionally be integrated with full-mission bridge simulators or various desktop simulators. Transformation of traditionally available kinematical squat data at constant-depth water into forces and storing them in a lookup-table is straightforward. At this stage of research, the reference position of sensing the actual depth shall conservatively be the ship's bow, that is the first point experiencing decrease in depth in the direction of movement. Such depth is next input to the lookup-table and considered the 'equivalent' constant depth in the varying seabed environment.

REFERENCES

Drobyshevskiy, Y.E. 2000. Ship Squat over a Stepped Bottom: Theoretical Model. *RINA Transactions*, 142: 59-75.

Edstrand, H., & Norrbin, N.H. 1978. Shallow Water Phenomena and Scale Model Research. Some Experience from the SSPA Maritime Dynamics Laboratory. *International Shipbuilding Progress* 25(no. 287/Jul): 181-195.

Ferguson, A.M. & Seren, D.B. & McGregor, R.C. 1982. Experimental Investigation of a Grounding on a Shoaling Sandbank. *RINA Transactions*, 124: 303-322.

Gourlay, T.P. 2003. Ship squat in water of varying depth. *International Journal of Maritime Engineering (RINA Transactions, Part A)* 145 (A1): 1-12.

Haatainen, P. & Lund, J. & Kostilainen, V. 1978. *Experimental investigation on the squat in changing depth conditions*. Report no. 14. Otaniemi: Helsinki University of Technology/Ship Hydrodynamics Laboratory.

Pawlowski, M. 1999. *Nonlinear Model of Ship Motions in Irregular Wave*. Technical Report no. 33/99. Gdansk: Polish Register of Ships (PRS) (in Polish).

Renilson, M.& Hatch, T. 1998. Preliminary investigation into squat over an undulating bottom. *The Naval Architect* (Feb): 36-37.

Tuck, E.O. 1966. Shallow-water flows past slender bodies. *Journal of Fluid Mechanics* 26(1): 81-95.

Safety of Transport

Safety of Transport
Safety of Marine Transport – Marine Navigation and Safety of Sea Transportation – A. Weintrit & T. Neumann (eds.)

State of Safety in the Polish Land Transport

J. Mikulski

Department of Transport, University of Economics in Katowice, Poland

ABSTRACT: Transport is the basis of every economy and the land transport plays here a special role. It allows an easy adaptation to the changing conditions of contracts for the carriage. In order to present issues of safety in road transport in the article sets out the factors determining the transport, and these are transport network and vehicles.

1 GENERAL

Transport is the basis of every economy and the land transport plays a special role, because it allows an easy adaptation to changing conditions of orders to carry people and goods.

The care of safety accompanies daily transportation processes and the modern technology finds its significant application in the transport safety improvement.

2 SAFETY AND HAZARDS IN THE ROAD TRANSPORT

To present closer the issues of safety in the road transport the factors determining this transport should be defined and they comprise the road network and the number of motor vehicles.

In recent years the road network was subject to substantial changes and also the amount of roads with improved pavement has increased. Also the motorways network has developed in those years (now there are more than 1500 km of them).This means that per 1000 km^2 of the country area the length of motorways is 4.7 km, while there are 3.5 km of them per 100,000 population (this is one of the lowest indices in the European Union).

The technical condition of road pavements is significant from the road traffic safety point of view. For road users the ruts and the smoothness and skid resistance during braking (anti-skid properties) are parameters which are most noticeable and at the same time affecting the road traffic safety.

Figure 1 presents the assessment of the national roads network condition. The comparison shows an improvement to the technical condition of road pavements (good condition +3.4%, unsatisfactory condition -2.4%, bad condition -1.0%), although the general condition of roads is still not satisfactory (Regulations… 1996).

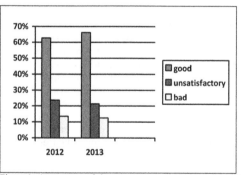

Figure 1. Assessment of the pavements technical condition of national roads network
Source: own study based on (Report … 2014)

The presented assessment of the technical condition shows that more than 37% of roads qualify to repairs of varying degree of urgency and nearly 1/3 of that require immediate repairs.

Factors that have the greatest influence on a poor technical condition of roads in Poland include many years of under investing, the lack of repairs and of current maintenance of roads, a systematically growing road traffic volume, non-adaptation of roads designed in the past to current loads and the lack of an effective system for elimination of overloaded vehicles from the traffic (Krystek 2009).

The number of motor vehicles and the resulting motorization index are important features characterizing the road transport and affecting the safety. For around twenty years a quick and systematic increase in the number of registered vehicles has been seen in Poland. In 2013 there were 504 cars per 1000 inhabitants in Poland.

Poland is a country of more than 38 million population, of total public roads network exceeding 40,000 km and more than 25 million registered vehicles. This dynamic growth of motorization in Poland brings a number of threats.

Dangerous behavior of road traffic participants belong to the main problems on Polish roads. This applies both to drivers and pedestrians and is related to disregard of road traffic regulations and to the lack of respect for rights of other traffic participants. Bad drivers' habits include speeding, forcing the right of way, incorrect maneuvers and to a large extent - alcohol.

Only in 2013 there were 35,847 road accidents on Polish roads and 44,059 persons were injured, not to mention notable material losses amounting to approx. PLN 30 billion per year. Although the data shows a decline in the number of accidents (by 10.5% as against 2011 and by 3.2% against 2012) and a decrease of fatalities number (by 19.9% and 6%, respectively), but anyhow these are alarming figures (Police Headquarters 2014).

Taking into consideration the experience of countries with a high level of safety the Polish vision consists in striving for elimination of fatalities. This programme is implemented in stages and every few years numerical targets are determined defining the authorities plans in relation to further reduction of fatalities in transport.

The programme assumes improving the road traffic safety system, forming safe behavior of road traffic participants, protection of pedestrians and cyclists, ensuring a safe road infrastructure and environment and reducing consequences of road accidents (Liberadzki & Mindur 2006).

3 SAFETY SITUATION AND HAZARDS ON RAILWAYS

The rail transport is a significant link of the transport system and the railway is considered an effective and environmentally friendly and at the same time safe means of transport.

Modernization and development of the railway network, involved in transporting people and goods, may contribute to the development of the country. When evaluating the level and situation of safety on the railway it is necessary to consider the technical condition of railway infrastructure and of the rolling stock, the railway traffic organization, the technology of shipments performance, including hazardous goods shipments, professional qualifications and due performance of duties by employees directly related to traffic management and safety on railway lines and driving rail vehicles as well as supervision of railway operating performance (Krystek 2009).

The railway network in Poland was subject to significant changes, that is it has substantially reduced in recent years. The same applies to freight and passenger traffic.

However, a poor condition of the railway infrastructure is a special hazard in Polish rail transport. Organizational transformations and financial problems make that the current condition of the railway lines infrastructure is unsatisfactory and subject to systematic degradation threatens an effective and safe traffic. The technical condition of tracks is presented in Figure 2. Although as it is visible the track length with good assessment makes around 43% of the total track length and this means an increase as against previous years (the result of maintenance - repair works carried out recently), but this result continues to be unsatisfactory (Office of Rail Transport ... 2011).

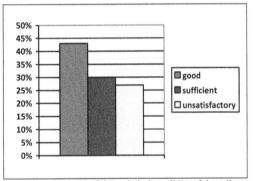

Figure 2. Assessment of the technical condition of the railway infrastructure at the end of 2013.
Source: Own study based on: (Annual Report ...2013)

Also the condition of the rolling stock used for passenger and freight traffic is not satisfactory, both due to its age and structure. Basic problems consist of obsolete designs and wearing out of most passenger cars, and especially electric multiple-unit sets (on average they are 26 years old), resulting in high service and maintenance costs, as well as the lack of locomotives adapted to speeds exceeding 160 km/h and a very small number of locomotives for speeds of 140-160 km/h. A low share of cars adapted to modern transport technologies and a substantial average age of cars are the basic problems in the freight traffic (Master Plan 2008).

The analysis of occurred railway accidents as well as of their causes and effects is one of main determinants to evaluate the situation of railway traffic safety. Accidents include in particular collisions, derailments, occurrences with involvement of persons caused by a moving railway vehicle, a fire of a railway vehicle and occurrences on level crossings.

It is not easy to categorize railway accidents and occurrences, because there are various reasons for them. Accidents occur as a result of a poor infrastructure or rolling stock condition, devastation or theft of railway traffic control equipment. In addition there is also a human factor in the form of suicidal incidents, imprudence of persons moving in railway areas as well as road vehicles passing through level crossings.

Analyses show that the human factor plays a prevailing role in the origination of railway accidents. The largest number of accidents with involvement of people occur on level crossings and there is a larger and larger number of derailments and collision because of that.

The threat with railway vandalism, resulting in the theft of significant elements of railway traffic control or in train robberies to steal the transported goods, becomes very important, affecting the railway traffic safety. Trains are then substantially delayed or it is necessary to arrange traffic as non-organized running to bypass an obstacle on the track. The costs of equipment theft or devastation incurred by the infrastructure managers due to that are very high and exceed expenditures allocated to repairs of the railway traffic control equipment.

Railway carriers, infrastructure managers, users of industry tracks as well as rolling stock manufacturers, rolling stock main repair shops, manufacturers of the railway traffic control equipment and bodies carrying out technical certification of the equipment are involved in ensuring safety in the rail transport.

4 LEVEL CROSSINGS – DANGEROUS POINTS OF THE LAND TRANSPORT

A level crossing is a railway line crossing with a road on the same level. The practice shows that road intersections, including also level crossings, introduce a big element of threat (increased risk) to traffic participants. Under specific traffic conditions, even at a minimal non-adaptation to those conditions and disregard to traffic rules, a possibility of vehicles collision occurs. A collision causes always notable material losses and frequently has tragic implications (injured persons and fatalities). It constitutes also, depending on the freight carried, a hazard to the environment and people present in the vicinity. A collision on a level crossing creates a special threat to the carried passengers and freight. Taking into account a mass nature of passenger and

freight traffic such accidents can result in numerous casualties and material losses of disastrous size. For railways, every accident on a level crossing, irrespective of its dimensions, results in impediments in the carried out traffic. It causes train delays, resulting from changes of traffic organization on the route, on which an accident occurred, as well as results in costs of removing the effects of the occurred accident.

On the network of the largest Polish railway infrastructure manager, Polish Railway Lines /PLK/, there are level crossings of various types, divided into the following categories (Regulations …1996):

- category A - public level crossings with barriers or public level crossings without barriers, on which the traffic on the road is controlled by signals given by railway employees,
- category B - public level crossings with automatic light signals and half-barriers,
- category C - public level crossings with automatic or operated by railway employees light signals,
- category D - public level crossings without barriers and half-barriers and without automatic light signals, and
- category E - public pedestrian crossings,
- category F - non-public level and pedestrian crossings.

The classification of a level crossing to a specific category depends on the traffic (railway - road) conditions and directly affects the equipment to be used for its protection. Level crossings may be protected by barrier drives, road traffic signals, lights on longitudinal barriers, acoustic warning devices, which actively warn road traffic participants about a closed traffic on the level crossing.

To improve the traffic safety on level crossings outer distant signals have been additionally introduced, providing engine drivers with information on the level crossing protection. Outer distant signals are placed ahead of a level crossing at a braking distance for the fastest train travelling on this railway line.

The Polish Railways /PKP/ specifications show that the largest number of level crossings have category D, without barriers and half-barriers and without automatic light signals, marked only with road signs. These are level crossings with the highest risk of accident or collision.

When studying the situation of safety on level crossings the occurred accidents and collisions should be considered. Road users are perpetrators of definite majority of accidents on level crossings, who disregard the highway code provisions and who do not exercise special caution at level crossings. Attention should be also drawn to the work of some railway employees operating level crossings, who were not closing barriers in the prescribed time before the approaching train or were opening them after the passage of the first train and just before the train moving on the adjacent track in the opposite direction.

5 LEVEL CROSSINGS FROM NEW TECHNICAL SOLUTIONS POINT OF VIEW

Looking for methods of improving safety on level crossings it is necessary to consider a possibility to use for that new technological solutions based on telematic solutions. The main issue is to control the dangerous zone on a level crossing or more precisely its occupancy by road vehicles.

The work on using such equipment has been carried out on the railways for a few years, where inter alia such devices as the following are tested (Pikus 2004):
– TV systems with digital image analysis,
– IR sensors,
– inductive loops,
– radar scanners.

The carried out tests show that radar scanners work best, eliminating observation of the level crossing via CCTV by an employee operating the level crossing. It is of particular importance when the level crossing is operated by a movements inspector. The 'line clear' signal is given once the barriers have been closed, which are activated by stating by the radar scanner that the dangerous zone on the level crossing is free. On level crossings protected with automatic signalling equipment there is interlocking between level crossing signals and signals given for the train moving towards the level crossing. In such case the 'line clear' signal for the train approaching the level crossing is given when the barriers have been closed (longitudinal barriers in the final position) and the radar scanner has found that the dangerous zone on the level crossing is free (Pikus 2004).

There is ongoing work on developing solutions adapting the level crossing protection equipment to implemented modern traffic control systems as well as on systems for signal repetition on the locomotive. In parallel there is a process of harmonisation of requirements applicable to displays, which will show the information about level crossings to engine drivers.

When analysing modern solutions in the field of level crossing protection technology one cannot forget about safe control systems on a level crossing in the form of automatic level crossing signals. The introduction to a common application on the PKP network of modern solutions in level crossing protection systems and the use of developed IT techniques will reduce the accident rate on level crossings.

6 IMPORTANCE OF ACCIDENTS IN ROAD TRANSPORT EXAMINATION

To improve safety in the road transport it is necessary to have a possibility of gathering specific data (see Mikulski 2015), which analyzed will allow to state the most frequent reasons for accidents and their effects. In Poland there is a central system for road accidents and collisions registration. A few other local databases based on the data collected in this system have been created in Poland to report and analyze accidents and transport occurrences.

The work carried out in the field of safety improvement allowed identifying main problems of road traffic safety in Poland. Also a National Programme for Road Traffic Safety 2013 - 2020 was created in Poland, based on a set of actions, which include (National Programme 2013) the application of such technical solutions, which increase the road safety and make that roads 'forgive' human errors, as well as technical solutions in vehicles protecting drivers, passengers and other not protected traffic participants and reducing possible damages. These actions come down to one - to protection of road traffic participants life and health. Three basic factors affecting the situation in the road traffic safety are specified among systemic analyses of road traffic safety. These are:
– human - being the weakest link of the whole safety system, whose behavior on the road is unpredictable and frequently results in accidents,
– road infrastructure - its errors and condition belong to the main reasons for the road accidents occurrence,
– vehicle - its equipment and technical condition contribute to accidents occurrence and their gravity.

The main targets, which should contribute to improving the whole safety system, include (National Programme ... 2013):
– construction and modernization of modern network of roads and motorways,
– implementation of ITS measures in the road traffic supervision and management (information about the traffic situation and warning, increased application of VMSs, traffic supervision systems in hazard zones, new systems on intersections, speed supervision, road incidents detection),
– modernization of automatic speed supervision systems,
– development of databases including improvement in techniques of road incidents location,
– equipment of vehicles with modern safety devices,
– development of the e-call system,
– common use of GPS navigation systems in vehicles.

7 MANAGEMENT AND SUPERVISION OF SAFETY IN THE RAIL TRANSPORT

The rail transport, because of its complexity in various structures of its operation, requires special procedures in the field of safety. Special organizations were established to determine such procedures and to determine directions of action in the field of safety improvement in the rail transport:
- Office of Rail Transport, which primarily supervises the railway traffic safety (apart from technical supervision of operation and maintenance of railway lines and of the rolling stock),
- State Commission for Railway Accidents Investigation, which investigates accidents, including accidents and incidents, without deciding about guilt and responsibility,
- Transport Technical Inspection, which supervises the equipment installed in the railway area and in the railway vehicles.

Also a whole system for safety management was established, to which all entities providing services in the field of rail transport must submit.

To eliminate an unnecessary risk, related to the operation and management of traffic in the rail transport, reliability tests are carried out for railway vehicles, which exactly in the same way as road vehicles participate in traffic and are directly exposed to occurrence of faults and breakdowns, which can contribute to origination of dangerous situations. But in the rail transport one of the main traffic safety aspects, having a huge importance for safe traffic management, consists of reliability tests of railway traffic control equipment, because it due to its reliability may to a large extent contribute to avoiding dangerous situations. The railway traffic control equipment designed now must meet the highest safety standards because of the intended elimination of human errors.

8 DIRECTIONS OF LEVEL CROSSING PROTECTION CHANGES IN POLAND

In accordance with (Regulation ... 1996) the infrastructure manager is responsible for protection of level crossings. The way of level crossing protection depends or will depend on an average daily traffic flow on a level crossing, taking into account the traffic of road and railway vehicles. However, the number of occurring accidents provokes thoughts, whether everything operates properly in the whole system of safety on level crossings.

The data published by the infrastructure manager, Polish Railway Lines, on the accidents occurred on level crossings shows that participants of the road traffic are responsible for more than 95% of accidents due to a widespread breaking of road traffic regulations related to safety and behaviour on level crossings. Obviously the road traffic participants and media, irrespective of the occurred accidents reasons, blame the railway for their occurrence (poor technical condition of level crossings, improper service of attended level crossings etc.) (Pikus 2004). To improve the bad situation a national social campaign Safe Level Crossing - 'Stop and live!' has started more than decade ago, which individual editions are continued up to date. Its overriding objective consists in raising the drivers awareness of originating hazards, resulting from improper behavior on a level crossing.

To mitigate and to reduce the frequency of accidents occurrence on level crossings PLK undertakes, together with its reporting structures responsible for the safety in the railway area, common actions consisting of (Assessment ... 2011):
- increasing the pace of level crossings modernization, improving the technical condition of equipment on level crossings or working with them,
- checking on a current basis conditions of 'visibility triangles' and in situations of permanent obstacles, restricting the visibility of approaching railway vehicles, appropriate adaptation of the approaching train speed,
- observations, including monitoring and increasing the inspections frequency of level crossings, where the number of binding regulations breaching is the largest,
- spreading on level crossings, exposed to the highest risk of accident occurrence, innovative warning systems on level and pedestrian crossings,
- marking especially dangerous level crossings with information signboards or billboards, like it is the case on roads (so-called accident black spot).

Now, in the era of permanently increasing number of road vehicles and permanently increasing vehicles traffic intensity, it is necessary to carry out traffic calculations on level crossings more often than every five years (Regulation ... 1996).

New calculations, taking into account new technical-traffic conditions on a level crossing, will result in changing the level crossing category. Additionally, the actions improving safety on level crossings should be expanded by a permanent necessity of investing and introducing new technologies for common use on the PKP network, based on the highest reliability and safety standards. In all places where it is possible, investment activities should be carried out consisting in cancelling the existing level crossing and building overpasses, which in a conflict-free (collision-free) way will take over the level crossings role. The situation may be improved by using CCTV systems, which will allow, like speed cameras at roads, to identify drivers or vehicles breaking regulations while passing the level crossing.

9 CONCLUSIONS

The growth in transport and increased road users are the cause of the related increase in the number of collisions.

Both, bodies supervising transport (including the police) and the scientists studying the land transport safety issues as the main causes of accidents perceive two factors, which are men and transport infrastructure. And so this situation forces the creation of new traffic control devices (both road and rail). Another aspect that has the effect of reducing the number of accidents and improve safety is the development of intelligent transport telematics technology.

REFERENCES

Krystek R. 2009. *Zintegrowany system bezpieczeństwa transportu /Integrated transport safety system/. vol. I. Diagnoza bezpieczeństwa transportu w Polsce /Diagnose of transport safety in Poland/.* Warszawa. WKŁ.

Liberadzki B.,& Mindur L. 2006. *Uwarunkowania rozwoju systemu transportowego Polski /Development conditions of transport system in Poland/.* Warszawa. Wydawnictwo Instytutu Technologii Eksploatacji.

Mikulski J. 2015. *Badania wpływu telematyki na bezpieczeństwo transportu/Surveys of the Influence of telematics on the land transport safety.* Gdynia (in press).

Pikus R. 2004. *Kierunki zmian w sposobach zabezpieczenia przejazdów kolejowych wynikające z badań systemów sygnalizacji przejazdowych /Directions of changes in methods for level crossings protection resulting from studies on level crossing signalling systems/. Rozwiązania skrzyżowań kolei z drogami w poziomie szyn /Level crossing solutions/.* Polish Scientific-Technical Conference. Częstochowa.

Police Headquarters. 2014. *Wypadki drogowe w Polsce w 2013 roku /Road accidents in Poland in 2013/,* Warszawa.

Ministry of Infrastructure. 2008. *Master Plan dla transportu kolejowego w Polsce do 2030 roku /Master plan for the rail transport in Poland by 2030/.* Warszawa.

Ministry of Transport. 2013. Construction and Maritime Economy. *National Programme of Road Traffic Safety 2013-2020.* A joint publication edited by the Secretariat of the National Road Traffic Council. Warszawa.

Report on the technical condition of the national roads network at the end of 2013. 2014. GDDKiA. Warszawa.

Regulation of the Minister of Transport and Maritime Economy of 26 February 1996 on technical conditions to be met by level crossings of railway lines with public roads. (Dz. U. of 1996, No 33, item 144 with later amendments).

Annual Report. 2013. PKP PLK S.A.. Warszawa.

Assessment of the railway traffic safety situation. 2013. Office of Rail Transport. *Transport i Komunikacja,* No 4, 2011.

Surveys of the Influence of Telematics on the Land Transport Safety

J. Mikulski

Department of Transport, University of Economics in Katowice, Poland

ABSTRACT: Safety is the most important factor concerning transport processes, and modern technology is used to improve transport safety. In order to present issues of safety in land transport article presents the results of studies on the influence of telematics on the land transport. In studies carried out impact of the modern technology on the safety situation on the roads and railways used method of survey, addressed to the respondents having a major impact on transport safety.

1 GENERAL

The ensuring of safety in transport is one of most important factors that determine its effective operation. It is especially important in the land transport, because its two modes in the form of road and rail transport have the highest share in the amount of transported goods and people.

In recent years the number of transport users has grown rapidly. As a result, the increase in number of accidents is unavoidable. The cause of accidents and collisions is often the improper and insufficient transport infrastructure (Barcik&Czech 2010). Issues related to the analysis and activities undertaken in the area of transport infrastructure (road and rail) are discussed by various authors, and it is a very timely topic (Bril&Łukasik 2013, Dąbczyński 2001, Luft &Kornaszewski 2012, Sitarz et al. 2012).

The study was aimed at drawing attention to issues of the current safety situation and to new technical capabilities, which can substantially affect the improvement to this situation. Surveys were carried out to verify theoretical considerations. The results obtained this way allowed to find a common denominator of the analyzed issues. The factors, which directly and indirectly influence the level of safety in the land transport were analyzed, that is the condition of the rail - road infrastructure, the protection of level crossings and modern ICT.

2 SURVEYS METHODOLOGY

The method of questionnaire surveys was applied in the studies (Widuch 2014) carried out on the influence of telematics on the improvement to the land transport safety and on the safety situation existing on roads and railways. Four questionnaire surveys were carried out, addressed to respondents having a major influence on the safety in transport. Because of the rail transport specific nature, in which a lot of factors contributes to the existing safety and a large group of people takes care of the safety, the respondents included railway operators (having an indirect influence on the safety), but primarily train dispatchers (running and managing the railway traffic) and drivers (directly participating in the railway traffic).

In the road transport the group of respondents comprised vehicle drivers, without breaking down to the size of driven vehicles.

3 QUESTIONNAIRE

The surveys carried out intended to find an answer to the question, whether and to what extent modern ICT systems used in the land transport have a positive effect on the safety improvement. The questions included in questionnaires were also directed towards checking the existing safety situation. A few questions were designed so as to find a common denominator, determining the improvement to the safety situation on level crossings. It is known that the safety systems used on the railways and on the roads differ, but some of them could work with each other or some could be used in the other mode of land transport.

Surveys comprised:
– railway operators (10 respondents) and three groups containing 100 respondents each:
– train dispatchers,
– train drivers,
– motorists,
and they were carried out at the turn of 2013 and 2014.

3.1 *Results of railway operators survey.*

The scope of questions addressed to railway car operators covered various areas affecting broadly understood safety in the rail transport.

The first question was related to the railway infrastructure used by operators carrying out their daily traffic. The results confirm considerations related to the situation of rail infrastructure condition in Poland (Mikulski 2015). As many as 70% of those surveyed considered the infrastructure condition as only sufficient and 30% as bad. These results show that actions to improve the rail infrastructure quality must be taken. Such a situation means applying continuously speed limits to eliminate hazards and hence a poorer quality of service provided.

The next questions covered the issue of motive power and the rolling stock condition. A modern rolling stock and in good working order is a safe rolling stock. The results illustrate the situation of the rolling stock moving on Polish tracks. 60% of those surveyed have stated that the condition of their rolling stock is good, but 40% that it is sufficient. This is not a bad situation, but to improve the safety this condition could be better than sufficient. 42% of respondents were in favor of withdrawing the old rolling stock and buying a new one and 58% in favor of modernization of the operated rolling stock. Such actions undoubtedly fill with optimism that the rolling stock moving on Polish tracks will be newer, technically improved and hence safer.

An important question was related to technical aspects of telematic systems, operating for an efficient and safe traffic management. The presented results show that 80% of respondents are in favour of using ICT systems to run a more efficient and safer transportation.

In the next question - presentation of systems used - respondents had a possibility to present their opinion and to provide more than one answer. A full set of answers covered a satellite navigation system GPS, but one of respondents (10% of those surveyed) has been using two systems, adding to the used GPS system also a mobile telephony system GSM-R, operating based on a GSM principle extended by a radio channel for transmission of data

and to implement functions anticipated for specialized applications for railways.

The next questions were related to safety on level crossings, occurred accidents, on what level crossing category there are most often collisions between motive power units and cars/lorries. This is this category of questions, which was included in two other groups of respondents.

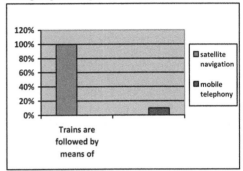

Figure 1. Graphical presentation of ICT systems used
Source: Own study[28]

The survey shows clearly that 90% of respondents carrying out passenger and freight traffic, being owners of the means of transport, experience accidents with cars/lorries on level crossings. The presented surveys of such occurrences frequency show that accidents on level crossings occur in all categories of occurrence frequency. 50% of respondents answered that very rarely, 40% that rarely and 10% that frequently.

The collected results reveal a bitter truth of level crossings and illustrate previous considerations of this issue (Mikulski 2015). According to those surveyed all accidents occur on level crossings of category C and category D, considered especially dangerous level crossings.

The last questions was aimed at obtaining answers on improvement to the safety on level crossings. It included a possibility to use telematic systems operating on level crossings and to utilize them to protect especially dangerous ones. 60% of carriers declared a willingness to make output signals of their ICT systems available to improve safety on level crossings (this signal could be e.g. a GPS signal of the locomotive, received by drivers in real time).

3.2 *Results of train dispatchers survey*

Train dispatchers are a professional group deserving a closer familiarization in relation to the railway transport safety. Their duties include primarily a safe management of trains traffic. The train dispatchers work is closely related to the railway traffic control

---

[28] All graphs were prepared based on (Widuch 2014)

equipment, thanks to which the traffic is carried out. Broadly understood safe running of railway traffic depends on its condition, current maintenance, defectiveness or modernity. The traffic safety depends also on the train traffic intensity, on systems supporting the train dispatcher work, traffic impediments during repair works and occurring human errors, which can be gradually eliminated due to modern solutions. The convention of questions and results will be continued in this and in the next surveys.

The scope of the first question covered the condition of railway safety control devices. The presented results show that the condition of railway safety control devices is not exactly good. 30% of respondents referred to a poor condition of those devices, 38% to sufficient, 21% to good and 11% to satisfactory condition.

The second question addressed the condition of local infrastructure, which should enable carrying out a safe traffic. The presented results confirm previous considerations included in (Mikulski 2015) and in the questionnaire survey of railway operators that the railway infrastructure, on which the railway traffic takes place, is not in the best condition. The largest number of answers referred to a poor (40% of respondents) and sufficient (44% of respondents) condition of infrastructure. It should be presumed that for the other answers - satisfactory (2%) and good (14%), the infrastructure was recently modernized or revitalized.

In recent years station control rooms have been equipped with IT systems - the Operating Performance Recording System (SEPE) and the System for Train Dispatchers Work Support (SWDR). The next question covered those systems functionality (monitoring of arriving trains, a graph of actual trains movement).

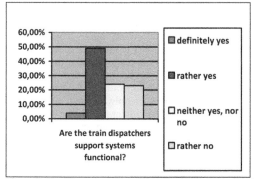

Figure 2. Graphic presentation of functionality assessment of train dispatchers support IT systems
Source: Own study

The distribution of respondents answers was not unequivocal. More than a half of those surveyed have stated that the supporting IT systems are exhaustive in their functions, but 25% of respondents did not have a specified opinion on those systems functionality and 22% of those surveyed stated that they were not exhaustive. Therefore this is an indication that these systems should be improved.

The traffic intensity has a major impact on the safety of trains traffic management. Based on questionnaire results it is possible to conclude that most of station control rooms have a high or medium intensity of the trains stream, which implies that a human operating the traffic is to a larger extent exposed to making an error. The traffic management during special impediments related to repairs or to current maintenance of infrastructure is a very important issue in the safe train traffic management. This is closely related to so-called line obstructions. These are particular situations, worth following (the next question).

A definite majority of surveyed persons have told that the train traffic during traffic impediments proceeds smoothly. However, in the surveyed group there were answers that certain problems and impediments did occur. For this last group, with the result of 11%, the occurring 'disturbances' may contribute to a slump in the level of safe train traffic management.

In the train traffic sometimes there are very dangerous situations, for which drivers should be blamed - passing a 'Stop' signal. The scope of the next questions covered such situations and possibilities of train dispatcher response by an emergency train stopping. The presented data shows that there exist cases of breaching elementary safety rules by train drivers, consisting in ignoring the 'Stop' signal. More than a half of respondents have told that such cases did not happen, however 46% of those surveyed stated that such situations occurred once or a few times.

Modern ICT systems, installed in new or in modernized motive power units, could be a panacea for such cases as well as a train dispatcher reaction by the 'Radio - STOP' button, situated on the train communication transceiver. Respondents statements on this button use prove that once or a few times (17% of those surveyed) decisions were made on using this button and thereby preventing the occurrence of threat to the railway traffic safety or even of an accident or disaster.

The last question of the questionnaire addressed the issue of improving a safe traffic management and the work organization at the station. Trains traffic management supported by already existing ICT systems operating in station control rooms and extended by telematic systems could definitely substantially contribute to a radical safety improvement. The question was worded as follows: *Would the real time trains monitoring by modern telematic systems (incl. GPS) and its presentation on a monitor screen, presenting the current situation not only on adjacent lines but also on the more remote ones, improve the situation of traffic management safety?*

The presented results show a positive attitude to the stated issue, because as many as 83% of those surveyed expressed their willingness to introduce new technologies to improve the work and a safe train traffic management.

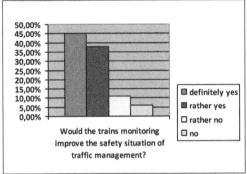

Figure 3 Graphical presentation of train dispatchers preferences related to the introduction of new support systems
Source: Own study

3.3 *Results of train drivers survey*

Train drivers, like train dispatchers, are a professional group directly affecting the level of railway traffic safety. A safe train driving depends on their skills and experience. However, it should be remembered that a safe train driving is strictly related to the condition of the railway infrastructure, working order and modernity of the on-board equipment, failure-free operation of railway traffic control equipment and technical conditions of railway lines. All these aspects were covered by the carried out survey and questions related to accidents and dangerous railway incidents could not be omitted, which in a way provide a measure of the existing safety level.

The first question, like in the previous questionnaires, was related to the condition of the railway infrastructure, because train drivers experience it directly.

Again, the presented results confirm previous considerations and questionnaire surveys of the railway infrastructure condition. As many as 64% of respondents consider that this condition is sufficient, but 21% of those surveyed that it is bad. The results are very similar to results of train dispatchers survey. Only 15% of respondents stated that the infrastructure condition was good, which is definitely too little to consider the railway infrastructure in Poland safe.

The defectiveness of railway traffic control equipment has a significant influence on the safety

situation of the carried out railway traffic. In such situations it is necessary for train dispatchers to dispatch trains based on so-called special-purpose signal. These are extremely dangerous situations, which must by accompanied by a doubled caution and a proper securing of the train route. The second question to respondents was asked in this relation.

Based on the answers provided it is possible to state that situations, where a train must go based on a special-purpose signal occur almost always. Only 16% of those surveyed stated that they encountered such situation once for a few shifts. This situation makes considering that the operated railway traffic control equipment, frequently obsolete, shows a high defectiveness, which can contribute to origination of dangerous situations.

The third question addressed the on-board equipment of motive power units - that installed on-board and taking care of safe driving of a motive power unit.

The picture of results in the surveyed group of respondents is nearly unequivocal. 99% of those surveyed stated that only the automatic train stopping devices (SHP) take care of safe train driving. Instead, at 1% these devices include also a system for automatic speed restriction - a system that does not allow to exceed the preset speed. It shows that most of motive power units are still not equipped with devices featuring a possibility of independent interference in a situation, where the driver has not adapted the speed to the situation.

However, there are devices that in the case of missing reaction of the driver to the occurred hazard would implement train braking. Therefore the next question referred to the effectiveness of such devices.

The results obtained in the survey show the condition of safety vigilance devices. The SHP and the active automatic warning system (CA) are generally in good working order: 77% of answers - always in good working order, 18% - always at least one of the systems is in good working order. However, it is necessary to take into account that 5% of surveyed persons stated a frequent poor working order of those systems, which can result in a situation threatening the traffic safety. Unfavorable technical conditions of railway lines, affecting the railway traffic safety, under which the visibility of signals may be reduced, play an important role. This topic was covered in the next questions.

The presented data shows that the weather conditions do not have an overly high share in deterioration of motive power units driving, because this visibility is mainly good - 11% or sufficient - 63%, however 26% of those surveyed stated insufficient visibility. The situation looks worse because of natural objects and structures situated close to the railway line.

Results for the question in the survey show clearly that in 98 out of 100 cases the observation of signals by train drivers is difficult, which is a worrying effect proving a lack of control bodies action in this field.

The questions covered also technical aspects of level crossings as well as behavior of vehicle drivers passing the most dangerous point of contact of two modes of land transport.

The answers of respondents were alarming, because 100% of those surveyed stated that vehicle drivers pass unmanned level crossings despite an approaching train. It is necessary to consider, whether social campaigns carried out on a wide scale in the field of improving safety on level crossings have a positive effect or whether perhaps it would be justified to support reversal of this dangerous situation with modern technology and influence the drivers behavior in a different way.

Questions addressed also the issue of protecting level crossing by caution signals (TOP), informing train drivers about the situation of level crossings protection. In the respondents opinion level crossing are well protected by caution signals, because it was expressed by as many as 91% of those surveyed. A small percentage of respondents - 9% - have an opposite opinion, which may contribute to the originating hazards and accidents on level crossings.

The scope of last questions covered participation of train drivers in railway occurrences specified as accidents, incidents or disasters. Such occurrences may have different forms - as occurrences with outcomes (substantial material losses, casualties/fatalities) or occurrences without outcomes (traffic impediments or operational breaks in the trains traffic).

The collected information reveals that 34% of train drivers taking part in the survey were involved in railway occurrences with outcomes. Generally categories of such occurrences vary, but imply serious material effects in the infrastructure, transported goods or persons.

The presented results show the fact that 40% of surveyed train drivers participated in railway occurrences without outcomes, which means that the percentage of occurrences is not small. The elimination of those occurrences should become a priority for services responsible for improvement in the railway transport safety.

3.4 *Results of Motorists Survey*

The road transport has the highest share in persons and goods transport among all modes of transport. This is the mode of transport, which in recent years experienced the greatest progress in the introduction of modern technical solutions. This is strictly related to the improvement in the safety of driving vehicles and also in safe controlling of vehicles traffic. Car

manufacturers, apart from improvement to the passive safety, focused on the improvement to the active safety, which through appropriate actions of the driver may substantially contribute to the improvement in safety and to the protection of other participants of the road traffic. Whole telematic subsystems are frequently elements of active safety in vehicles, contributing to the entire system of safety in vehicles. During previous ICT considerations a high share of Telematics and its contribution to the road traffic controlling and to informing the drivers in real time about the hazards have already been proven. Therefore the last questionnaire survey of this group comprised the issue of vehicles, road traffic and level crossings safety.

Larger and larger share of vehicles in the road traffic means an increase in the occurring hazards and originating dangerous situations, therefore the first questions referred to the frequency of using cars, irrespective of the road conditions.

The presented data shows the fact that 69% of those surveyed have been using a car everyday, 20% less frequently (life issues, current matters), and 11% seldom. Results pretty unequivocally determine that for a large group of motorists, because for as many as 80%, changing road conditions, dependent on the season of the year, do not have a major importance. It should be emphasized that the remaining 20% of respondents do not withdraw from the road traffic during the autumn and winter.

The next question was devoted to the issue of traffic flow on intersections controlled by road traffic control systems. Respondents statements provided to the question asked confirmed that there are still certain gaps and deficiencies in traffic control systems, because 38% of those surveyed indicated the lack of traffic flow, and 26% referred to traffic impediments for non-privileged directions. However there is certain optimism, because 36% of those surveyed consider that there is a full traffic flow on intersections.

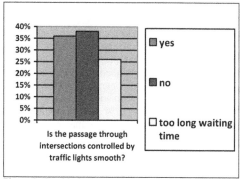

Figure 4. Graphical presentation of opinions on the passage flow through controlled intersections
Source: Own study

279

The next question in the questionnaire survey covered the active safety systems installed by vehicle manufacturers and used by drivers.

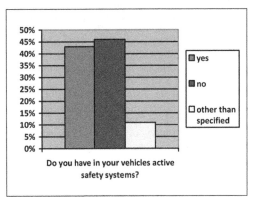

Figure 5. Graphical presentation of the share of various active safety systems in vehicles
Source: Own study

Respondents were asked about having or the lack of various active safety systems in the vehicles they use (Automatic Stability Requirement, Electronic Stability Control, Adaptive Cruise Control, Anti-Collision System etc.). They could choose those systems from the question content or indicate others not specified in the question. More than a half of those surveyed have responded that they had systems specified in the question or had others. Results have shown that in the majority of cases this was one of specified systems - 39%, 26% declared two systems, 15% had more than two, while 20% of respondents stated having entirely different systems. The last group of surveyed persons may be not quite reliable, because technically advanced new systems can comprise the already existing ones. Nevertheless cars become technically more and more advanced means of transport.

Another aspect of safety on roads is real time reliable driver information about the existing hazards or traffic impediments provided via VMSs by road traffic control centers. Reliable and genuine information provided to drivers can very effectively affect the improvement in the road traffic safety. The information subject of messages addressed to drivers was the scope of the next question.

It results from the obtained answers that as many as 58% of surveyed persons stated a high effectiveness of the information provided to drives through VMSs. For 27% of respondents the provided information is of no major importance, while 15% considered that it made the traffic more difficult. It is possible to conclude that VMSs controlled by ICT systems fulfill well their functions.

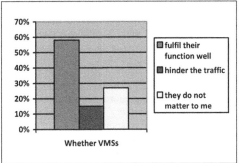

Figure 6. Graphical presentation of VMSs functionality
Source: Own study

The next questions were devoted to the issues of safety on level crossing and the information messages via the GPS satellite navigation.

Level crossings are dangerous places of two modes of transport - rail and road - contact. A very large number of accidents occurs there, generally resulting in huge material, but primarily human, losses therefore the survey could not miss car drivers with respect to this topic.

Questions were related to respondents opinion on the frequency of passing through level crossings and to what extent those crossings should be protected.

Presented results (graph 7) show that 28% of drivers frequently pass level crossings, 45% sometimes, and 27% rarely. Level crossings encountered in their journeys were frequently protected in various ways. Therefore a question could not be missed, which specified subjective impressions and suggestions about the way of level crossings protection.

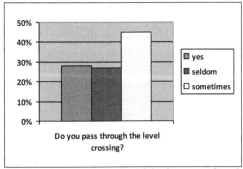

Figure 7. Graphical presentation of the frequency of passing through level crossings
Source: Own study

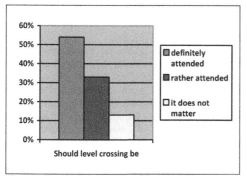

Figure 8. Graphical presentation of the way of level crossings protection
Source: Own study

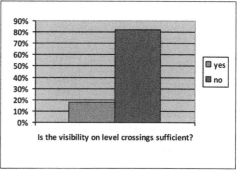

Figure 10. Graphical presentation of visibility on level crossings
Source: Own study

A definite majority of surveyed respondents were in favor of attended level crossings. These are level crossings serviced by authorized railway employees, defined in the railway nomenclature as category A level crossings (Regulations … 1996). Taking into account previously quoted statistics and respondents opinions, these level crossings are considered the safest level crossings among all their categories.

To a large extent the safety on level crossings depends on the drivers themselves. Widespread breaching of road traffic regulations in this area frequently results in the accidents occurrence. The largest number of such occurrences happens on the most dangerous category of level crossings - unmanned category D. Therefore the next question addressed to respondents referred to the issue of breaching road traffic regulations in force on level crossings.

The data obtained as a result of survey reveal a dangerous trend to break road traffic regulations by drivers. Despite the fact that 59% of respondents stated that they have never broken the regulations, the fact that 41% of those surveyed passed an unmanned level crossings without stopping, ignoring the 'Stop' sign, may be worrying.

Figure 10 presents responses related to sufficient visibility on level crossings. Level crossings are frequently situated in the area, which is inconvenient for observation. It frequently happens that the visibility of the 'safety triangle' is hindered. The results distribution may be worrying, because as many as 82% of respondents considered that the visibility on level crossings is insufficient, which can threaten the traffic safety.

The last two questions were addressed to finding benefits resulting from the GPS satellite navigation. Figure 11 presents results of the survey in the area of using the GPS satellite navigation.

The data shows that every fourth driver has been regularly using the satellite navigation (26% of those surveyed), 35% use it sometimes, while 13% very seldom. The presented answers are undoubtedly closely related to the frequency of vehicle driving, because from a logical point of view drivers that travel infrequently do not feel the need of using the satellite navigation.

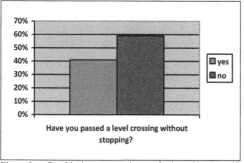

Figure 9. Graphical presentation of breaching traffic regulations in force on level crossings
Source: Own study

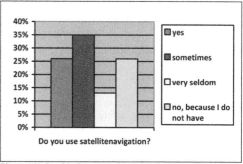

Figure 11. Graphical presentation of GPS navigation using by drivers
Source: Own study

The last question was aimed at finding the answer in relation to improving the safety on level crossings through a modern technique in the form of GPS satellite navigation. The obtained results are presented in Figure 12.

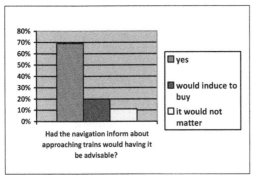

Figure 12. Graphical presentation of respondents opinion on supporting safety on level crossings by a satellite technology Source: Own study

Respondents answers clearly indicated a high willingness to use the satellite technology (69%), or - if it was missing - to purchase it (20%) to improve the road traffic safety on level crossings. Organizational structures of railway operators and railway infrastructure managers would undoubtedly have to be involved in this project, to integrate the railway systems with the satellite technology, aimed also to serve motorists in this area.

4 CONCLUSIONS

The carried out surveys were aimed at obtaining information on the situation of safety in the land transport and on telematic technologies involved in this field. The groups of respondents were not chosen at random, because the obtaining of reasonably reliable results was motivated by the aim to reach persons directly involved in driving or in managing the traffic. Questions for individual survey groups were presented in similar thematic blocks, which from various points of view presented similar issues in the field of assessment of:

− existing safety situation in the land transport,
− telematic systems used in vehicles to improve safety,
− protection of level crossings and the GPS navigation system.

The obtained results, taking into account the assessment of existing safety situation in the road and railway transport, have confirmed considerations presented in (Mikulski 2015), that the situation of railway and road infrastructure as well as of used vehicles, mainly the older ones, is not the best (but it should be admitted that not the worst).

Answers to questions on ICT systems have confirmed that they have most developed and have most been used in the road transport. Increasingly more complex safety systems introduced for use in cars deserve attention. It is possible to conclude from the questionnaire survey that after the introduction to use of new technologically developed devices, ICT based, the safety of railway traffic control could increase.

To provide an additional protection of level crossings VMSs could be situated on unmanned level crossings, providing drivers with real time information on a train approaching the level crossing. A condition fulfilling this solution would be the installation by the infrastructure managers of appropriate track side equipment, working with VMSs, installed by the road manager, in whose area the level crossing is situated.

Another solution to improve the safety on level crossings could consists of information provided in real time to drivers about an approaching train, by means of GPS satellite navigation. It is well known that the satellite navigation is mainly used for objects positioning, so for moving objects there is some information delay - the object is already in another place. However, despite that the information provided from a receiver could undoubtedly influence the imagination of many drivers.

Having analyzed results of the carried out questionnaire surveys it is possible to state that telematic systems in the land transport have already well settled and some of them are simply indispensable.

REFERENCES

Barcik J. & Czech P. 2010. Wpływ infrastruktury drogowej na bezpieczeństwo ruchu /The influence of road infrastructure on safety of road traffic/. *Zeszyty Naukowe PŚl* No 69, 2010

Bril J. & Łukasik Z. 2013. Bezpieczeństwo transport /Transport safety/. *Autobusy* No 3, 2013

Dąbczyński Z. 2001. Bezpieczeństwo ruchu drogowego /Road safety/. SITK, Kraków, 2001

Luft M. & Kornaszewski M. 2012. Bezpieczeństwo ruchu na przejazdach kolejowych /Traffic safety at level crossings/. http://www.not.org.pl/not/files/2012/bezpieczenstwo-transport/prezentacje2/04.pdf

Mikulski J. 2010. Kierunki rozwoju telematycznych systemów transportowych w Polsce /Development directions of telematic transport systems in Poland/. Szczecin. University of Szczecin, *Problemy Transportu i Logistyki* No 10. 2010.

Mikulski J.: 2007. Obecny stan w dziedzinie telematyki systemów transportowych /Contemporary Situation in Transport Systems Telematics/. *Technika Transportu Szynowego* No 11, 2007.

Mikulski J. 2015. Stan bezpieczeństwa w transporcie lądowym /Situation of safety in the land transport/, Gdynia (in press).

Sitarz M. et al. Bezpieczeństwo w transporcie kolejowym Unii Europejskiej /Railway safety performance in the European Union/. *tts (Eksploatacja)* No 5-6, 2012

Widuch K. 2014. Telematyka i jej wpływ na bezpieczeństwo transportu lądowego /Telematics and its influence on the land transport safety/, B.Sc. Eng. Thesis, ŚWSZ, Katowice, 2014.

Regulation of the Minister of Transport and Maritime Economy of 26 February 1996 on technical conditions to be met by level crossings of railway lines with public roads. (Dz. U. of 1996, No 33, item 144 with later amendments).

Approaches and Regulations Regarding Significant Modifications in Transportation and Nuclear Safety

N. Petrek
Institute of Railway Engineering and System Safety, Technische Universität Braunschweig, Germany

H.P. Berg
Bundesamt für Strahlenschutz, Salzgitter, Germany

ABSTRACT: Any modifications on board of ship classed by a certification organization deviating from approved drawings or cause alterations of previously approved documents are regarded to have effect on the validity of class. For the European railway industry the European Commission implemented in 2009 a regulation which harmonizes the risk assessment process containing a process for judging the significance of modifications. In European air traffic control an excel-based approach is used for assessing the use of new technology and procedures regarding the effort which is needed to safely implement them in the existing system. In the area of nuclear safety such a common regulation does not exist to date. In Germany nuclear regulations are issued by the Federal Government. After introducing the different methods for evaluating modifications these methods are compared with regard to their structure and procedures, in particular in case of significant modification.

1 INTRODUCTION

The harmonization of the risk assessment process for European railways by the release of European Commission regulation 352/2009 (2009a) has led to a rather new process for judging the significance of a proposed modification. This process is part of the Common Safety Methods (CSM) which are defined within this regulation. In our paper we will point out the similarities and differences between this new approach used in the European railway sector and the approaches of further technology sectors. For this purpose the methods which are used in the area of maritime transportation and in European air traffic control will be discussed. Besides these three approaches used for the evaluation of modifications in transport systems we examine the procedure to classify modifications in nuclear power plants (NPPs) on the basis of the approach used in the German Federal State of Baden-Württemberg.

These four approaches are compared regarding their structure, the role of the proposer within each method and the relevant aspects which are used for the determination of the significance. The intent of the paper is not to compare the requirements for the transport of radioactive material via ship, train and airplanes but to provide insights in the processes regarding significant modifications applied in these three transport areas and the approach for NPPs.

2 SIGNIFICANT MODIFICATIONS IN MARITIME TRANSPORTATION

Modification is inevitable within any type of business and arises from the need to respond and adapt to varying conditions. This is also the case for maritime transportation. Modifications may be required to the equipment, operational policies, and organizational structure or personnel. Whenever a modification is made, the potential consequences of that modification should be assessed before implementation.

If a modification is technically inappropriate, poorly executed, its risks poorly understood, or management fails to ensure communication to key personnel, accidents or other undesired consequences can result. Thus, a formal and effective management of modification program plays a critical role in preventing accidents and losses (ABS 2013). It requires organizational support, assignment of necessary resources, and a clear, defined process. Therefore, guidance notes to the maritime and offshore industries are offered (e. g. ABS 2013) as a tool to aid in the development and implementation of an effective management of modification strategy to optimize existing safety and risk management efforts.

Modifications and conversions of qualifying ships are zero-rated under Group 8, items 1 and 2 of

VTRANS 120200 provided, after modification or conversion, the ship remains a qualifying ship (VTRANS 2010). This includes, for example:
- rebuilding or lengthening,
- updating or improvement of serviceable equipment,
- structural alterations.

It is important to note that this provision requires the ship to be qualifying before the modification work is started. This means that the modification or conversion of a non-qualifying ship is not zero-rated even if the modification or conversion results in a qualifying one. However, after conversion the ship will then be treated in the same way as any other qualifying ship for future supplies.

For example, the conversion of a trawler (gross tonnage of 20.72 tons) to a vessel designed for commercial scientific research would be zero-rated under Group 8, item 1. However, the services of modifying a 14 ton ship to be a 16 ton ship would not be zero-rated as the modification is not of a qualifying ship.

Where a contract to supply modification services across a fleet of ships is being undertaken it is permissible for parts being modified to be removed from one ship, be modified, and then installed in a sister-ship whose parts are similarly destined for another sister-ship after modification.

The interaction between ship repair and ship conversion is also discussed in (Senturk 2011).

There are various different kinds of conversions but no commonly defined definition does exist. Repairs in accordance with approved drawings and documents are not considered to be a conversion. A conversion includes, e.g., any modifications on board of a classed ship which deviates from the approved drawings or an increase of the maximum allowable draft.

Considering the various scope of conversion issues it is to be noted that some modifications may be regarded to be so-called major conversions which is comparable to significant modifications in other transportations systems and in the nuclear field.

The definition of a major conversion is to be distinguished from above definition of a conversion. The definition of a major conversion is individually provided in the applicable statutory instruments (SOLAS, Marpol etc.). However, a major conversion does include but is not limited to each modification which substantially alters the dimensions or carrying capacity or engine power of the ship as all these measures do normally imply new requirements which are to be observed. Major conversions normally do imply complete application of rules effective at the time of conversion. Moreover, any conversion which substantially alters the energy efficiency of the ship and includes any modifications that could cause the ship to exceed the applicable required energy efficiency design index

as set out in regulation 21 is a major conversion (MEPC 2011).

A significant modification of a ship is, for example, the change of the type of vessel, e.g., from a cargo ship to a passenger ship in order to carry more than 12 passengers. In this case, the applicable rules for the whole ship are to be applied as in the case of a new built. A simple replacement of the engine the regulation is not to be applied to the whole ship. The new components must comply with the latest regulations.

Another example would be an extension of a passenger vessel by a new inserted section. Basically, this might be seen as a significant modification; however, it has been agreed that the current rules as for a new built ship is applicable only for the new section.

As already indicated in the maritime transportation the ships are classified. The objective of ship classification is to verify the structural strength and integrity of essential parts of the ship's hull and its appendages, and the reliability and function of the propulsion and steering systems, power generation and those other features and auxiliary systems which have been built into the ship in order to maintain essential services on board (IACS 2011).

The purpose of a classification society is to provide classification and statutory services and assistance to the maritime industry and regulatory bodies as regards maritime safety and pollution prevention, based on the accumulation of maritime knowledge and technology. The different classification organizations have their own rules (e.g. DNV 2013 or DNV-GL 2014) which are not in any case identical.

Any modifications on board of a classed ship which deviates from the approved drawings or cause alterations of previously approved documents are regarded to be a conversion of the ship. Such modifications normally do have effect on the validity of class and in addition also on the statutory certificates issued by the classification society on behalf of the flag State Administration or by the flag State Administration itself. In so far such intended modifications are to be planned well in advance in order to maintain validity of class or validity of the corresponding statutory certificates, or even to ensure the issue of new additional statutory certificates which might be required after conversion.

The tasks and roles of the classification society and flag depend very much on the flag. Usually, the flag accepts the results of the investigation of the classification society, sometimes the flags partly check the modifications themselves.

For example in case of Iceland, no major modifications may be made to a ship, such as enlargement of the cargo spaces or superstructure, replacement of the main engine or modifications which affect the ship's measurements, seaworthiness and stability, safety and/or facilities of the crew, unless approval has been given by the Icelandic Maritime Administration, or another party authorized by the Administration. Modifications shall be carried out under the monitoring of the Maritime Administration, and the same rules apply concerning monitoring and notification, as in the case of the construction of a new ships (IMA 2003).

3 EVALUATION OF MODIFICATIONS REGARDING SIGNIFICANCE IN OTHER TRANSPORT SYSTEMS

3.1 Evaluation of modifications in the European railway system

In the area of European railways the revised CSM regulation 402/2013 (EC 2013) describes a harmonized risk assessment process where only safety relevant and significant changes have to be considered. In order to determine the significance of a change, the proposer has to apply the following six criteria which are described in Article 4 of the CSM regulation:

1 failure consequence: credible worst-case scenario in the event of failure of the system under assessment, taking into account the existence of safety barriers outside the system under assessment;
2 novelty used in implementing the change: this concerns both what is innovative in the railway sector, and what is new for the organization implementing the change;
3 complexity of the change;
4 monitoring: the inability to monitor the implemented change throughout the system life-cycle and intervene appropriately;
5 reversibility: the inability to revert to the system before the change;
6 additionality: assessment of the significance of the change taking into account all recent safety-related changes to the system under assessment and which were not judged to be significant.

While the criterion additionality is used for the consideration of changes in the past which were non-significant and are part of the same area as the proposed change, the regulation itself does not give any further guidance how to apply the five criteria (a) to (e). Due to the problems which exist with the correct application of the given process numerous approaches were developed by the industry, national authorities and also by scientific studies. These concepts and approaches as well as the CSM regulation and its criteria are discussed in detail in (Petrek 2014a). Furthermore, beside the discussion of the relevance of the criteria this document and

also (Petrek 2014b) explain how the application of multivariate statistics can be used to determine the weighting of the criteria and also a threshold above which a change is significant.

The framework for judging the significance of a change is given in Figure 1 and is also developed and explicitly discussed in (Petrek 2014a). Figure 1 shows that the consideration of the safety relevance and the analysis of the significance of the change represent only two elements of the whole process which are located within the light blue box. The whole process also covers an analysis if the change falls within the scope of the regulation. In addition, if the change is assessed as significant by applying the five criteria, the proposer has to consider the associated risk and how it will be managed.

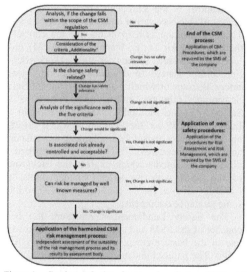

Figure 1. Caption of the framework of the CSM risk management process.

The process of Figure 1 finally classifies the proposed change into one of three different categories. A change which does not fall within the scope of the CSM regulation or which is not safety related only requires the application of QM-Procedures and the harmonized risk management process does not have to be applied. If a change is safety relevant but not significant, the proposer has to apply those own safety procedures which are required by the Safety Management Systems (SMS) of the company. Only for safety relevant and significant changes the proposer has to apply the harmonized risk management process. The main difference between the application of an own safety procedure and the application of the harmonized risk management process is the participation of an independent assessment body within the harmonized process. The independent assessment body checks the suitability of the chosen methods within the risk management process as well as the results. Therefore the judgment of the significance of a change within the European railway system does not only consider the risk-based criteria safety relevance and failure consequence but also the four qualitative criteria innovation, complexity, monitoring and reversibility. And these qualitative criteria consider, if the company already has any experience with the implementation of the change. Thus, a significant change and the need for an independent assessment is more likely, if the company has no or only little experience with the proposed change.

3.2 Evaluation of modifications in European air traffic management

Eurocontrol is the European Organization for the Safety of Air Navigation and was founded in 1960. This organization is responsible for the control of the European Air Traffic Management (ATM) and is located in Brussels. In the year 2004 a workgroup of Eurocontrol started to develop a tool for system safety assessment which was initially named Safety Screening Technique and is now known as the Safety Scanning Method (SSM). It is used in the European ATM for the evaluation which effort is needed in order to safely implement new material and new procedures. The SSM is described in different documents which explain the background of the SSM with its Safety Fundamentals, the EXCEL tool used for the assessment and in addition some guidance documents.

The Safety Fundamentals represent the basic concept of the SSM and such Fundamentals are also one of the basic elements in nuclear safety (IAEA 2006). The Safety Fundamentals are explained in the Eurocontrol document regarding Safety Fundamentals (2011a): "Safety Fundamentals describe basic requirements, which need to be in place at all times and at all lifecycle stages, presenting the objectives, concepts and principles of protection and safety and providing the basis for the safety requirements." The workgroup has identified such Fundamentals within the regulatory framework of different technology sectors like the nuclear and chemical industry as well as the aviation and railway sector. The workgroup has identified 21 of these Fundamentals where the origin of each fundamental is described in (Eurocontrol 2011a). These 21 Fundamentals are structured within the perspectives safety architecture, operational safety and safety management where each perspective consists of five to seven fundamentals. The fourth perspective considers the regulation framework with the existing regulatory principles and the requirements related to them. Those fundamentals are used to analyze a system within a safety analysis while their generic structure allows the analysis of different types of systems. For example the fundamental "Transparency" can either be applied to the interactions within an organization or to the functionality of a technical system.

The assessment of each perspective is done by answering multiple choice questions within the Safety Scanning EXCEL tool which guides the user through the assessment. Within this tool each fundamental possesses one guiding and several subordinate questions. The multiple choice questions possess the three answer options "yes", "no" and "possible" where each answering option is assigned to one certain point value. Additionally, each Fundamental has a weighting itself with regard to the aspect how easy respectively difficult it is to solve any arising problem concerning this Fundamental. If the user exclude an influence of the modification to a certain fundamental by answering the guiding question, there is no further consideration of this fundamental. However, if the answer of the guiding question indicates any influence on the fundamental, the user also has to answer all subordinate questions.

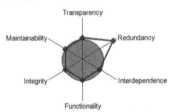

Figure 2. Example of a result diagram for the perspective Safety Architecture within the EXCEL tool of the SSM.

After all questions are answered, the EXCEL tool generates a circular diagram for each perspective which is exemplarily shown in Figure 2. These four diagrams consist of a grey circle where each fundamental of the given perspective represents one axis within this grey circle. The general idea behind the chosen representation of the results is that each of the diagrams shows the effort which is necessary to solve any arising safety-related problem within one of the four perspectives. In this regard dots outside the grey circle indicate a higher effort which is necessary to address arising problems and therefore an influence of the modification on safety whereas dots inside the grey circle allow the conclusion that the given modification or change has no significant impact on this Fundamental.

The application of the SSM tool should be done by groups representing all concerned stakeholders where the moderation of such an assessment is extensively described in the guidance documents of Eurocontrol (2011b).

The explanation of the SSM shows that this approach does not deliver a final result regarding the significance of the proposed modification but allows a deep understanding of this modification and its possible difficulties when it comes to the implementation. Furthermore, the SSM only makes use of qualitative considerations and analyzes the influence on the existing Fundamentals. Although this method is used for the assessment of safety-related modifications, the application of the SSM does not contain risk-based criteria itself.

4 PROCEDURE TO CLASSIFY MODIFICATIONS IN NUCLEAR POWER PLANTS

The nuclear technology is not monitored by European or other international authorities. The licensing and supervision of nuclear installations is perceived by the respective national authorities. However, the International Atomic Energy Agency provides fundamental principles, requirements and guidance with respect to nuclear safety which is not legally binding. One safety guide regarding classification of structures, systems and components (SSCs) has been recently issued (IAEA 2014).

At European level, so-called safety reference levels have been defined by the Western European Nuclear Regulators' Association which were recently updated (WENRA 2014). These reference levels should be adhered to by all member countries of the European Union (EU). In Part G of these reference levels, the safety classification of SSCs is described. The goal is to identify all safety related SSC and to classify them according to their importance for safety. Part Q reflects modifications to a nuclear power plant (NPP) and it is requested that no modification degrades the plant's ability to operate safely. National regulations and international agreements are supplemented by a directive of the European Commission (2009b) which the EU members have to be implement into the national law.

The licensing and supervision of nuclear installations in Germany is the responsibility of the Federal States (Länder) who are subject of expediency supervision of the Federal Ministry for the Environment, Nature Conservation, Building and Nuclear Safety (BMUB). In addition to the German Atomic Energy Act (Atomgesetz – AtG) (2013), there exist several national requirements which have to be considered in the approval and supervision process of nuclear facilities. A comprehensive update of the German sub-legal nuclear regulations has recently been issued (BMU 2013).

Basically, each NPP in Germany must have a valid operating license. An essential part of this approval is the condition of the AtG to show all planned modifications (plant, operation and organization) to the competent authority and to examine their safety relevance. Significant modifications of systems according to § 7 AtG are subject to approval by the supervisory authority. The implementation of authorizations below this level is subject to a graded supervisory control depending on their safety significance.

The approach in nuclear technology is that all important equipment of a NPP are classified in terms of their safety significance, accompanied by respective requirements and specifications. The procedural rules for the treatment of modifications are regulated in operating manuals which are specific for each NPP and part of the approval of the authority to start/continue the operation of the NPP.

In the following, the procedure to evaluate modifications in the Federal State of Baden-Württemberg is exemplarily described because this Federal State has developed a concept for regulatory supervision of NPPs - the last version is issued last year (UM BW 2013) - and a supervisory manual (UM BW 2011) with a separate detailed chapter describing the regulatory plant modification procedure (LEÄV) anticipating the safety requirements for NPPs (BMU (2013). In this manual of Baden-Württemberg modifications are subdivided into three categories, designated as category A, B or C where A contains the highest requirements.

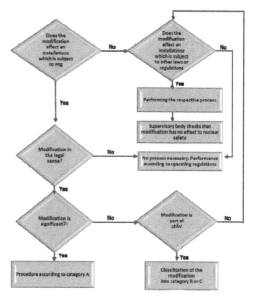

Figure 3. Procedure of classifying modifications of equipment in NPPs.

Figure 3 shows the procedure how to classify modifications. In a first step, it must be considered whether the proposed modification relates to equipment which is subject to the AtG. In that case it must be examined whether this modification is a

287

modification in the legal sense. If yes, one has to consider whether the modification is significant, i.e. the modification can have more than obviously insignificant effects regarding the safety level of the plant. In this case, the modification falls into category A.

Otherwise, it must be checked whether the modification is subject to the uniform modification process. In this case, the modification falls into the category B or category C. The decision regarding category B/C depends on the type of modification: is it only an exchange by equivalent equipment (components or systems) or leads the modification to a deviation in the approved specification. In the first case it would be a modification of category C, in the second one a modification of category B.

In the following it is explained which procedure depending on the classification of the modification in one of the three categories is required according to (UM BW 2011).

Category A: If the modification belongs to category A, an authorization procedure is required by § 7 of the AtG and pursuant to § 7 paragraph a permit application has to be submitted by the licensee to the supervisory body. The licensing procedure is conducted in line with the Nuclear Licensing Procedure Ordinance (2006). In addition, a probabilistic assessment is required to check the influence of the modification on the probabilistic safety analysis (PSA). It must be explained why the proposed modification has no effect on the PSA.

Category B: If the modification belongs to category B, it must be supervised by the authority under § 19 of the AtG. A notification of the planned modification must be submitted by the licensee to the supervisory authority. The implementation of the modification can only take place if the supervisory authority has provided its written supervisory opinion that the proposed modification is seen from the perspective of the supervisor as "without any concern or not subject to licensing." The supervisory authority and the technical support organisation monitor and accompany the implementation of the planned modifications. In addition, a probabilistic assessment is required, as in case of a modification of category A, showing the influence of the modification on the PSA. It must be explained why the proposed modification has no effect on the PSA.

Category C: If the modification belongs to category C, it must be monitored by the authority under § 19 of the AtG. An examination of the modifications in category C by an expert according to § 20 of the AtG is required. The implementation of the modification can only start after the expert has finalised his supervisory report and the authority does not make any objections until the modification starts. It is, however, no probabilistic evaluation required as in categories A and B.

Modifications that are not subject to the procedure described above are performed by the operator according to its internal written operating procedures without participation of the authorities and experts. The supervisory authority will perform random checks on the basis of the documentation provided by the operator, if he has correctly classified the modifications.

The importance of a modification has to be assessed by comparing the actual plant state with the state after having performed the modification. The decisive factor is whether and to what extent the modification (e.g. of SSCs) has safety relevance.

5 COMPARING THE APPLIED PROCEDURES IN THE DIFFERENT AREAS

At first the comparison of all four approaches shows that only the area of European railways has a common method which is valid for all changes within the European railway system. In contrast, modifications in nuclear power plants as also modifications in maritime transportation are not controlled by such a European regulation like the CSM, whereas the European legislator has shown some efforts to create a common framework in the area of nuclear safety as a reaction to accidents at the NPPs of Fukushima-Daiichi in March 2011. For the area of nuclear safety the licensing and supervision falls within the national authority. Concerning modifications in maritime transportation, here exist international guidance notes from the ABS, the Maritime Environment Protection Committee (MEPC) and also from the International Association of Classification Societies (IACS). So these three organizations have a comparable role to the IAEA in the area of maritime safety by providing guidelines and creating general rules. Moreover, the classification societies are an essential part within the assessment of modifications in maritime transportation and, in addition, the characteristic element in the area of maritime safety. However, the responsibilities of the internationally active classification societies and the applicable requirements regarding modifications depend on the flag of the ship. Regarding changes in European ATM, the SSM used in this area is no method which is required by the European legislator. This method was developed by Eurocontrol itself and is applied prior to proposed safety relevant modifications in order to determine all relevant aspects concerning the given modification, while the supervision of the air navigation services falls within the responsibility of the national supervisory authority (see EC 2005).

The comparison of the four methods themselves shows that each approach has an explicit or implicit definition which modifications are relevant and have to be considered. Within the CSM approach the proposer has to check, if the proposed change falls within the scope of the regulation and in the ongoing if the change is safety relevant. In case of the LEÄV of the nuclear sector, the proposer has to check if the modification affects any installation which is subject to the AtG and in the next step if this change is a modification in the legal sense. Regarding modifications in maritime transportation, there does not exist any common definition of the term modification. However, the remarks in Section 2 illustrate that relevant technical modifications are described by the term conversion. This term covers all modifications on board of a classed ship which deviate from approved drawings or represent an increase of the maximum allowable draft. In this respect, repairs in accordance with approved drawings and documents are no conversions. This approach resembles the respective aspect of the CSM, where the application of existing guidelines or the simple substitution of technical components is no relevant change for the CSM regulation and the risk management process. The SSM of the European ATM does not contain any precondition concerning relevant modifications, however, the guidance document states that the method shall be applied for proposed changes which could possibly have any effect on safety.

Basis of the judgment of any modification within the LEÄV are deterministic and risk-based considerations where qualitative criteria do not have any relevance (Petrek & Berg 2014). In contrast, the characteristic element of the CSM approach are the qualitative criteria which are used in combination with a risk-based consideration. Regarding the consideration of modifications in maritime transportation, the experience of the company either has no relevance for the judgment of the modification. In this area, the distinction between a major and normal conversion is done by the consideration if any substantial aspects of the ship are altered by the proposed modification. Within the SSM, the consideration of modifications is done by evaluating the influence of the modification on each of the 21 Safety Fundamentals. This evaluation contains qualitative aspects of the change including the experience of the company with the given modification (Petrek 2014). In contrast to the other three approaches, the SSM does not include any risk-based consideration and, in this respect, does not provide a final result concerning the significance of the modification unlike the other three approaches. The CSM approach distinguish between changes which are safety relevant but not significant and significant changes. The application of the LEÄV provides one out of three different classifications for the given modification determining the further process. Modifications of ships are distinguished in two groups: modifications which are conversions and others. Conversions in turn are either normal or major conversions where the applicable regulations and further processes depend on the classification of the modification in these three groups.

Regarding the possible maneuvers for the assessment and the implementation of the proposed modification, the procedure of the Federal State Baden-Württemberg for the classification of equipment in NPPs is much more restrictive and also more prescriptive in comparison to the CSM regulation of European railways (Petrek & Berg 2014). All three possible criteria of modifications in the nuclear field require an authorization procedure or a supervision by the authority, while in the field of European railways only safety relevant and significant modifications require an independent assessment of the chosen risk management process and its results. In the area of maritime transportation the level of participation of the authority depends on the flag of the ship. Normally the flag accepts the results of the investigations of a proposed modification performed by the classification society, but there are also some flags where no major modifications are permitted without the approval of the responsible Administration. Regarding modifications within the European ATM, the SSM does not provide a final classification of the given modification. Consequently a statement concerning the restrictiveness of the method and the possible maneuvers for the proposer is not possible because the further proceedings do not result from the application of the SSM itself. However, the SSM allows the proposer understand the possible consequences of the modification and which effort is needed to safely implement it. In the ongoing, the proposer of the modification respectively the affected stakeholders of the company have to discuss the modification with theirs national supervisory authority on the basis of the SSM results.

6 CONCLUDING REMARKS

The comparison in the previous Section 5 has illustrated the fundamental differences between the approaches and regulations of the four different technology sectors. Firstly, this affects legislation and supervision, e. g. if national or European institutions are responsible. For both, the European ATM and the European railway system, the applicable requirements directly result from European regulations, whereas licensing/supervision is the duty of national authorities. Within the European railway system, the proposer of a significant change has to consult an independent assessment body and the licensing and supervision is within the responsibility of the national safety agency. For the European ATM, the supervision of modifications falls within the duty of the national supervisory authority (EC 2005).

In contrast, the nuclear technology is subject to national legislation and the respective acts like the German AtG. The licensing and supervision is still task of national authorities (in the case of Germany by the federal states). In the maritime sector the applicable requirements depend on the flag of the ship and any European regulations or requirements do not exist. It also depends on the flag, if in the case of an intended modification the results of the investigation done by the classification society are accepted or if the flag and its authorities check the modifications and the results of the investigation.

There are also fundamental differences regarding the considerations within the approaches. The basis of the LEÄV are deterministic and risk-based considerations, for instance, if any possible influence by the modification on the existing PSA can be excluded. Qualitative aspects of the modification are not relevant within this approach. A risk-based consideration is also part of the CSM approach whereas also the qualitative aspect, the experience of the organization with the implementation of the change, is an essential part of this approach. Therefore, the classification of the change depends not only on the change itself but also on the organization and its experience. In contrast, the assessment of modifications in maritime transportation does not contain such qualitative considerations. The final judgment depends on the assessment if any substantial aspects of the ship are altered and therefore on the basis of somehow risk-based consideration. But this considerations are less deterministic and straightforward in comparison to the considerations within the LEÄV which possesses clear classifications of modifications on the basis of their type into the three different categories. In maritime transportation, the requirements are less prescriptive and therefore the judgment of modifications allows more room for maneuvers. Unlike the other three approaches, the SSM does not include any risk-based consideration and represents an approach which is only qualitative.

The partially significant differences within the management of modifications and changes between the different technology sectors raises the question if a harmonization of the proceedings and approaches could be reasonable. Additionally, this leads to the central and general question, if the differences described and analyzed within this paper are acceptable against the background of similar tools and fundamentals used for the demonstration of safety within the different technology sectors.

REFERENCES

American Bureau of Shipping (ABS). 2013. *Guidance notes on management of change for the marine and offshore industries*, Houston, Texas, USA, February 2013.

Act on the peaceful utilisation of nuclear energy and the protection against its hazards (Atomic Energy Act, Atomgesetz – AtG), 23 December 1959, promulgated on 15 July 1985, last amendment of August 28, 2013.

Det Norske Veritas (DNV). 2013. *Conversion of ships*, April 2013.

DNV-GL. 2014. *Rules for classification*, DNVGL-RU-0050, General Regulations, Edition October 2014.

Eurocontrol. 2011a. *Safety Fundamentals for Safety Scanning*, SRC Document 46 Annex A, June 2011.

Eurocontrol. 2011b. *Guidance for Moderating a Safety Scanning Event*, SRC Document 46 Annex B, June 2011.

European Commission (EC). 2005. *Commission regulation (EC) NO 2096/2005 of 20 December 2005 laying down common requirements for the provision of air navigation services*, December 2005.

European Commission (EC). 2009a. *Commission regulation (EC) NO 352/2009 of 24 April 2009 on the adoption of a common safety method on risk evaluation and assessment as referred to in Article 6(3) (a) of Directive 2004/49/EC of European Parliament and of the Council*, April 2009.

European Commission (EC). 2009b. *Directive 2009/71/EURATOM establishing a community framework for the nuclear safety of nuclear installations*, June 25, 2009.

European Commission (EC). 2013. *Commission Implementing regulation (EU) No 402/2013 of 30 April 2013 on the common safety method for risk evaluation and assessment and repealing Regulation (EC) No 352/2009*, April 2013.

Federal Ministry for the Environment, Nature Conservation and Nuclear Safety (BMU). 2013. *Safety requirements for nuclear power plants of 22 November 2012*, Federal Gazette, AT January 24, 2013 B3.

International Association of Classification Societies (IACS). 2011. *Classification societies – what, why and how?* June 2011.

Icelandic Maritime Administration (IMA). 2003. *Ship Survey Act No 47/2003*.

International Atomic Energy Agency. 2006. *Fundamental Safety Principles*. Safety Standards No. SF – 1, 2006.

International Atomic Energy Agency. 2014. *Safety classification of structures systems and components in nuclear power plants*. Specific Safety Guide No. SSG-30, IAEA, Vienna, May 2014.

Maritime Environment Protection Committee (MEPC). 2011. *Resolution MEPC.203 (62), MEPC 62/24/Add., Annex 19*, July 2011.

Ministry of the Environment, Climate Protection and the Energy Sector Baden-Württemberg (UM BW). 2011. *Regulatory Plant Change Procedure (Landeseinheitliches Änderungsverfahren – LEÄV)*, September 2011.

Ministry of the Environment, Climate Protection and the Energy Sector Baden-Württemberg (UM BW). 2013. *Concept for regulatory supervision of nuclear power plants in Baden-Württemberg*. June 2013.

Ordinance on the Procedure for Licensing of Installations under §7 of the Atomic Energy Act (Nuclear Licensing Procedure Ordinance - Atomrechtliche Verfahrensverordnung - AtVfV) of 18 February 1977, promulgated on 3 February 1995, last amendment of 9 December 2006.

Petrek, N. 2014a. Konstruktion eines Verfahrens zur Signifikanzbewertung von Änderungen im europäischen Eisenbahnwesen. *Ph.D.-Thesis of the Department of*

Architecture, Civil Engineering and Environmental Science of the Technische Universität Braunschweig, August 2014.

Petrek, N. 2014b. A New Approach for Judging the Significance of Changes in European Railways. *FORMS/ FORMAT 2014 – 10ᵗʰ Symposium on Formal Methods*, September 2014.

Petrek, N. & Berg, H. P. 2014. *Comparing the two methods for judging changes in European railways and in European nuclear safety. In* Nowakowski et al. (eds), Safety and Reliability: Methodology and Applications: 1649 – 1654. London: Taylor & Francis Group.

Senturk, Ö. U. 2011. The interaction between the ship repair, ship conversion and shipbuilding industries, *OECD Journal:General Papers*, Vol. 2010/3, August 2011.

VTRANS120200. 2010. Repair, maintenance, modification and conversion of ships and aircraft and their parts: Repair, maintenance, modifications and conversions, Draft November 2010.

Western European Nuclear Regulators' Association (WENRA). 2014. Report WENRA safety reference levels for existing reactors, update in relation to lessons learned from TEPCO Fukushima-Daiichi accident, 24th September 2014.

Safety of Transport
Safety of Marine Transport – Marine Navigation and Safety of Sea Transportation – A. Weintrit & T. Neumann (eds.)

Optimization of the Transport Service of Fishing Vessels at Ocean Fishing Grounds

S.S. Moyseenko & L.E. Meyler
Baltic Fishing Fleet State Academy of the Kaliningrad State Technical University, Kaliningrad, Russia

ABSTRACT: Improving the efficiency of the fishing fleet through the use of methods and models for optimization of transport services is the relevant task. The transportation service consist of the delivery on the fishery technical and technological supplies, food, products for crews' life support, fuel, fresh water, as well as unloading of finished products from fishing vessels and its transportation to the ports of destination, etc. A project-based approach and methods of the operations research are used as the methodological basis for designing and planning the complex of the fleet service. The paper presents mathematical and heuristic methods which allow to optimize the design solutions and plans of the fishing vessels service at the fishing grounds. According to experts estimations the use of the project-based approach in the management allows to improve significantly the quality of transport services of the fishing fleet as well as the efficiency of fishery and safety.

1 INTRODUCTION

1.1 *The statement of the problem*

The efficiency of the fishing fleet operation at fishing grounds is largely depend on the level of transport service organization. In this regard, improving the efficiency of the fleet through the use of methods and models of the transport service optimization is the relevant task (Moyseenko 2009). Increasing the time when vessels operate on fishery and the probable volume of their catch depends on the speed of handling (servicing) the vessels at sea. In this regard, it is important to define an order of the vessels approach the servicing vessel (a floating fish factory or a transport vessel). The main condition is to minimize the total handling time of the group of fishing vessels as well as the downtime of the servicing vessel. A methodological basis for designing and planning the fishing fleet servicing are the systematic approach and methods of operations research.

2 A TASK OF FISHING VESSELS TRANSPORT SERVICING

2.1 *Conditions of vessels relocation*

A group of fishing vessels operates at a fishing ground. Their fish products or fish raw materials have to be discharged on the servicing vessel: the refrigerated fish transport or the fish carrier (further – the transport vessel). Vessels operate at several fishing squares within the fishing ground and have different distances to the transport vessel.

Let us suppose that vessels are located at the fishing square in a certain sequence: 1, 2, 3, ... N. Each fishing vessel with its catch follows to the transport vessel for handling. After that the vessel returns to the fishing ground to continue fishery. Schematically the whole cycle performed by the vessel is shown in Figure 1.

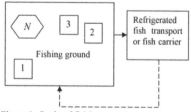

Figure 1. Cycle of fishing vessels servicing

It is necessary to perform the above-described conditions for minimizing the handling time of fishing vessels.

Let t_i = the time of the i-th fishing vessel trip to the transport vessel; τ = the handling time of the i-th vessel; and x_L = the downtime of the transport vessel waiting the approach of the next fishing vessel for handling.

There are the following recurrence relations:

$$
\left.
\begin{aligned}
&X_1 = t_1 \\
&X_2 = \max\left(t_1 + t_2 - \tau_1 - X_1, 0\right) \\
&X_1 + X_2 = \max\left(t_1 + t_2 - \tau_1, t_1\right) \\
&X_3 = \max\left(\sum_{i=1}^{3} t_i - \sum_{i=1}^{2} \tau_i - \sum_{i=1}^{2} X_i, 0\right) \\
&X_1 + X_2 + X_3 = \max\left(\sum_{i=1}^{3} t_i - \sum_{i=1}^{2} \tau_i, \sum_{i=1}^{2} t_i - \tau_1, t_1\right)
\end{aligned}
\right\} \quad (1)
$$

and then by an induction:

$$
\left.
\begin{aligned}
&\sum_{i=1}^{N} X_i = \max, K_n \\
&K_n = \sum_{i=1}^{n} i - \sum_{i=1}^{n-1} \tau_i
\end{aligned}
\right\} \quad (2)
$$

$$1 \le n \le N$$

It is required to find such relocation of fishing vessels realization of which would minimize the expression (2). The algorithm for solving this problem was developed by the dynamic programming method (Ackoff & Sasieni 1971, Bellman & Kalaba 1969].

2.2 Solution of the problem by the dynamic programming method

Let us denote:

$$\varphi(t_1, \tau_1, t_2, \tau_2, \ldots, t_n, \tau_n, T)$$

as the time required to make a full cycle for the N number of fishing vessels. There is the condition: the handling process starts at the time T after the i-th fishing vessel begins the trip to the transport vessel. Here t_i is the trip time, τ_i is the handling time of the i-th vessel at the optimum relocation. Pairs of the vessels are denoted by symbols ij in order to identify their relocations. The functional equation (3) can be obtained if the i-th vessel makes the first trip to the transport vessel for handling:

$$
\phi(t_1, \tau_1, t_2, \tau_2, \ldots, t_n, \tau_n, T) =
$$
$$
= \min\left[
\begin{aligned}
&t_i + \phi(t_1, \tau_1, t_2, \tau_2, \ldots, 0, 0, \ldots, t_N, \tau_N, \tau_i + \\
&+ \max(T - t_i, 0))
\end{aligned}
\right] \quad (3)
$$

The pair (0, 0) replaces the pair (t_i, τ_i) in the expression (3). The optimal variant of the relocation is obtained from the expression (3) by replacing two

vessels, i.e. the i-th vessel follows first to the transport vessel and then it makes the j-th vessel.

$$
\begin{aligned}
&\phi(t_1, \tau_1, t_2, \tau_2, \ldots, t_T, \tau_T, T) = t_i + t_j + \\
&+ \phi(t_1, \tau_1, 0, 0, \ldots, t_N, \tau_N, T_{ij})
\end{aligned} \quad (4)
$$

where

$$
\begin{aligned}
T_{ij} &= \tau_j + \max\left[\tau_i + \max(T - t_i, 0) - t_j, 0\right] = \\
&= \tau_j + \tau_i - t_j + \max\left[\max(T - t_i, 0), t_j - \tau_i\right] = \tau_j + \\
&+ \tau_i - t_j + \max\left[T - t_i, t_j - \tau_i, 0\right] = \tau_j + \tau_i - \\
&- t_i - t_j + \max\left[T, t_i + t_j - \tau_i, t_i\right] = \tau_j + \tau_i - t_j - \\
&- t_i + \max\left[T, \max(t_i + t_j - \tau_i, t_i)\right]
\end{aligned} \quad (5)
$$

It is seen from the expression (5) that:

$$\max(t_i + t_j - \tau_i, t_i) < \max(t_i + t_j - \tau_j, t_j) \quad (6)$$

and it makes sense to relocate the i-th and the j-th vessels. Their relocation is reasonable when:

$$\min(\tau_i, t_j) > \min(\tau_j, t_i) \quad (7)$$

2.3 An algorithm for determining the optimal vessels' relocation

The obtained results described by expressions 6 and 7 allow to determine the optimum relocation of the fishing vessels using the following algorithm:
1. to obtain the information about the state of fishing vessels and to fill Table 1;
2. to determine the list of fishing vessels to be handled by the transport vessel;
3. to define values of parameters t_i and τ_i according to data in Table 1 and put them in Table 2;

Table 1. Data on the location and catches of fishing vessels

Parameters		Data			
Hull number of the vessel	N	N_1	N_2	...	N_n
Type of the vessel		A	B	...	...
Coordinates of the vessel	Lat. φ_i	φ_1	φ_2	...	φ_n
	Long. λ_i	λ_1	λ_2	...	λ_n
Distance to the transport vessel S_n	S_i, miles	S_1	S_2	...	
Speed of the vessel	V_i, knots	V_1	V_2	...	V_n
Catch	Q_i, tons	Q_1	Q_2	...	Q_n

4. to find the least among the values of t_i and τ_i;
5. if the least value would be one of the value t_i, this vessel begins first to make the trip to the transport vessel;
6. if the least value would be one of the value τ_i, the transport vessel handles this vessel the last;
7. to delete values of t_i and τ_i in Table 2;
8. to repeat this process with the (2n -2) remaining values;

Table 2. Values of fishing vessels parameters

Parameters		Data			
Hull number of the vessel	N_i	N_1	N_2	...	N_n
i		1	2	...	N
t_i		T_1	T_2	...	T_n
τ_i		τ_1	τ_2	...	τ_n

9 to select the vessel with the lower number of priority if there are the several minimum values; if $t_i = \tau_i$ to put vessels in order on the value of t_i;

10 to calculate the schedule of vessels handling using the form below (Table 3).

Table 3. Form of the schedule of fishing vessels handling by the transport vessel

Parameters			Data			
№ of the order			1	2	...	n
Hull number of the vessel	N_i		N_1	N_2	...	N_n
Catch	Q_i, tons		Q_1	Q_2	...	Q_n
Period of working time	Start T_{si}		T_{s1}	T_{s2}	...	T_{sn}
	Finish T_{fi}		T_{f1}	T_{f2}	...	T_{fn}
Working time	τ_i, hours		τ_1	τ_2	...	τ_n

2.4 A practical example of the logistic solution for planning the transport service of fishing vessels

Let us assume that the transport vessel operates at sea with six fishing vessels. The fishing conditions are relatively stable. The state of the vessels group is determined at the beginning of the operating period by the parameters given in Table 4.

Table 4. Dislocation of fishing vessels at the fishing ground

Parameters		Data					
Hull number of the vessel	N	1	2	3	4	5	6
Type of the vessel		A	B	C	D	E	F
Coordinates of the vessel	Lat. φ_i	φ_1	φ_2	φ_3	φ_4	φ_5	φ_6
	Long. λ_i	λ_1	λ_2	λ_3	λ_4	λ_5	λ_6
Distance to the transport vessel	S_i, miles	7	10	5	12	5	6
Speed of the vessel	V_i, knots	7	9	10	10	7	10
Catch	Q, tons	5	10	7	15	8	12

Parameters t_i and τ_i for each fishing vessel are defined and Table 5 can be filled.

It is possible to obtain the optimal order of the fishing vessels approach for handling by the transport vessel following further to the above algorithm (Table 6).

Table 5. Stepwise solution of the problem of determining the order of fishing vessels handling by the transport vessel

i	t_i, hours	τ_i, hours
1	1.0	0.7
2	1.1	1.0
3	0.5	0.9
4	1.2	1.1
5	0.7	1.2
6	0.6	1.0
On the first step		
3	0.5	0.9
1	1.0	0.7
2	1.1	1.0
4	1.2	1.1
5	0.7	1.2
6	0.6	1.0
On the second step		
3	0.5	0.9
6	0.6	1.0
1	1.0	0.7
2	1.1	1.0
4	1.2	1.1
5	0.7	1.2

Table 6. The optimum order of the fishing vessels approach to the transport vessel

i	t_i	τ_i
3	0.5	0.9
6	0.6	1.0
5	0.7	1.2
4	1.2	1.1
2	1.1	1.0
1	1.0	0.7

Then the schedule of vessels handling is calculated (Table 7).

Table 7. The schedule of vessels handling

Parameters		Data					
№ of the order		1	2	3	4	5	6
Hull number of the vessel, N_i		3	6	5	4	2	1
Catch Q_i [tons],		7	12	8	15	10	5
Period of handling time,							
	Start T_{si}	0^{00}	0^{54}	1^{54}	3^{06}	4^{12}	5^{12}
	Finish T_{fi}	0^{54}	1^{54}	3^{06}	4^{12}	5^{12}	5^{54}
Handling time τ_i, hours		0.9	1.0	1.0	1.1	1.0	0.7

An analysis of the solution sensitivity allows to determine the limits of the parameters variation. The obtained strategy (order) of fishing vessels handling by the transport vessel remains optimal in these limits (Moyseenko et al. 2011). Thus, in the above example an increase of the catch of 20% does not change the order but the handling schedule requires adjustment on time.

3 CONCLUSION

In conclusion, it is necessary to note that the algorithm for determining the optimal handling of fishing vessels was tested in practice. The efficiency of the proposed method was verified by comparing the decisions obtained by the traditional method and based on the use of the algorithm for solving the problem by dynamic programming.

Thus, the result of 20-th realizations (under approximately the same conditions of the fleet operation), shows that the use of the optimization algorithm of the fishing vessels priority service allows to minimize by 15-20% the downtime of the transport vessel and the loss of fishing time by fishing vessels (Moyseenko & Meyler 2011).

The fishing process has a probabilistic character; however the proposed method allows to find approximate strategies even in more complex models.

REFERENCES

Ackoff R.L.& Sasieni M.W. 1971. Fundamentals of Operation Research. Moscow: Mir (in Russian)

Bellman P. & Kalaba P. 1969. Dynamic programming and modern control theory. Moscow: Nauka (in Russian)

Moyseenko S.S. 2009. Management of the fleet operation. Kaliningrad: BFFSA Publ. House (in Russian)

Moyseenko S.S. & Meyler L.E. 2011. Optimal management of fleet relocation at deep-sea fishing grounds. Green ships, Eco shipping, Clean seas; Proc. 12-th Annual General Assembly of the International Association of Maritime Universities (IAMU) AGA12. Gdynia, June 2011.

Moyseenko S.S., Meyler L.E. & Semenkov V.N. 2011. Organization of Fishing Fleet Transport Service at Ocean Fishing Grounds. Maritime logistics in the global economy. Current trend and approaches; Proc. Hamburg intern. conf. on logistics (HICL 2011). Hamburg, 10-12 September 2011. EUL Verlag GmbH

Safety of Transport
Safety of Marine Transport – Marine Navigation and Safety of Sea Transportation – A. Weintrit & T. Neumann (eds.)

Selected Transport Problems of Dangerous Goods in the European Union and Poland

G. Nowacki
Military University of Technology, Warsaw, Poland

C. Krysiuk & A. Niedzicka
Motor Transport Institute, Warsaw, Poland

ABSTRACT: The paper refers to threat assessment of dangerous goods (DG) in transportation of the European Union and the Republic of Poland. Dangerous goods in the European Union are carried by inland waterways, rail and road. In Poland 88% of DG have been carried by road and 12% by rail. DG can cause an accident and lead to fires, explosions and chemical poisoning or burning with considerable harm to people and the environment. There is not monitoring system in Poland to control in real time road transportation of dangerous goods. Proposition of National System of Monitoring Dangerous Goods in Poland was presented. Realization of mentioned system may significantly contribute to improving safety of people and environment.

1 INTRODUCTION

Dangerous goods (DG) have known in more commonly as hazardous materials, (abbreviated as HazMat). Dangerous goods include materials there are flammable, explosive, radioactive, corrosive, oxidizing, asphyxiating, toxic, pathogenic, or allergic.

Dangerous goods transported by road can cause accidents and lead to fires, explosions and chemical poisoning or burning with considerable harm to people and the environment, not only in Poland but in other countries in European Union and all over the world.

Poland ratified the European Agreement concerning the International Carriage of Dangerous Goods by Road - ADR in 1975. Even, if the ADR agreement is ratified, unfortunately there are still problems in the transport sector, especially problems regarding the transport of dangerous goods, that offers extensive deficiency at streets. The most numerous group of dangerous goods include items of class 3 (liquid, flammable materials), especially liquid fuels.

The European Union has passed numerous directives and regulations to avoid the dissemination and restrict the usage of hazardous substances, important ones being the restriction of Hazardous Substances Directive. There are also long-standing European treaties, that regulate the transportation of hazardous materials by road (ADR, 2013), rail (RID, 2013), river and inland waterways (AND, 2013).

There is not monitoring system to control in real time dangerous goods vehicle in Poland. The aim to develop a cooperative system for dangerous goods vehicles (DGV) through route monitoring, re-routing (in case of need) enforcement and driver support, based upon dynamic, real time data, in order to minimize threats related to movements of DGV.

2 CHARACTERIZATION OF DANGEROUS GOODS

2.1 General characterization

Hazardous material is a material or object which, is not be accepted for carriage by road, or is approved for such carriage under the conditions laid down in those provisions (ADR, 2013; RID, 2013; AND, 2013). There are nine classes of dangerous goods as follows:

- Class 1. Explosives (1.1. Substances and articles (SA) which have a mass explosion hazard, 1.2. SA which have a projection hazard but not a mass explosion hazard, 1.3. SA which have a fire hazard and either a minor blast hazard or a minor projection hazard or both, but not a mass explosion hazard, 1.4. SA which present only a slight risk of explosion in the event of ignition or initiation during carriage, 1.5. Very insensitive substances having a mass explosion hazard which are so insensitive that there is very little probability of initiation or of transition from

burning to detonation under normal conditions of carriage, 1.6. Extremely insensitive articles which do not have a mass explosion hazard);
- Class 2. Gases (2.1. Flammable gas, 2.2. Non flammable, non Toxic gas, 2.3. Toxic gas);
- Class 3. Flammable liquid;
- Class 4. Flammable solids (4.1. Flammable solid, 4.2. Spontaneously combustible substance, 4.3. Substance which emits flammable gas in contact with water);
- Class 5.1. Oxidizing substances (5.1. Oxidising substance, 5.2. Organic peroxide);
- Class 6. Toxic substances (6.1. Toxic substance, 6.2. Infectious substance);
- Class 7. Radioactive material;
- Class 8. Corrosive substances;
- Class 9. Miscellaneous dangerous substances and articles.

2.2 Percentage share of dangerous goods delivery

The transport of dangerous goods in the EU-28 slightly increased from 78 billion tonne-kilometres in 2009 to 80 billion tonne-kilometres in 2012, but has decreased by 3.9% in 2013 and was just over 77 billion tonne-kilometres.

Between 2009 and 2013, most Member States have observed a fall in the transport of dangerous goods. The highest fall was recorded in Greece (-62%), followed by the Netherlands (-44%) and Portugal (-34%). On the other side, very high increases of transport of dangerous goods were registered in countries like Estonia (+99%), Luxembourg (+95%) and the United Kingdom (+72%)

Figure 1 shows the types of dangerous goods in road transport in 2013 (the first group was flammable liquids - 58.2% of the total, the second gases – 13.6% and third corrosives – 10.3%.

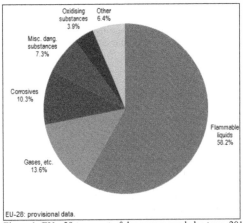

Figure 1. EU - 28 transport of dangerous goods by type, 2013 (% in tonne-kilometres), http://ec.europa.eu/eurostat/statistics, 2013

This represents very little change compared with previous years showing a very similar distribution between product groups.

The transport of dangerous goods by road, rail or inland waterway (within or between Member States) presents a considerable risk of accidents. Measures should therefore be taken to ensure that such transport is carried out under the best possible conditions of safety (Directive 2008/68/EC).

The South Eastern Europe (SEE) area is a sea and river transit space of vessels carrying hazardous freight which constitutes many potential environmental risks for coasts and inland waterways.

Economic development and a strong growth of transport and increased traffic in the SEE area aggravate the already increased threats of pollution and thus require a good management and high performance of observation, communication and monitoring response systems (figure 2).

Figure 2. South East Europe marine and river information systems, http://www.seemariner.eu

Therefore the SEE MARINER (South Eastern Europe Marine and River Integrated System for Monitoring and Transportation of Dangerous Goods) project was developed (Squillante, 2013) from February 2011 to December 2013 (overall project budget: 2.188.000,00 €).

It was focusing on mitigating environmental risks arising from the transportation of dangerous goods in marine areas and rivers by applying an integrated system for the joint prevention and response procedures, enhanced monitoring of maritime and river traffic and increased coordination capacity for the mobilization of the relevant authorities and stakeholder groups. SEE Mariner project results:
- Improved coordination, harmonization and availability of data on the transportation of dangerous cargoes;
- Enhanced managerial skills and equipment for handling dangerous cargoes;

- Developed and tested common management structures and tools for the monitoring of dangerous goods transportation;
- Streamlined procedures and protocols for emergency situations or disasters caused by the transportation of dangerous goods.

According to statistics (Eurostat) dangerous goods in European Union in 2004 were carried by transport in three manners as follows (see figure 3):
- by Inland Waterways – 6.8% (ADN),
- by Rail – 27.4% (RID),
- by Road – 65.8 % (ADR).

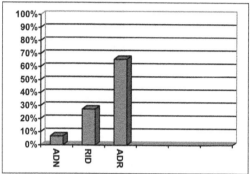

Figure 3. Transport of Dangerous Goods in European Union, http://ec.europa.eu/eurostat/statistics, 2005

According to Commission Recommendation (C, 2011) Member States should follow the guidelines set out in the Annex when completing the annual report on checks concerning the transport of dangerous goods by road as follows:
- Number of transport units checked on the basis of the contents of the load (and ADR),
- Number of transport units not conforming to ADR Number of transport units immobilised,
- Number of infringements noted, according to risk category (Risk category I, II, III);
- Number of penalties imposed, according to penalty type (Caution, Fine, Other).

In Poland, according to statistics (CSO, 2014), 87.2% of dangerous goods are carried by road transport and only 12,8% by rail in 2013.

Table 1. Percentage share of the dangerous goods transported (Grzegorczyk & others, 2007, 2011; Rożycki, 2013)

Class	Share	Percentage
Class 1	Explosive substances and articles	0.95%
Class 2 25.17%	Gases, including compressed, liquefied and vapours	
Class 3	Flammable liquids	66.19%
Class 4.1	Flammable solids, self-reactive substances	1.50%
Class 8	Corrosive substances	1.62%
Class 9	Miscellaneous dangerous substances and articles	2.93%
Rest	Class 4.2, 4.3, 5.1, 5.2, 6.1, 6.2, 7	1,64%

Table 2. Percentage share arranged according to the form of transport (Grzegorczyk & others, 2007, 2011; Różycki, 2013)

Transport means	Percentage
Tankers	79%
Containers	20%
Goods in bulks	1%

According to data provided by M. Różycki and P. Grzegorczyk the largest amounts transported concern liquid materials (table 1, 2).

To make a right choice of packaging for the transport of dangerous goods, some materials classified as hazardous classes ADR, may be included in the so-called packing groups according to the degree of danger they present.

In most cases, the degree of risk is assessed on a three-stage scale:
- Packing group I: Substances presenting high danger,
- Packing group II: Substances presenting medium danger,
- Packing group III: Substances presenting low danger.

3 ORGANIZATION OF CARRIAGE OF DANGEROUS GOODS

3.1 Responsibility of dangerous goods in Poland

Road transport of dangerous goods, means any movement of dangerous goods by the vehicle on public road or other generally accessible roads, including stops required during the transportation and activities related to this haulage (OJ. 2011, No 227, pos. 1367).

Road transport of dangerous goods is a complex process, which is why many sectors are responsible for its implementation (OJ. 2011, No 227, pos. 1367; OJ. 2000, No 50, pos. 601; OJ. 2000, No 122, pos. 1321):
- the minister responsible for transport, supervises the transport of dangerous goods and the entity executing tasks associated with this transport with the exception of the armed forces vehicles,
- Minister of National Defence, supervises the transport of dangerous goods by the transport means belonging to the armed forces,
- Minister responsible for the economy, deals with the matters of technical conditions and testing of packaging of dangerous goods,
- Minister responsible for health, deals with the matters of the conditions of carriage of infectious substances,
- Provincial Inspector of Road Transport - in the matters of the road transport safety inspections of hazardous materials,

– President of the National Atomic Energy Agency, on in the matters concerning conditions of carriage of radioactive material.

3.2 *Control services of dangerous goods*

The participant of carriage of dangerous goods is required to send a copy of the annual report on its activities in the carriage of dangerous goods and related activities, hereinafter referred to as "annual report" before the February 28 of each year following the year covered by the report, to the voivodship road transport inspector appropriate for the seat or place of residence for the participant of dangerous goods carriage.

If, a serious accident or breakdown occurred in connection with the carriage of dangerous goods, within the meaning of ADR, the transport participant, within 14 days from the day of the event, is to submit the report:

– to the appropriate, for the place of the event, Regional Road Transport Inspector - in the case of road transport of dangerous goods,

– Head of the Armed Forces Support Inspectorate - in the case of transport of dangerous goods by the transport means belonging to the armed forces or means of transport, for which the armed forces are responsible.

Information about a serious incident or accident in the carriage of dangerous goods is transferred to the Minister responsible for transport by these authorities, immediately upon receipt by them of the accident report (OJ. 2011, No 227, pos. 1367; OJ. 2000, No 50, pos. 601; OJ. 2007, No 42, pos. 276).

The inspection of the transport of dangerous goods is conducted by:

– Road Transport Inspectorate officer – on the roads, parking lots and at the place of business of the participant in the carriage of dangerous goods,

– Police officers - on the roads and parking lots,

– Border Guard officers - on the roads and parking lots,

– Customs officers - on Polish territory,

– Military Police Soldiers – with respect to the carriage of dangerous goods performed by the armed forces.

In carrying out inspection, the officers work together, to the extent necessary, with the authorized representatives of:

– Nuclear regulatory bodies - on the conditions of carriage of radioactive material,

– Transport Technical Supervision - on the conditions of carriage of dangerous goods,

– Inspectorate for Armed Forces Support and the Military Technical Inspection – on the carriage of dangerous goods performed by the armed forces,

– Inspection of Environmental Protection – on the matters relating to compliance with environmental regulations.

Road Transport Chief Inspector reports serious or repeated infringements, jeopardizing the safety of the transport of dangerous goods, carried out by the vehicle or company from another Member State of the European Union, to the competent authorities of the Member State of the European Union, in which the vehicle or the company is registered.

According to the competence possessed, the Road Transport Chief Inspector, provides the Minister responsible for transport matters the information, before March 31 of each calendar year, on penalties imposed for violations relating to the carriage of dangerous goods and the number of checks on the transport of dangerous goods, observed breaches of regulations, relating to the carriage of dangerous goods.

Road Transport Chief Inspector conveys to the European Commission each calendar year, and not later than 12 months from the end of this year, a report on the inspection of road transport of dangerous goods which contains the following information:

– If possible, the actual or estimated volume of dangerous goods by transported by road (in the tons transported or in ton-kilometres),

– The number of checks carried out,

– The number of vehicles checked at registration location (vehicles registered in this country, in other EU Member States or other countries),

– The number and types of infringements.

3.3 *Problems of dangerous goods monitoring*

Road transport of goods within the EU, including Poland is growing constantly, as is evidenced by the data presented in table 3.

According to the of Central Statistical Office (CSO), about 10 percent of cargo transported by trucks on the Polish roads are dangerous goods.

In 2013, it was 155.3 million tons of dangerous goods transported by road and only 23.3 million tons by rail, often representing lethal threat. 155.3 million tons per year, is 425 thousand tons a day - to carry the load on standard semi-trailers with a capacity of 18 tons, it takes 23.6 thousand trucks per day.

Table 3. Transport of dangerous goods (CSO, 2014)

Year	Road transport weight (ton)	Rail transport weight (ton)
2005	107 976 100	26 955 300
2010	149 125 300	23 456 800
2012	149 338 600	23 087 800
2013	155 305 000	23 259 600

According to the Road Transport Inspection data, in 2011, inspectors checked more than 16 thousand vehicles carrying hazardous materials. The most common violations consisted of by passing restrictions on drivers' driving times and mandatory rest periods, lack of fire-fighting equipment in vehicles, poor labelling of goods and lack of required transport certificates and documents.

Similar data can be found in the report of the Supreme Audit Office (SAO). Irregularities lead to a situation in which entrepreneurs, advisors for the safety matters and drivers are not adequately prepared to organize and carry out transport of dangerous goods. Hazardous materials are transported in Poland often during peak traffic hours, near public buildings and green areas. There are more and more frequent accidents and crashes involving their transport. Provincial governors and marshals are not aware of the potential risks, and the persons directly responsible for the transport of these materials are poorly prepared for that.

Every day on the Polish roads, one can meet thousands of trucks carrying explosives, corrosive, or radioactive materials. They drive virtually unattended.

Improper handling of them can result in death, the huge material losses and environmental contamination.

According to statistics (Michalik & others, 2009), threats of DG delivery in Poland were gained serious number from 2005 to 2007 (see table 4).

Under the ADR agreement, in the case of vehicles carrying dangerous goods at high risk, there should be monitoring devices used for dangerous goods (telemetry systems, tracking devices for movement of goods), effectively prevent the theft of vehicles and cargo.

In addition, following the tragic events of September 11, 2001 in New York, and March 11, 2004 in Madrid, the EU resolution was adopted, which drew attention to the possibility of terrorist attacks, including the use of dangerous goods, which are subject to the obligation of monitoring (Resolution, 2010).

Therefore, to ensure the monitoring of dangerous goods in Poland, it is necessary to design and implement a national dangerous goods monitoring system.

Table 4. Local threats No of DG delivery in Poland from 2005 to 2007 (Michalik & others, 2009)

Category	Chemical threats	Ecological threats	Total (No)
Man deliberate activity.	3	3	6
Defects and improper usage.	6	1	7
Improper storage of DG.	6	2	8
Man unintentional activity.	4	7	11
Defects of mechanical devices.	6	15	21
Improper operation of transport means.	5	20	25
Unidentified.	25	66	91
Defects of transport means.	37	78	115
Lack of condition monitoring of DG.	67	87	154
Unlawful of Road Transport Safety.	104	446	550
Total.	263	725	988

4 PROPOSITION OF NATIONAL SYSTEM OF MONITORING DANGEROUS GOODS VEGICLES - NSMDGV

4.1 *Structure of system*

To decrease the risk during transportation, there is a complex information system which will monitor the oversized and dangerous goods in real time, and which will be interconnected on-line with an integrated emergency system. To determine this kind of system there should be taken into consideration the following architecture qualities: modularity, flexibility, possibility to be used in heterogeneous, environment, interoperability, use of open standards, performance, language independency, system reliability, information security and safety, user interface usability, service intelligence (Kršák, Herkt, 2012).

A consortium composed of: the Motor Transport Institute, Institute of Communications, Roads and Bridges Research Institute, Elte Company and Pimco Company intends to carry out the following tasks:

– Analysis of legislative, international, EU and national documents on the transport of dangerous goods,
– Analysis of operating and implemented ITS solutions in the EU countries and Poland, associated with the process of monitoring dangerous goods,
– Analysis of the functional, communication and physical (transport layer) structure of the systems monitoring vehicles carrying dangerous goods,
– Identifying the needs for interoperability, reliability, security, and mobility of the ITS solutions at different user levels,
– Identification of monitoring systems for vehicles carrying dangerous goods in the world and the EU,
– Developing a model of the national monitoring system for vehicles carrying dangerous goods - NSMDGV (see figure 4),
– Developing technical specifications for the demonstrator of the national monitoring system for vehicles carrying dangerous goods – NSMDGV,
– Producing the national monitoring system demonstrator for vehicles carrying dangerous goods - NSMDGV.

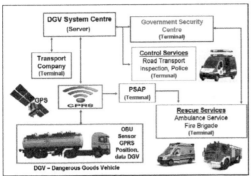

Figure 4. Scheme of NSMDGV (own study)

The final effect of project will be model and demonstrator of the National Monitoring System of Dangerous Goods Vehicles consists of following elements:
- demonstrator of system centre,
- five demonstrators of on-board unit (OBU) for DGV,
- five demonstrators of terminals for Crises Management Centre, Control Services, PSAP (Public Safety Answering Point), Rescue Services and Transport Company,
- demonstrator of innovation sensor for detecting chemical and gases threats.

Demonstrator of System Centre will be perform as a form of communications server with interface, that provides:
- communications with objects,
- random configuration of objects,
- monitoring of objects on map,
- configuration of alarm state of objects,
- creating of standard rotes.

Demonstrator of OBU for DGV can provide some functions as follows:
- localization of DGV,
- communications to System Centre,
- identification and configuration of OBU,
- data collection from sensors, meters and logical inputs.

Demonstrator of terminal will be perform as a form of client applications connected to communications server. The applications provide as follows:
- monitoring of actual positions of objects on map,
- configuration of alarm state of objects,
- creating of standard rotes.

4.2 Technical characteristics of system

The use of GPS and GSM technology, supported by a specialized software package enables the location of vehicles on the Polish territory as well as the entire Europe. This solution not only enables the precise location of the vehicle, but allows:

- monitoring the cargo, its physical and chemical state, which substantially affect the safety of those involved in transport as well as members of the public
- localising the vehicle transporting dangerous goods and other vehicles on the road,
- more efficient management of the fleet of transport companies, which has a direct impact on reducing the cost of transport,
- remote immobilisation of the vehicle, in case of e.g. theft,
- acquisition of a vehicle operating data,
- acquisition of the prevailing meteorological data from the vehicle,
- maintaining constant communication, vehicle – base, and sending messages,
- in the event of a breakdown or disaster automatic notification of the appropriate crisis management centre and emergency services,
- selection of the optimal and economical routes (defining route and maximum deviation from it for safety reasons such as traffic, weather conditions and surface condition).

The accidents of transport dangerous goods are caused mainly by changes in the tankers and containers environment during transportation (such as temperature, humidity, pressure, etc.) or a mixture of goods caused by a chemical reaction and lead to combustion, explosion, toxic gas leaks and so on. Therefore, it has great significance of improving the safety of road tankers and container in the real-time state can be monitored during whole the transportation of dangerous goods. Early and accurate detection, characterization and warning of a chemical and gas event are critical to an effective response. To achieve these objectives, an integrated system of sensors is needed and a supporting information technology network.

Demonstrator of NSMDGV (National System of Monitoring Dangerous Goods Vehicles) should be in line with all specifications (environmental, physical and electromagnetic compatibility) determined by EU directives and standards defined by CEN, ISO and ETSI. The main of them are following:
- Council Directive 73/23/EEC of 19 February 1973 on the harmonization of the laws of Member States relating to electrical equipment designed for use within certain voltage limits,
- Council Directive 89/336/EEC of 3 May 1989 on the approximation of the laws of the Member States relating to electromagnetic compatibility,
- Directive 1999/5/EC of the European Parliament and of the Council of 9 March 1999 on radio equipment and telecommunications terminal equipment and the mutual recognition of their conformity,
- PN-ETS 300 135:1997/A1:1999. Radio Equipment and Systems. Angle-modulated

Citizens' Band radio equipment. Technical characteristics and methods of measurement,
- PN-ETS 300 673;2005. Radio Equipment and Systems RES. Electromagnetic Compatibility (EMS),
- PN-ETSI EN 300 433-2 V1.1.1:2003. PN-ETSI EN 301 489-1 V1.6.1:2006. PN-ETSI EN 301 489-13 V1.2.1:2003. Electromagnetic compatibility (EMC) and Radio Spectrum Matters,
- PN-EN 60950-1:2007/A1:2011. Information technology equipment – Safety.

5 CONCLUSIONS

Transport of dangerous goods can cause an accident and lead to fires, explosions and chemical poisoning or burning with considerable harm to people and the environment.

There is not monitoring system in Poland to control in real time road and rail transportation of dangerous goods.

Developing the model and the implementation of the national monitoring system demonstrator for the vehicles carrying dangerous goods - NSMDGV can significantly contribute to:
- improving the safety of people and the environment,
- developing methods to minimize the damages and costs,
- improving the exchange of information between the centres of production, transport, collection and rescue,
- developing methods of cooperation at the breakdown site.

REFERENCES

Commission Recommendation of 21.2.2011 on reporting of checks concerning the transport of dangerous goods by road. Brussels, C(2011) 909.

Directive 2008/68/EC of the European Parliament and of the Council of 24 September 2008 on the inland transport of dangerous goods. Official Journal of the European Union, L 260/13, 30.9.2008,

Grzegorczyk, K., Hancyk, B., Buchcar R., Dangerous Goods in Road Transport. ADR 2007-2009. BUCH-CAR. Błonie 2007.

Grzegorczyk, K., Buchar, R, Dangerous Goods. Transport in practise. ADR 2011-2013. BUCHCH-CAR. 1 Issue 2011.

Grzegorczyk, K., Changes in Dangerous Goods Transport. Dangerous Goods (Towary Niebezpieczne), No 1/ 2011.

Grzegorczyk, K., Transport of Dangerous Goods. Dangerous Goods (Towary Niebezpieczne), No 3/ 2011.

Kršák, E., Herkt, P., Technical Infrastructure for Monitoring the Transportation of Oversized and Dangerous Goods. Proceedings of the Federated Conference on Computer Science and Information Systems. 2012.

Michalik, J., Gajek, A., Grzegorczyk K. & others. Threats reasons of DG in road transport. Bezpieczeństwo Pracy. Nauka i Praktyka nr 10/2009. CIOP, Warsaw 2009.

Różycki, M., Safety Transport of Dangerous Goods. PPHU Moritz Marek Różycki, 2009.

Regulations concerning the International Carriage of Dangerous Goods by Rail (RID), concluded at Vilnius on 3 June 1999, http://www.cit-rail.org/en/rail-transport-law/cotif. 01.01.2013.

Resolution of Council of the European Union on preventing and combating road freight crime and providing secure truck parking areas. Brussels, 9 November 2010.

Squillante, P., South Eastern Europe Marine and River integrated Monitoring System for the Transportation of Dangerous Goods, Italy, 31/1/13, http://www.southeast-europe.net/en/projects/approved_projects/?id=147.

The act from 15 November 2011. OJ. 2011, No 227, pos. 1367.

The act from 15 November 1984. Transport Law. OJ. 2000, No 50, pos. 601.

The act from 29 November 2007. Atomic Law. OJ. 2007, No 42, pos. 276.

The act from 21 December 2000 refers Technical Inspection. OJ. 2000, No 122, pos. 1321.

The European Agreement concerning the International Carriage of Dangerous Goods by Road (ADR), concluded at Geneva on 30 September 1957, 1 January 2013, www.unece.org/trans/danger/publi/adr/adr_e.html.

The European Agreement concerning the International Carriage of Dangerous Goods by Inland Waterways (ADN), concluded at Geneva on 26 May 2000, 1.01. 2013, http://www.unece.org/trans/danger/publi/adn/adn2011/13fil es_e.html, 1 January 2013.

The statistics of Central Statistical Office from June 2014. Concise Statistical Yearbook of Poland. Warsaw 2014.

AUTHOR INDEX

Printed and bound by CPI Group (UK) Ltd, Croydon, CR0 4YY

24/10/2024

01778293-0004